G-W PUBLISHER

CNC | Manufacturing Technology

Rick Calverley

CNC Manufacturing Technology helps prepare learners for entry-level CNC technician positions in today's advanced manufacturing industry.

Be Digital Ready on Day One with EduHub

EduHub provides a solid base of knowledge and instruction for digital and blended classrooms. This easy-to-use learning hub delivers the foundation and tools that improve student retention and facilitate instructor efficiency. For the student, EduHub offers an online collection of eBook content, interactive practice, and test preparation. Additionally, students have the ability to view and submit assessments, track personal performance, and view feedback via the Student Report option. For instructors, EduHub provides a turnkey, fully integrated solution with course management tools to deliver content, assessments, and feedback to students quickly and efficiently. The integrated approach results in improved student outcomes and instructor flexibility.

Michael Jung/Shutterstock.com

eBook

The EduHub eBook engages students by providing the ability to take notes, access the text-to-speech option to improve comprehension, and highlight key concepts to remember. In addition, the features enable students to customize font and color schemes for personal viewing.

Objectives

Course objectives at the beginning of each eBook chapter help students stay focused and provide benchmarks for instructors to evaluate student progress.

eAssign

eAssign makes it easy for instructors to assign, deliver, and assess student engagement. Coursework can be administered to individual students or the entire class.

Monkey Business Images/Shutterstock.com

Assessment

Self-assessment opportunities enable students to gauge their understanding as they progress through the course. In addition, formative assessment tools for instructor use provide efficient evaluation of student mastery of content.

Reports

Reports, for both students and instructors, provide performance results in an instant. Analytics reveal individual student and class achievements for easy monitoring of success.

	🖶 Print	⬇ Export
Score	**Items**	
100%	●	●
80%	●	●
100%	●	●
80%	●	●
100%	●	●
100%	●	●

Instructor Resources

Instructors will find all the support they need to make preparation and classroom instruction more efficient and easier than ever. Lesson plans, answer keys, and PowerPoint® presentations provide an organized, proven approach to classroom management.

Learn more about EduHub at www.g-w.com/eduhub

Guided Tour

The instructional design includes student-focused learning tools to help students succeed. This visual guide highlights the features designed for the textbook.

Chapter Outline provides a preview of the chapter topics and serves as a review tool.

Learning Objectives clearly identify the knowledge and skills to be obtained when the chapter is completed.

Key Terms list the terms to be learned in the chapter.

Tech Tips provide you with advice and guidance that is especially applicable for on-the-job.

Safety Notes alert you to potentially dangerous materials and practices.

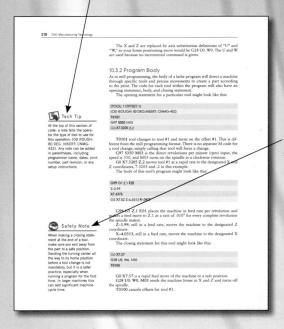

Illustrations have been designed to clearly and simply communicate the specific topic.

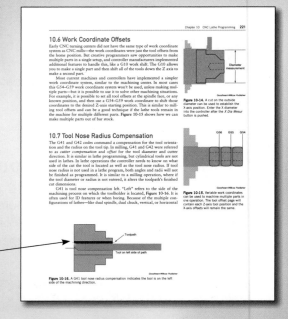

From the Shop presents additional information related to the topic being discussed to deepen understanding.

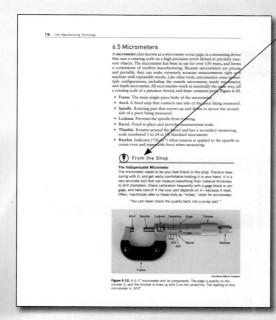

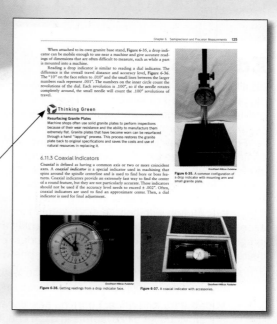

Thinking Green notes highlight key items related to sustainability, energy efficiency, and environmental issues.

Summary feature provides an additional review tool for you and reinforces key learning objectives.

Know and Understand questions allow you to demonstrate knowledge, identification, and comprehension of chapter material.

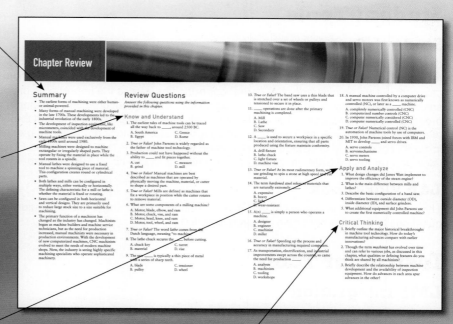

Apply and Analyze questions extend your learning and help you analyze and apply knowledge.

Critical Thinking questions develop higher-order thinking and problem-solving skills.

TOOLS FOR STUDENT AND INSTRUCTOR SUCCESS

EduHub

EduHub provides a solid base of knowledge and instruction for digital and blended classrooms. This easy-to-use learning hub provides the foundation and tools that improve student retention and facilitate instructor efficiency.

For the student, EduHub offers an online collection of eBook content, interactive practice, and test preparation. Additionally, students have the ability to view and submit assessments, track personal performance, and view feedback via the Student Report option. For the instructor, EduHub provides a turnkey, fully integrated solution with course management tools to deliver content, assessments, and feedback to students quickly and efficiently. The integrated approach results in improved student outcomes and instructor flexibility. Be digital ready on day one with EduHub!

- **eBook content.** EduHub includes the textbook in an online, reflowable format. The eBook is interactive, with highlighting, magnification, note-taking, and text-to-speech features.
- **Vocabulary activities.** Learning new vocabulary is critical to student success. These vocabulary activities, which are provided for all key terms in each chapter, provide an active, engaging, and effective way for students to learn the required terminology.
- **eAssign.** In EduHub, students can complete online assignments—including text review questions, activities from the Lab Workbook, and assessments—as specified by their instructor. Many activities are autograded for easy class assessment and management.

Student Tools

Student Text

CNC Manufacturing Technology helps to prepare learners for entry-level CNC technician positions in today's advanced manufacturing industry. Early foundational chapters create an awareness of career opportunities, stress the importance of shop safety, and show how math is used in machining. Subsequent chapters develop a broad base of knowledge of basic machining skills such as print reading and the use of measuring instruments. The author employs a building-block approach to present in-depth coverage of CNC equipment, cutting tools and inserts, CNC mill and CNC lathe programming, CAM, multiaxis programming, and additional advanced topics.

Lab Workbook

The Lab Workbook includes lab activities for hands-on learning and study guide questions to help learners gain required knowledge.

Instructor Tools

LMS Integration

Integrate Goodheart-Willcox content in your Learning Management System for a seamless user experience for both you and your students. Contact your G-W Educational Consultant for ordering information or visit www.g-w.com/lms-integration.

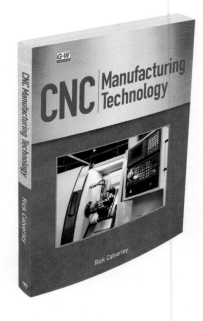

Instructor Resources

Instructor Resources provide all the support needed to make preparation and classroom instruction easier than ever. Available in one accessible location, you will find Instructor Resources, Instructor's Presentations for PowerPoint®, and Assessment Software with Question Banks. These resources are available as a subscription and can be accessed at school, at home, or on the go.

Instructor Resources One resource provides instructors with time-saving preparation tools such as answer keys, editable lesson plans, and other teaching aids.

Instructor's Presentations for PowerPoint® These fully customizable, richly illustrated slides help you teach and visually reinforce the key concepts from each chapter.

Assessment Software with Question Banks Administer and manage assessments to meet your classroom needs. The following options are available through the Respondus Test Bank Network:

- A Respondus 4.0 license can be purchased directly from Respondus, which enables you to easily create tests that can be printed on paper or published directly to a variety of Learning Management Systems. Once the question files are published to an LMS, exams can be distributed to students with results reported directly to the LMS gradebook.
- Respondus LE is a limited version of Respondus 4.0 and is free with purchase of the Instructor Resources. It allows you to download question banks and create assessments that can be printed or saved as a paper test.

G-W Integrated Learning Solution

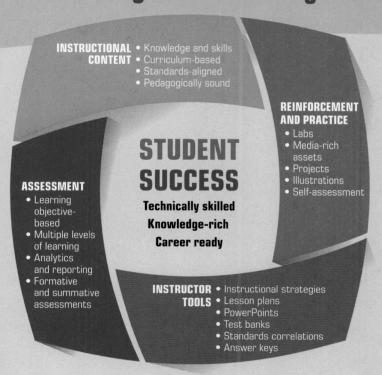

INSTRUCTIONAL CONTENT
- Knowledge and skills
- Curriculum-based
- Standards-aligned
- Pedagogically sound

REINFORCEMENT AND PRACTICE
- Labs
- Media-rich assets
- Projects
- Illustrations
- Self-assessment

STUDENT SUCCESS
Technically skilled
Knowledge-rich
Career ready

ASSESSMENT
- Learning objective-based
- Multiple levels of learning
- Analytics and reporting
- Formative and summative assessments

INSTRUCTOR TOOLS
- Instructional strategies
- Lesson plans
- PowerPoints
- Test banks
- Standards correlations
- Answer keys

The G-W Integrated Learning Solution offers easy-to-use resources that help students and instructors achieve success.

▶ **EXPERT AUTHORS**
▶ **TRUSTED REVIEWERS**
▶ **100 YEARS OF EXPERIENCE**

EMPLOYABILITY SKILLS · TECHNICAL SKILLS · ACADEMIC KNOWLEDGE · INDUSTRY RECOGNIZED STANDARDS

CNC | Manufacturing Technology

First Edition

by

Rick Calverley
Director of Education
Lincoln College of Technology
Grand Prairie, Texas

Publisher
The Goodheart-Willcox Company, Inc.
Tinley Park, IL
www.g-w.com

Manufactured in the United States of America.

Library of Congress Catalog Card Number 2019020816

ISBN 978-1-63563-883-7

1 2 3 4 5 6 7 8 9 – 21 – 24 23 22 21 20 19

Image Credits. Front cover: Allgusak/Shutterstock.com; Timofeev Vladimir/Shutterstock.com
From the Shop: nexusby/Shutterstock.com

Library of Congress Cataloging-in-Publication Data

Names: Calverley, Rick, author.
Title: CNC manufacturing technology / by Rick Calverley.
Other titles: Computer numerical control manufacturing technology
Description: First edition. | Tinley Park, IL : The Goodheart-Willcox
    Company, Inc., [2021] | Includes index.
Identifiers: LCCN 2019020816 | ISBN 9781635638837
Subjects: LCSH: Machine-tools--Numerical control.
Classification: LCC TJ1189 .C38 2021 | DDC 671.3/5--dc23 LC record available at https://lccn.loc.gov/2019020816

Preface

The technology of the manufacturing industry is progressing at a pace that has not been seen before. As the manufacturing industry grows and the previous generation of machinists continue to retire, companies are experiencing difficulties filling skilled positions. This is a great time to be entering the field of manufacturing!

In the past, manufacturing facilities were filled with manual lathes and mills, a skilled machinist crafting parts at each machine. Today, manufacturing facilities are filled with CNC machines, and CNC programmers and CNC operators use these machines to produce parts at a higher quality, a higher rate of production, and a higher level of consistency than was attainable in the past.

A new educational approach is required to prepare tomorrow's workforce. Students must develop the relevant skills needed for career success. *CNC Manufacturing Technology* has been designed to meet these needs. This book will prepare you to begin a career in CNC manufacturing. Early foundational chapters create an awareness of career opportunities and required soft skills, stress the importance of shop safety, and show how math is used in machining. Subsequent chapters develop a broad base of knowledge of basic machining skills such as print reading and the use of measuring instruments. The author employs a building-block approach to present in-depth coverage of CNC equipment, cutting tools and inserts, CNC mill and CNC lathe programming, CAM, multiaxis programming, and additional advanced topics.

CNC Manufacturing Technology makes extensive use of illustrations and photographs to support and extend beyond the text explanations of concepts. Code examples are provided as appropriate to help students understand the basics of CNC programming.

Advanced manufacturing offers an excellent variety of career opportunities. *CNC Manufacturing Technology* will help you begin along a rewarding and fulfilling career path.

About the Author

Rick Calverley is the Director of Education at Lincoln College of Technology in Grand Prairie, Texas, where he designed and implemented Lincoln's first program in CNC Machining and Manufacturing Technology. As a third-generation machinist, a trained mold maker, tool and die maker, and CNC multiaxis programmer, Mr. Calverley has seen this industry go through radical changes. He started his career in 1982, running manual machinery in his father's shop. During his 30+ years of making parts, he has produced parts that make turbochargers, transmissions, plastic injection molds, aircraft parts, and even parts for the International Space Station. Before joining Lincoln, Mr. Calverley worked as a CNC programmer for several manufacturing firms, including Solidiform, Inc., a defense contractor in the aerospace industry.

Mr. Calverley holds an AS degree in marine technology from the College of Oceaneering, plus he has completed additional college-level industry and academic programs. He holds all twelve NIMS Machining Level 1 certifications, as well as three Associate Level Mastercam certifications. Mr. Calverley has served on the national councils for the Fabricators and Manufacturers Association, Mastercam Educational Development, and the Haas Technical Education Center. He also established the first National Tooling and Machining Association student chapter.

Reviewers

The author and publisher wish to thank the following industry and teaching professionals for their valuable input into the development of *CNC Manufacturing Technology*.

David Black
Anderson W. Clark Magnet
    High School
Glendale, California

Doug Bowman
Vincennes University
Vincennes, Indiana

Daniel Colquitt
Baker College of Flint
Flint, Michigan

Jeffrey C. Friend
Savannah Technical College
Savannah, Georgia

Jeff Hines
College of Lake County
Grayslake, Illinois

Tom Larson
Dunwoody College of
    Technology
Minneapolis, Minnesota

Jordan Lesinski
Wayne State University
Detroit, Michigan

Warren Palmer
Red River College
Winnipeg, Manitoba, Canada

Garret Parker
Randolph Community College
Asheboro, North Carolina

Todd Sanders
Danville Community College
Danville, Virginia

Clinton Smith
Lincoln College of Technology
Indianapolis, Indiana

Xavier M. F. Smith
Jamestown Community
    College
Olean, New York

Leonard Walsh
Goodwin College
East Hartford, Connecticut

Chris Waterworth
Greater Lawrence Technical
    School
Andover, Massachusetts

Acknowledgments

The author and publisher would like to thank the companies, organizations, and individuals listed below for their contribution of resource material, images, or other support in the development of *CNC Manufacturing Technology*.

The author and publisher extend special appreciation to Madalyn Belle Photography for substantial contribution to the photography program.

Accuform®
Apex Tool Group
CNC Software, Inc.
DC Turbo
Haas Automation, Inc.

Iscar
L.S. Starrett Company
Masterlock
MSA Safety Inc.

Rick Calverley thanks his wife, Gina, and his five sons—Jacob, Zech, Sam, Gabe, and Israel—for the support they provided throughout the process of writing this textbook.

Brief Contents

Chapter 1: **Progressing to CNC Manufacturing** 2

Chapter 2: **The Skills to Compete** 20

Chapter 3: **Machine Shop Safety** 46

Chapter 4: **Machining Math** 64

Chapter 5: **Print Reading** 86

Chapter 6: **Semiprecision and Precision Measurements** 110

Chapter 7: **CNC Machinery** 132

Chapter 8: **Cutting Tools and Inserts** 156

Chapter 9: **CNC Mill Programming** 178

Chapter 10: **CNC Lathe Programming** 210

Chapter 11: **Computer-Aided Manufacturing** 242

Chapter 12: **Multiaxis Programming** 264

Chapter 13: **Subprogramming and Probing** 286

Chapter 14: **Automated and Additive Manufacturing** 312

Contents

CHAPTER 1

Progressing to CNC Manufacturing 2

1.1 The Earliest Machine Tools 4
1.2 Early Measuring Tools 5
1.3 Basic Manual Machines 6
1.4 Evolution of Machinists 14

CHAPTER 2

The Skills to Compete 20

2.1 The Modern CNC Environment 22
2.2 Supplied Industries 24
2.3 Careers in CNC Manufacturing 28
2.4 Applying for a Position 30
2.5 Succeeding in the Workplace 36
2.6 Technical Training 41

CHAPTER 3

Machine Shop Safety 46

3.1 General Shop Safety 48
3.2 Fire Safety 53
3.3 Hazardous Materials Identification
System 56
3.4 Safety Data Sheets 57
3.5 Safety Aids 59

CHAPTER 4

Machining Math 64

4.1 Introduction to Machining Math 66
4.2 Fractions 66
4.3 Geometric Shapes 70
4.4 Angles 73
4.5 Trigonometry 74
4.6 Cartesian Coordinate System 77
4.7 Calculating Speeds and Feeds 79

CHAPTER 5

Print Reading 86

5.1 Introduction to Print Reading 88
5.2 Drawings and Prints 88
5.3 Title Blocks 92
5.4 Auxiliary and Sectional Views 93
5.5 Line Usage 95
5.6 Dimensions 97
5.7 Tolerances 101
5.8 Holes 102
5.9 Arcs 104
5.10 Threads 105
5.11 Geometric Dimensioning and Tolerancing . . 106

CHAPTER 6

Semiprecision and Precision Measurements . . . 110

6.1 Introduction to Precision and
Semiprecision Measurements 112
6.2 Steel Rules 112
6.3 Protractors 113
6.4 Calipers 114
6.5 Micrometers 116
6.6 Comparison Gages 118
6.7 Height Gages 120
6.8 Gage Blocks 121
6.9 Sine Bars 122
6.10 Telescoping Gages 123
6.11 Indicators 123
6.12 Advanced Gaging and Inspection Tools . . 126

CHAPTER 7

CNC Machinery 132

7.1 Introduction to CNC Machinery 134
7.2 CNC Machining Centers 134
7.3 CNC Turning Centers 140
7.4 Mill-Turn Machines 146

7.5 Swiss Machines 147
7.6 CNC Grinders 147
7.7 Water Jet 149
7.8 Plasma Cutters 150
7.9 CNC Lasers 150
7.10 CNC Routers 151

CHAPTER 8

Cutting Tools and Inserts 156

8.1 High-Speed Machining 158
8.2 Inserts 158
8.3 Drills 164
8.4 Taps 166
8.5 Countersinks and Counterbores 167
8.6 End Mills 167
8.7 Face Mills 168
8.8 Toolholders 169
8.9 Machinability Ratings 173

CHAPTER 9

CNC Mill Programming 178

9.1 Introduction to CNC Mill Programming . . 180
9.2 Program Planning 180
9.3 Program Format 182
9.4 Address Codes 184
9.5 Contouring 187
9.6 Radius and Angle Milling 191
9.7 Pocketing 195
9.8 Canned Cycles 197
9.9 Tool Setting 202
9.10 Work Coordinate Offsets 205

CHAPTER 10

CNC Lathe Programming 210

10.1 Introduction to CNC Lathe Programming . . 212
10.2 Program Planning 212
10.3 Program Format 215
10.4 Address Codes 217
10.5 Tool Setting 220
10.6 Work Coordinate Offsets 221
10.7 Tool Nose Radius Compensation 221
10.8 Profile Cutting 222

10.9 Drilling Canned Cycles 231
10.10 Threading Cycles 235
10.11 Parting Off 237

CHAPTER 11

Computer-Aided Manufacturing 242

11.1 Introduction to Computer-Aided
Manufacturing 244
11.2 CAD versus CAM 244
11.3 Geometry Creation 247
11.4 Geometry Import 251
11.5 Toolpath Generation 251
11.6 Associated Geometry 256
11.7 Job Setup 257
11.8 Post-Processing: From Screen to G Code . . 259

CHAPTER 12

Multiaxis Programming 264

12.1 Multiaxis Machining 266
12.2 Multiaxis Turning 267
12.3 Multiaxis Milling 270
12.4 Mill-Turn Machines 280

CHAPTER 13

Subprogramming and Probing 286

13.1 Programming Efficiency 288
13.2 Main Programs and Subprograms 288
13.3 Lathe Subprogramming 295
13.4 Implementation of Probing Systems 298

CHAPTER 14

Automated and Additive Manufacturing 312

14.1 Introduction to Automated Manufacturing . 314
14.2 Robotics 314
14.3 Cellular Manufacturing 318
14.4 Flexible Manufacturing Systems 321
14.5 Additive Manufacturing 322
14.6 The Future 327

Reference Section *332*
Glossary *352*
Index *362*

Feature Contents

From the Shop

Hardened Steel 13

Success in the Shop 36

Protect Yourself at All Times 49

Feeds and Speeds 80

Bolt Hole Circles 100

The Indispensable Micrometer 116

Building Our Future 146

Check the Data 158

G41: Left-Hand Compensation 189

How Fast Can You Tap? 201

Walking the Path 222

Is My Surface Solid? 251

Dynamic Work Offsets 277

Our Robot Future 314

3D Printing 327

Thinking Green

Green Manufacturing 17

Green Fire Safety 54

Electronic CAD Drawings 89

Resurfacing Granite Plates 125

Reducing Chips 141

Recycling Cutting Tools 167

Recycling Tungsten Carbide Drills 200

Health and Environmental Impacts of Cutting Fluids 234

Government Programs as Green Catalysts 246

Leadership in Energy and Environmental Design (LEED) 281

Proactive Maintenance 297

Additive Manufacturing 325

1

Progressing to CNC Manufacturing

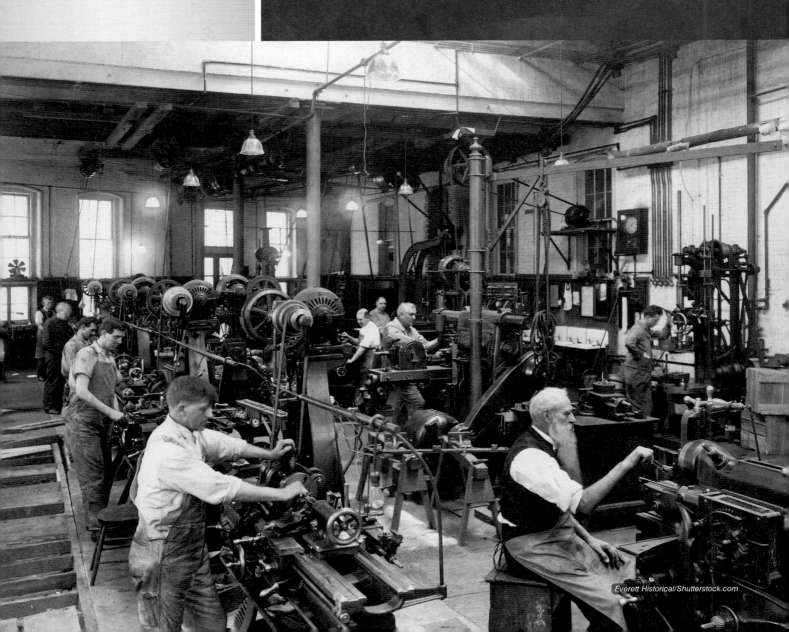

Everett Historical/Shutterstock.com

Chapter Outline

1.1 The Earliest Machine Tools

1.2 Early Measuring Tools

1.3 Basic Manual Machines

1.3.1 Mills

1.3.2 Lathes

1.3.3 Saws

1.3.4 Drill Presses

1.3.5 Grinders

1.4 Evolution of Machinists

1.4.1 Machine Builders

1.4.2 Machine Service Technicians

1.4.3 Manual Machinists

1.4.4 CNC Machinists

1.4.5 Machining Specialists

Learning Objectives

After reading this chapter, you should be able to:

- Summarize the earliest forms of machining.
- Explain the importance of James Watt and John Wilkinson in early lathe design.
- Describe how early measuring devices mirrored machinery development.
- Define manual machining.
- Identify major components of manual milling machines.
- Identify major components of manual lathes.
- Describe workholding in a lathe and mill.
- Identify multiple types of saws and describe their functions.
- Discuss possible uses of a drill press.
- Name three different types of grinders and describe their functions.
- Describe how machining as a profession has evolved over time.
- Describe the advent of CNC technology.

Key Terms

band saw

chuck

CNC machinist

drill fixture

drill press

grinding

hardened steel

horizontal band saw

inside diameter (ID) grinder

lathe

machine vise

machining specialist

machinist

manual machine

manual machinist

micrometer

milling machine

numerical control (NC)

outside diameter (OD) grinder

surface grinder

turret

vertical band saw

workpiece

1.1 The Earliest Machine Tools

The earliest accounts of machine tools can be traced all the way back to Ancient Egypt, around 2500 BC, when logs were spun by human power while a crude tool was used to remove the bark and attempt to make the log round. After that, we have examples of pottery wheels and bow drills being used to form shapes and make holes. These machines were man- or animal-powered and usually involved some hand tool that was held in place by man. Today, we generally consider machining as an automated process where a machine drives a cutting tool through material to create a desired shape.

There has always been a need for precision manufacturing, but it has taken us thousands of years to design and produce equipment to meet those needs. As our need for manufacturing grew, so did our ingenuity to build better machines. In 1763, James Watt was working at the University of Glasgow when he was asked to repair a steam engine. He immediately noted inefficiencies in its design and began to consider possible design changes. He was successful with many design changes, including adding a condenser system to remove heat from the piston and a chamber and seal on the top of the system that was fully enclosed.

Watt also noted that he needed a tight-fitting seal and a closer tolerance piston and chamber to further improve design and increase performance. In the early 1770s, he commissioned John Wilkinson to build a machine that could bore out diameters and produce parts more effectively. In 1774, Wilkinson developed what is considered to be the first lathe. This early lathe was a boring machine that produced round, close tolerance cannon bodies. Wilkinson's lathe had a fixed bar that was moved manually to set cutting depths, and the cast-iron cannons were turned by water power. After each cut, the bar was unbolted from the machine and repositioned by hand for the next cut. The accuracy of this machine was no better than 1/16th of an inch. Watt eventually used this machine to bore cylinders for his steam engines, and Wilkinson continued to make advancements in iron metalworking and machinery.

Another important inventor was born just before Watt's steam engine went into production. Henry Maudslay was born in 1770 and is widely regarded as the father of machine tool technology. He began working at age 12 in a weapons arsenal, filling cartridges with black powder. He also worked as a carpenter and eventually a blacksmith. Maudslay was involved in design improvements for the hydraulic press and modifications to various machine tools, but it was in 1800 that he dramatically changed production and assembly. That year, Maudslay developed the first thread-cutting lathe, and henceforward fastening assemblies together would never be the same. Up until this point, threads were cut only in external rods with hand tools like a file and chisel, and making a fastening nut was impossible. Most screws or bolts were used to fasten metal pieces into wood by driving the screw into the wood. Now it was actually possible to make repetitive, accurate external threads that could be screwed into a nut for joining.

Although Maudslay's standardization of screw threads was extremely important, it is not his most significant contribution. If production of parts was to continue to expand, the industry needed to have a tool that could inspect parts consistently at a level higher than the machine tools' accuracy. With this in mind, he invented an accurate measuring tool we now call the ***micrometer***. The word *micrometer* comes from the Greek words "micro," meaning small, and "metrons," meaning measure. True to its name, it was invented to make small measurements. Maudslay's bench micrometer could measure to 0.0001″ (one ten-thousandths of an inch). Now, it was possible to machine a piece and check its accuracy relative to all other pieces produced.

There are many different machine builders credited with making various forms of what we refer to as milling machines. There are records of many types of machines using rotary files or fixed workpieces with rotating tools, but many consider Eli Whitney's 1818 invention to be the first true milling machine. There is also evidence that Robert Johnson of Middletown, Connecticut, used a milling machine to create firearm pieces at an earlier date. The addition of a machine that could make square or rectangular as well as round pieces truly opened new doors in manufacturing.

1.2 Early Measuring Tools

Before the industrial revolution, Wilkinson's boring machine, Whitney's milling machine, and other early machinery, there was an inherent issue with production. All parts to be assembled had to be made one at a time and made to fit other existing parts. For example, all wagon wheels and axles were made one at a time to fit together. There was no "standard" size or even a reliable way to measure one piece to make another piece to fit. Imagine today if getting a new wheel on a car required taking it to a machine shop so they could cut a wheel to fit the car's axle. No two wheels would be exactly the same. Instead, imagine that a wheel maker knew exactly what size axle was on every wagon or car. That same wheel maker could make hundreds of wheels that could be sold as soon as they were needed.

In the late 1790s the world began to change with the beginning of what is known as the industrial revolution. Skilled craftsmen who worked from their homes or shops began to be replaced by factories and machinery that could produce products in large volume and at lower costs. First seen in the textile industry, this sweeping change in the manufacturing process rapidly took over most manufacturing. This new industrialization provided for many jobs and improved the quality of life for all, as more products became available for the common household at an affordable cost.

Consistent and reliable manufacturing relies on both machines and measuring tools. Large-scale production could not have happened without the ability to measure and fit pieces together and then exactly repeat that process many times over. Without the implementation of measuring tools and the ability to produce repeatable parts, the industrial revolution could not have occurred. Today, we still judge machinery by its repeatability and accuracy or, in other words, by its ability to repeat to the same measurable position consistently over a given period of time.

Compare the historical implementation of measuring tools to the invention and improvement of machines:

- In 1772, Watt invented a tabletop micrometer that was limited to a bench-mount-only design.
- In 1775, Wilkinson invented the boring machine that would eventually make Watt's steam engine components.

- In 1800, Maudslay invented the gear-driven, screw-cutting lathe. With the addition of screw threads, Maudslay then invented the tabletop-mounted micrometer that became the measuring device used to inspect all his production pieces.
- In 1818, Whitney created the horizontal milling machine.
- In 1830, the sewing machine was invented.
- In 1838, Robert Buchanan, a manufacturing engineer, invented the radial boring machine.
- In 1876, the combustion engine was invented.
- In 1877, Victor Machine advertised its first micrometer in *American Machinist Magazine*.
- In 1877, L. S. Starrett began production of his first measuring tool, the adjustable square.

1.3 Basic Manual Machines

Manual machines are best described as machines that are operated by physically moving the machine, material, or cutter to shape a desired part. *Manual machinists* are highly skilled and dexterous operators of this equipment. The use of basic manual machines dominated the machining industry from the beginning of manufacturing until around 1980. The most common manual machine is the *milling machine*, **Figure 1-1**.

1.3.1 Mills

Milling machines, also called *mills*, fix the material to be worked, or *workpiece*, in position while the cutter rotates to remove material. The workpiece is secured to the mill table, either directly, by use of a machine fixture, or more often in a *machine vise*, **Figure 1-2**. A machine vise is a

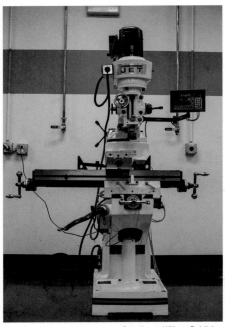

Goodheart-Willcox Publisher

Figure 1-1. A manual milling machine.

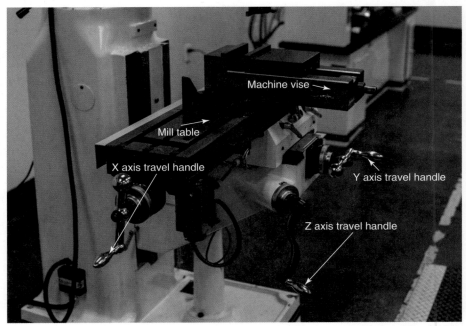

Goodheart-Willcox Publisher

Figure 1-2. Components of a manual milling machine, including machine vise, mill table, and travel handles.

heavy clamping system mounted on a mill table that secures material by means of a screw or lever.

There are several types of milling machines, and they are most often categorized by their spindle orientation. A vertical mill has a vertical spindle, as pictured in **Figure 1-1**. A horizontal mill has a spindle that is perpendicular to the table, as in **Figure 1-3**. Vertical or horizontal refers to the direction of the cutter. A milling machine travels in the X, Y, and Z directions, and these directions refer to the spindle direction, with the Z axis heading down the centerline of the spindle or cutter.

Figure 1-4 shows a vertical milling machine using a machine vise to hold a piece of aluminum while it is cut. This is a vertical mill because the centerline of the cutting tool is oriented vertically. The mill consists of several components:

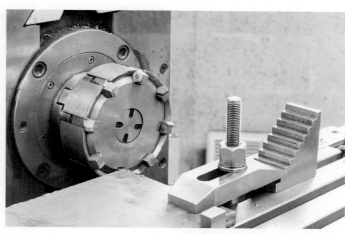

AnnaElizabeth photography/Shutterstock.com

Figure 1-3. Horizontal milling machine with a face mill cutter in a spindle.

- The motor provides power to the tool head.
- The tool head creates rotation to the spindle and tool and secures the tool in place.
- The ram attaches the tool head to the column and can be moved forward or backward to change the work envelope.
- The column attaches the ram and knee and allows the workpiece to move vertically.
- The knee holds the mill table and allows travel in the X, Y, and Z directions.
- Table traverse cranks are handles that incrementally move the mill table into the correct cutting position.

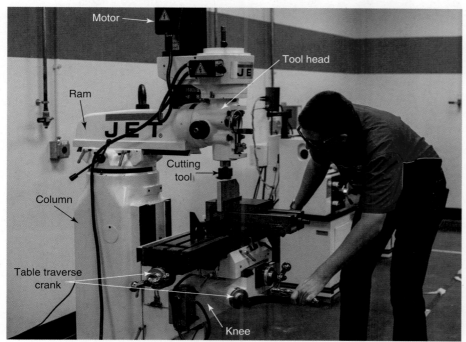

Goodheart-Willcox Publisher

Figure 1-4. The major components of a manual milling machine, including the motor, tool head, ram, cutting tool, column, table traverse crank, and knee.

 **Tech Tip**

Workholding refers to any device that is used to secure a workpiece against the forces of machining.

All these components work in unison to move a cutting tool or workpiece into its desired position and remove material. The milling machine can drill, tap, ream, and mill straight lines and angles. With the use of creative workholding and tooling, this machine has the ability to create a multitude of shapes and parts.

There are a variety of configurations and models for the milling machine, but they all feature a fixed workpiece with a rotating spindle. The spindle can be driven into the workpiece, or the table can be driven to move the work into the spindle. The milling machine is easily the most versatile manual piece of manufacturing equipment based on its ability to hold a variety of shapes and sizes in multiple configurations.

1.3.2 Lathes

Another common manual machine is the **lathe**. The word *lathe* comes from the Dutch language, meaning "to turn." Often we call lathe operations "turning" because the lathe spins or turns the workpiece while the cutter is fixed or driven through the material. Most often, the raw material is secured in a set of jaws or **chuck**, but material can also be secured in a collet or mounted to a fixture plate. The most common lathe is the horizontal engine lathe, **Figure 1-5**, also called a *turret lathe*. **Turret** refers to the tool holder in a lathe, which allows multiple tools to be used and indexed into position for repetitive processes.

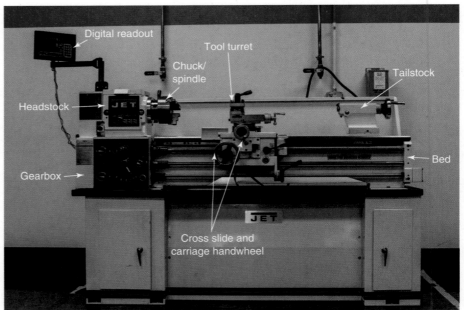

Goodheart-Willcox Publisher

Figure 1-5. A manual lathe and its major components, including chuck/spindle, tool turret, tailstock, bed, cross slide and carriage handwheel, gearbox, headstock, and digital readout.

Goodheart-Willcox Publisher

Figure 1-6. A three-jaw scroll chuck used in a manual lathe application.

Goodheart-Willcox Publisher

Figure 1-7. A four-station tool turret used to secure and quickly index between cutting tools.

The lathe consists of multiple components that turn the stock piece into a desired shape:

- The chuck secures material before cutting, **Figure 1-6.**
- The tool turret secures the cutting tool in place, **Figure 1-7.**
- The gearbox adjusts speeds and feeds and allows for threading, **Figure 1-8.**

Lathes, like mills, can have multiple configurations, including the vertical lathe, center lathe, Swiss lathe, and even a wood lathe—but they all use spinning material and fixed tools. Lathes make parts that are generally round or cylindrical in shape. With the addition of a tailstock, shown in **Figures 1-9A** and **1-9B**, the lathe can also drill holes on the centerline of a spinning workpiece. Because the part is spinning, only holes in the center can be drilled. Floating taps can be used to create internal threads, and specialty-formed tools can be used to create external threads or grooves.

Goodheart-Willcox Publisher

Figure 1-8. A multi-position lathe gearbox. The gearbox has multiple settings to allow for changes in spindle speed and tool feed.

A

B   *Goodheart-Willcox Publisher*

Figure 1-9. Tailstock attachment. A—With live center. B—With drill chuck.

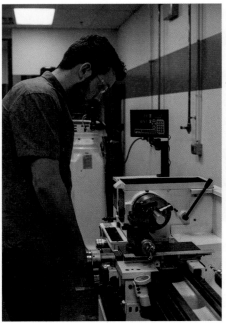

Figure 1-10. A lathe operator moving the tool into a cutting position.

The lathe is one of the earliest machine tools, and in the hands of a skilled operator, it can be an excellent machine for creating parts in multiple industries, **Figure 1-10.**

1.3.3 Saws

Material is often purchased in long lengths of bar, either round or rectangular, and must be cut closer to size before beginning machining operations. Cutting operations are completed using a saw. The saw is typically a blade or thin piece of metal with a series of sharp teeth attached to a handle. The saw is a vital complementary machine tool in manufacturing because it is nearly impossible to load extended lengths of material into a machine. The saw, like most other machine tools, comes in multiple configurations. Saws can be completely manually operated or use power-driven blades to cut material into sections.

The *band saw* has been a staple in metal cutting machine shops for decades and is still used as the primary cutting machine because of its speed and accuracy. The band saw uses a thin blade that is stretched over a set of wheels or pulleys and tensioned to secure it in place. The band saw comes in a vertical or horizontal configuration. The terms *vertical* and *horizontal* refer to blade orientation. A **vertical band saw**, **Figure 1-11**, has a blade oriented perpendicular to the floor. A **horizontal band saw**, **Figure 1-12**, uses a blade oriented parallel to the floor or base.

Using a vertical band saw requires the operator to push material through the fixed blade point. To operate this machine safely, a push bar or guide must be used to move material through the blade. Never attempt to hold material by hand when operating in proximity to a rotating blade. A vertical band saw is often used with thin, sheet-type material and easily allows an operator to cut angles or handle thin material. Stops and guides can be used to create multiple part cuts at given dimensions.

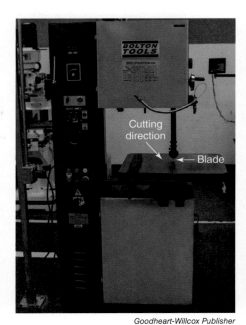

Figure 1-11. A vertical band saw.

Figure 1-12. A horizontal band saw.

Safety Note

Saw blade teeth are very sharp and curved to improve cutting performance. Always be careful when installing, removing, or handling blades. Make sure all blade guards are in place, and never try to clamp a piece in place or handle material while the saw blade is turning or operating.

The horizontal band saw is the most common saw used in manufacturing. Material is secured in place in vise-style jaws with the desired cut length of material extending beyond the saw blade. Once the material is secured, the blade motor can be turned on to activate the blade. The saw blade is lowered using a hydraulic (fluid) or pneumatic (air) cylinder. This creates a safe work environment for the operator because the blade is not engaged until the material is secured. Using the cylinder to control the cutting speed of the blade increases accuracy and blade life. The horizontal band saw is a safe and accurate sawing method for most applications. The blades come in a variety of tooth configurations for different material types and should be considered when choosing the correct blade for the application.

sNike/Shutterstock.com

Figure 1-13. A standard drill press used for secondary operations.

1.3.4 Drill Presses

Machining often requires secondary operations. Secondary operations are done after the primary machining is completed to add detail, remove sharp edges, or perform operations not easily done in the initial operation. A *drill press* is commonly used in secondary operations. A drill press can also be used to drill single holes in a workpiece that do not require the accuracy of a milling machine. A drill press resembles a manual milling machine, but it does not have the same versatility or power. It is simply a drill motor with a table, **Figure 1-13**, which can be a great asset when used appropriately.

Drill fixtures are still used today as an accurate and efficient method for creating multiple holes in production parts. A drill fixture is used to securely locate and orient a workpiece to ensure that all parts produced maintain conformity. A series of stops are used to locate a piece repetitively, and a drill bushing or guide directs the drill to a precise location. Fixtures can be used for rectangular plate, **Figure 1-14**, or round stock, **Figure 1-15**.

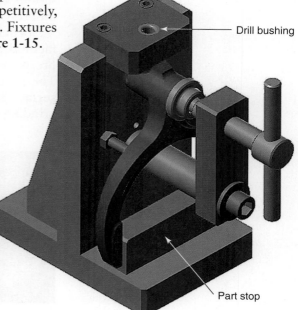

Drill bushing

Part stop

Goodheart-Willcox Publisher

Figure 1-15. A simple drill fixture for centered hole drilling.

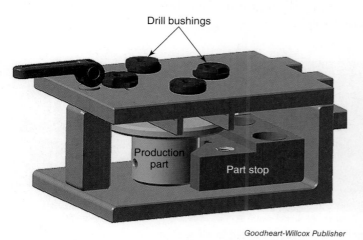

Drill bushings

Production part

Part stop

Goodheart-Willcox Publisher

Figure 1-14. A drill fixture using mounted drill bushings and part stops.

Figure 1-16. A multi-head drilling machine.

Multiple drill press heads can be mounted onto a table to create a bank of drills or taps of various sizes to machine more complex parts, **Figure 1-16**. The relatively low cost and high versatility of this equipment makes it a valuable tool. Before the advent of CNC mills with multiple tool holder locations, the use of multiple drills or taps was a common practice in high-production manufacturing.

1.3.5 Grinders

The mill, lathe, saw, and drill press all remove metal through a cutting process. A cutting process removes material from a workpiece by forcing a hardened, sharpened edge through the material with an action similar to a knife. Whether the material is steel, brass, aluminum, or even plastic, the basic process is the same. An entirely different process, *grinding*, is the process by which most of these sharp cutting tools are made.

In the context of machining, *grinding* means to wear smooth or sharpen by abrasion or friction. In the most rudimentary form, grinding operations involve spinning a stone at high speed to remove material through friction. Grinding may be the oldest form of machining, but grinding and grinders are still used today. In fact, grinding remains one of the most precise techniques for metal removal. Precision grinders can achieve sizes as close as 0.00005″, or 60 times smaller than a human hair, and can grind steel as smooth as glass. Bench grinders or pedestal grinders, **Figure 1-17**, can be used to sharpen drills, remove sharp edges, or create semi-precise angles and radii. Just as the name indicates, these grinders are mounted on a benchtop or secured to the floor with a pedestal. The part being ground is held and manipulated manually by the operator. This process can have serious safety concerns, so always use safety glasses, a safety shield, and the tool rest to secure the part in position.

Bench and pedestal grinders are not meant to be precision grinding methods, but when used properly they can produce at least a semi-precision result. However, another group of grinders is highly precise. The outside diameter (OD) grinder, inside diameter (ID) grinder, and surface grinder are all precision grinding machines used primarily in the grinding of *hardened steel*.

Figure 1-17. A—Bench grinders are mounted to a fixed workbench and commonly used for tool sharpening or smoothing off sharp edges. B—Pedestal grinders are mounted to a freestanding base.

From the Shop

Hardened Steel

The term *hardened steel* refers to materials that are naturally extremely wear-resistant or, through a process of heat treatment, made very wear-resistant. Hardened steel is commonly used in bearings, gears, cutting blades, and tools.

Most steel is machined in a natural state, which is more machinable, then made harder through a process of heating and cooling. Occasionally you will have to machine hardened steel, which will require some special tools and smaller cuts. Common tools you might see that are made from hardened steel are knives, files, and punches.

Outside diameter (OD) grinders apply a precisely balanced grinding wheel to the outside of a spinning cylindrical part to slowly remove material from the workpiece. See **Figure 1-18**. Both the wheel and the part are spinning at a relatively high speed, measured in revolutions per minute (rpm). This process is intended for small metal removal rates. Typically, a grinder does not remove more than 0.010″ of material. Because the part is not being cut, OD grinding produces an extremely smooth surface.

Inside diameter (ID) grinders are similiar to OD grinders, with the obvious distinction of grinding inside a workpiece. The grinding wheel is mounted on a shaft, and as the part spins, the wheel reciprocates in and out of the part to remove material. See **Figure 1-19**. There must be an existing hole in the piece for the grinding stone to enter. Like outside diameter grinding, inside diameter grinding is a slow process with very small metal removal rates. Although it is considered a manual machine, most machines are mounted with a hydraulic cylinder to move the stone back and forth into the part. A series of "stops" are used to control the motion, similar to a piston.

The third precision grinding machine is the *surface grinder.* The surface grinder differs from OD and ID grinders in that it does not rotate the workpiece. Its primary uses are to grind surfaces flat or make custom cutting tools. A variety of attachments can be added to a surface grinder, but the most common configuration is a mounted grinding wheel that

oYOo/Shutterstock.com

Figure 1-18. Outside diameter (OD) grinder used to perform a finished grind operation.

Vadim Ratnikov/Shutterstock.com

Figure 1-19. Inside diameter (ID) grinder performing internal grinding operation.

Vadim Ratnikov/Shutterstock.com

Figure 1-20. Surface grinding operation on a magnetic table.

spins while a workpiece is held stationary on a magnetic surface plate. See **Figure 1-20**.

The surface grinding process can create precisely flat and parallel parts. The term *flat* refers to a surface lying within a single plane, and *parallel* refers to the measurement from one plane to a second plane. When a surface is flat, it does not fluctuate from a single plane. The grinding wheel can be used to grind the table extremely flat with a superior finish, so the resulting piece is ground to near exacting parallelism. This is also a delicate process that is typically expected to remove no more than 0.002″ per pass of the wheel over the part.

The machines listed in the preceding sections by no means constitute a comprehensive list of manual machines. Even so, all these machines have a wide variety of applications, attachments, and configurations. Manual machinists are some the most highly skilled and versatile craftsmen in the world. It is truly remarkable that machinists developed and built these machines so they could either build more machines or create amazing parts and pieces for use in everyday life. Consider that every car, train, plane, and any other machine built from the early 1700s until the 1960s was made by hand on some type of equipment listed in this chapter. Literally, the world as you know it today was built by machinists using manual machines.

1.4 Evolution of Machinists

A *machinist* is simply a person who operates a machine. However, that definition is nowhere close to comprehensive. Actually, what it means to be a machinist has grown and evolved for more than 200 years. Being a machinist has specific meaning in the historical context of the manufacturing industry. When the term was first coined during the industrial revolution, *machinist* meant someone who built machines. Then, the meaning changed to someone who could repair machines. During the 1920s and beyond, a machinist worked at a factory making pieces in a production environment. In the late 1970s, this term expanded to include a new acronym, CNC machinist, which refers to someone who could operate computer numerical controlled machinery. Today, the term *machinist* refers to a highly technical and qualified *machining specialist*, or someone with advanced training who operates specific machinery for industries such as medical, aerospace, or automotive.

1.4.1 Machine Builders

As the industrial revolution swept from Europe to the United States in the mid-1800s, there was a tremendous movement away from manual operations in textiles and mills to machine operations. Increasing speed and accuracy in manufacturing required machinery and new workers with expertise in metals, castings, and early mills and lathes. These machinists, known as machine builders, began to build machines that would drive the industrial revolution.

These early industrial machines were created with crude drawings that had no exact specifications. As the machines were built, they were tested for function and redesigned as needed. Early machinists were often machine designers and problem solvers for complicated issues never seen before. Adding to the difficulty, none of these machines were alike because they were all made one part at a time.

1.4.2 Machine Service Technicians

Early machine builders birthed the next generation of machinists, machine service technicians, who were arguably the first real machining tradespeople. As machines began to be produced, they inevitably broke down and needed repair. Since most of these machines were custom made, there were no replacement parts. Even if parts existed, it could take days or weeks to get them. A machinist, or machine service technician, would be called in to fix a broken machine, **Figure 1-21**.

These machinists had to be experts in all forms of machine repair. They were knowledgable about various steels, early welding practices, proper use of lathes and mills, and machine building functions. They were called out to factories to fix existing pieces or make new ones. They were experts in cutting threads, welding, making steel sleeves, bushings, and correcting early design flaws that led to failures. Equipment was usually too large to move and had to be repaired on-site. All these factors required machine service technicians to be creative and highly skilled in all principles of machine operations.

Jannarong/Shutterstock.com

Figure 1-21. A transmission gear train being disassembled for inspection.

1.4.3 Manual Machinists

As transportation, electrification, and other industrial improvements swept across the country, so came the need for production machinists. These tradespeople made great strides in creating interchangeable parts to meet the industrial needs of a growing country. Now parts could be made

Figure 1-22. A manual machinist operating a manual machine in a factory.

in large quantities and shipped as needed to repair or build machines. With the production of guns, cars, trains, ships, farming implements, and factory equipment, came the need for repetitive machining. Although most machinists were still experts in multiple machines and operations, many became specialists in lathe, mill, or grinding machines, **Figure 1-22.** These manual machine operators dominated manufacturing from 1920 until around the early 1990s.

1.4.4 CNC Machinists

The rapid growth and development of manufacturing during the last 50 years can be largely attributed to the development of numerical controls. *Numerical control (NC)* is the automation of machine tools by use of a computer. This is vastly different from the manual machines covered earlier in this chapter because there are no handwheels or levers to create the machine motion. Computer numerically controlled (CNC) machines allow tools to move through the use of servo drive motors and encoders.

In the mid-1940s, a group led by John T. Parsons began using the first computers to solve complex mathematical machining problems involving the accurate interpolation of curves in helicopter rotor blades. This was a huge machining issue because the blades consisted of a constant twisting shape. The manufacturing of these blades was labor intensive and slow. By 1948, the Parsons Corporation was receiving contracts to create the challenging tapered wings for military aircraft.

In 1950, Parsons joined forces with IBM and MIT to develop servomechanisms and servo drives. The computer calculations were only one piece of the solution, and servo drives, or drive motors, needed to be developed to physically drive the machine to the coordinates calculated. In 1952, Parsons patented the servo control and Automatic Programmed Tool (APT) programming language. In January 1958, Parsons had finally married his math, servo motors, and a computer to create the first true numerically controlled machine. For this achievement, John Parsons is widely regarded as the "Father of CNC."

In the early 1960s, the world entered the technology age. With the race to space and early computer development, machinists evolved again to meet the needs of production. The development of computer-controlled machining began.

In the late 1970s, a new type of machine was introduced. It was a manual machine controlled by a computer drive and servo motors. This new machine was first known as a numerically controlled (NC) machine, then later as a computer numerically controlled (CNC) machine. These machines were very expensive, which made them virtually impossible for the average machine shop to own. That changed in 1988, when Gene Haas released his first VF-1 CNC milling machine at a fraction of the cost. Haas's innovation started the next phase in machining, which required a new machinist: the **CNC machinist**. CNC machinists have traditional machining skills as well as the necessary skills to operate a technologically advanced computer-driven system, **Figure 1-23**.

Monkey Business Images/Shutterstock.com

Figure 1-23. Modern CNC machinists work on technologically advanced machines and must understand how CNC machines are programmed as well as traditional principles of machining.

1.4.5 Machining Specialists

Today, machining is still evolving at a rapid pace. Machinists now work in climate-controlled facilities, operating machinery with high-speed capabilities. The new machinist is a specialized, highly sought-after operator of machinery who is capable of producing thousands of parts in a day. The precision and speed of production is astronomically higher than it was just 20 years ago, and the value of machine specialists is at an all-time high. Although we no longer see the "general" machinist, this new specialist is going to be asked to push the envelope of manufacturing into the future, even further than was ever imaginable.

Thinking Green

Green Manufacturing

Green manufacturing describes a twofold trend in manufacturing: manufacturing green products and making the process of manufacturing green. Green products are environmentally friendly in their use, and they are durable, free of toxic compounds, and often biodegradable. As part of green manufacturing, green products are produced with less waste and less pollution than with traditional manufacturing practices and are often made from renewable or recycled materials. International attention on environmentally friendly manufacturing using renewable resources has driven a drastic increase in green manufacturing worldwide. "Greening" the manufacturing process by reducing waste and pollution offers economic and environmental benefits and could also provide avenues for innovation and job growth.

Chapter Review

Summary

- The earliest forms of machining were either human- or animal-powered.
- Many forms of manual machining were developed in the late 1700s. These developments led to the industrial revolution of the early 1800s.
- The development of inspection equipment, like micrometers, coincided with the development of machine tools.
- Manual machines were used exclusively from the early 1800s until around 1980.
- Milling machines were designed to machine rectangular or irregularly shaped parts. They operate by fixing the material in place while the tool rotates in a spindle.
- Manual lathes were designed to use a fixed tool to machine a spinning piece of material. This configuration creates round or cylindrical parts.
- Both lathes and mills can be configured in multiple ways, either vertically or horizontally. The defining characteristic for a mill or lathe is whether the material is fixed or rotating.
- Saws can be configured in both horizontal and vertical designs. They are primarily used to reduce large stock size to a size suitable for machining.
- The primary function of a machinist has changed as the industry has changed. Machinists began as machine builders and machine service technicians, but as the need for production increased, manual machinists were necessary in production environments. With the development of new computerized machines, CNC machinists evolved to meet the needs of modern machine shops. Now, the industry is seeing highly specific machining specialists who operate sophisticated machinery.

Review Questions

Answer the following questions using the information provided in this chapter.

Know and Understand

1. The earliest tales of machine tools can be traced all the way back to _____, around 2500 BC.

   A. South America
   B. Egypt
   C. Greece
   D. Rome

2. *True or False?* John Parsons is widely regarded as the father of machine tool technology.

3. Production could not have happened without the ability to _____ and fit pieces together.

   A. cut
   B. grind
   C. measure
   D. mill

4. *True or False?* Manual machines are best described as machines that are operated by physically moving the machine, material, or cutter to shape a desired part.

5. *True or False?* Mills are defined as machines that fix a workpiece in position while the cutter rotates to remove material.

6. What are some components of a milling machine?

   A. Motor, blade, elbow, and ram
   B. Motor, chuck, vise, and ram
   C. Motor, head, knee, and ram
   D. Motor, tool, wheel, and ram

7. *True or False?* The word *lathe* comes from the Dutch language, meaning "to machine."

8. The lathe chuck secures the _____ before cutting.

   A. chuck key
   B. material
   C. turret
   D. vise

9. The saw _____ is typically a thin piece of metal with a series of sharp teeth.

   A. blade
   B. pulley
   C. tensioner
   D. wheel

10. *True or False?* The band saw uses a thin blade that is stretched over a set of wheels or pulleys and tensioned to secure it in place.

11. _____ operations are done after the primary machining is completed.

    A. Mill
    B. Lathe
    C. Saw
    D. Secondary

12. A _____ is used to secure a workpiece in a specific location and orientation, ensuring that all parts produced using the fixture maintain conformity.

    A. drill fixture
    B. lathe chuck
    C. light fixture
    D. machine vise

13. *True or False?* At its most rudimentary form, we use grinding to spin a stone at high speed and cut material.

14. The term *hardened steel* refers to materials that are naturally extremely _____.

    A. expensive
    B. heavy
    C. light
    D. wear-resistant

15. A(n) _____ is simply a person who operates a machine.

    A. designer
    B. engineer
    C. machinist
    D. miller

16. *True or False?* Speeding up the process and accuracy in manufacturing required computers.

17. As transportation, electrification, and industrial improvements swept across the country, so came the need for production _____.

    A. analysts
    B. machinists
    C. tooling
    D. workshops

18. A manual machine controlled by a computer drive and servo motors was first known as numerically controlled (NC), or later as a _____ machine.

    A. completely numerically controlled (CNC)
    B. computerized number controls (CNC)
    C. computer numerically considered (CNC)
    D. computer numerically controlled (CNC)

19. *True or False?* Numerical control (NC) is the automation of machine tools by use of computers.

20. In 1950, John Parsons joined forces with IBM and MIT to develop _____ and servo drives.

    A. servo controls
    B. servomechanisms
    C. servo meters
    D. servo tooling

Apply and Analyze

1. What design changes did James Watt implement to improve the efficiency of the steam engine?

2. What is the main difference between mills and lathes?

3. Describe the basic configuration of a band saw.

4. Differentiate between outside diameter (OD), inside diameter (ID), and surface grinders.

5. What additional equipment did John Parsons use to create the first numerically controlled machine?

Critical Thinking

1. Briefly outline the major historical breakthroughs in machine tool technology. How do today's manufacturing advances compare with earlier innovations?

2. Though the term *machinist* has evolved over time and can refer to various jobs, as discussed in this chapter, what qualities or defining features do you think are shared by all machinists?

3. Briefly describe the relationship between machine development and the availability of inspection equipment. How do advances in each area spur advances in the other?

2 The Skills to Compete

Chapter Outline

2.1 The Modern CNC Environment

2.2 Supplied Industries

2.2.1 Automotive Industry

2.2.2 Medical Technology

2.2.3 Aerospace Applications

2.2.4 Defense

2.2.5 Computing and Technology

2.2.6 Woodworking

2.2.7 Sheet Metal Fabrication

2.3 Careers in CNC Manufacturing

2.3.1 CNC Machine Operator

2.3.2 CNC Machine Setup Operator

2.3.3 CNC Programmer

2.3.4 QC Inspector

2.3.5 Engineer

2.3.6 Manager

2.3.7 Trainer

2.4 Applying for a Position

2.4.1 Your Résumé

2.4.2 Letter of Application

2.4.3 Job Application Forms

2.4.4 The Job Interview

2.5 Succeeding in the Workplace

2.5.1 Health and Hygiene

2.5.2 Work Habits

2.5.3 Time Management

2.5.4 Attitude on the Job

2.5.5 Professional Behavior

2.5.6 Decision-Making and Problem-Solving

2.5.7 Communication Skills

2.5.8 Ethical Workplace Behavior

2.5.9 Interpersonal Skills

2.5.10 Staying Safety Conscious

2.6 Technical Training

2.6.1 Apprenticeships

2.6.2 On-the-Job Training

2.6.3 NIMS

2.6.4 Society of Manufacturing Engineers

2.6.5 National Tooling and Manufacturing Association

2.6.6 Education

Learning Objectives

After reading this chapter, you should be able to:

- Describe the modern CNC environment.
- Explain the impact of manufacturing on gross domestic product (GDP).
- List some of the industries supported by manufacturing.
- Give examples of the multiple career paths in CNC machining.
- Identify and prepare job application materials.
- Describe desirable traits for succeeding in the workplace.
- List common soft skills.
- Describe the technical skills needed to compete in manufacturing.
- Discuss apprenticeships.
- Explain the benefits of on-the-job training.
- Outline the certification processes of NIMS, SME, and NTMA.

Key Terms

aerospace

apprenticeship

attitude

CNC laser

CNC programmer

computer numerically controlled (CNC) machines

ethical behavior

gross domestic product (GDP)

journeyperson

leadership

machine operator

machine setup operator

manufacturing engineer

mechanical engineer

micromachining

National Tooling and Manufacturing Association (NTMA)

negotiation

NIMS

nonverbal communication

on-the-job training (OJT)

punctual

quality control (QC) inspector

reference

résumé

self-motivation

sheet metal fabrication

Society of Manufacturing Engineers (SME)

soft skills

team

verbal communication

water jet

2.1 The Modern CNC Environment

Chapter 1 described the history of the manufacturing industry and traditional machinery. This industry was built with hard work in harsh environments. Old machine shops were often dark, hot sweatshops where machinists worked long hours in dangerous conditions. Early machinists endured overheated shops where they were pelted with hot metal shavings, oil, and other debris from the machines they ran. Heavy steel bars or plates were lifted by hand on and off machines. Unsurprisingly, these were difficult jobs resulting in frequent and often severe injuries. From these beginnings, however, modern manufacturing has come a long way. Where is machining today, and what industries employ machinists?

Chuck Rausin/Shutterstock.com

Figure 2-1. A modern machine shop in operation.

Today, machinists work in an entirely different environment, albeit one that still requires an immense amount of skill and pride, **Figure 2-1**. Modern *computer numerically controlled (CNC) machines* are controlled by computer drives and servo motors. They are technologically advanced pieces of equipment operated by trained and highly skilled technicians. The complexity of the parts and the tolerances to which they are machined require CNC machines to be more specialized than traditional machinery. CNC machines are susceptible to failure in hot or dusty environments, so most machine shops are climate controlled to some degree. The increasingly small tolerances produced also require constant temperature and humidity control.

The most significant improvement in machining is safety, **Figure 2-2**. Since machine operators do not have to move machines manually, almost all CNC machines

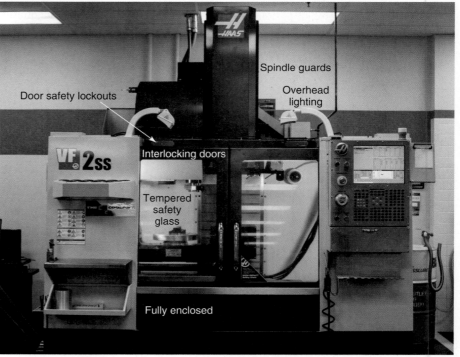

Goodheart-Willcox Publisher

Figure 2-2. A Haas VF-2 with noted safety features.

are enclosed with sheet metal sides and safety glass doors. The enclosures prevent metal shavings from being thrown from the machine and injuring operators, and the doors also lock to prevent an operator from accidentally putting extremities in the cutting area. Today, the oil used to reduce friction has been replaced with water-soluble or synthetic coolants that are safe for the environment as well as machinists. Machines have built-in lighting to give machinists better visibility, eyes are protected by machine enclosures, and machining chips are removed by an auger directly into a bin for recycling.

The current manufacturing job market is very strong with high earning potential. In 2017, there were 12.4 million workers in the manufacturing industry in the United States, accounting for 8.5% of the US workforce. Manufacturing occurs in every state, though some states have a significantly higher percentage of manufacturing jobs than others, **Figure 2-3**. Not only are jobs plentiful, but skilled workers in manufacturing also earn comparatively high salaries. In 2017, average annual compensation for manufacturing jobs was $84,832—well above the average of $66, 847 for nonfarm businesses. A primary reason for the pay differential is the impact of manufacturing on *gross domestic product (GDP)*, or the monetary value of all goods and services produced by residents of a nation during

Percentage of Workforce Employed in Manufacturing

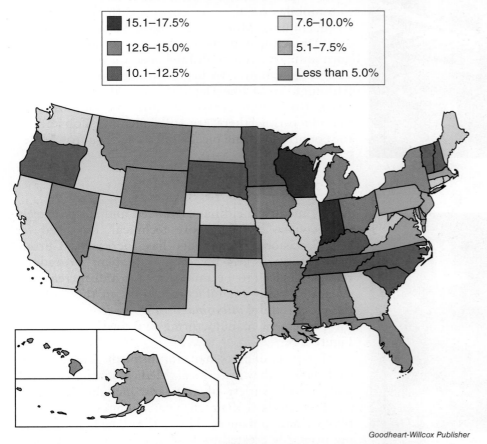

- 15.1–17.5%
- 12.6–15.0%
- 10.1–12.5%
- 7.6–10.0%
- 5.1–7.5%
- Less than 5.0%

Goodheart-Willcox Publisher

Figure 2-3. Manufacturing has an industry presence nearly everywhere, including all states in the United States. However, some areas have a larger manufacturing workforce than others. This map shows the percentage of the state's workforce that is employed in the manufacturing industry, based on data from the Bureau of Economic Analysis.

a specific time period. In 2017, manufacturing alone accounted for over $2 trillion in US GDP. Manufacturing is responsible for nearly 12% of the total output of the United States.

2.2 Supplied Industries

The manufacturing industry has massive impacts on national prosperity, individual workers, and any industry that uses machined parts, tools, or precision fits and tolerances. Nearly all industries rely on machinists. Some of the largest supplied industries include the automotive industry, medical technology, aerospace applications, defense, computing and technology, woodworking, and sheet metal fabrication.

2.2.1 Automotive Industry

The automotive industry is probably the most recognizable manufacturing field. The historically powerful Detroit-based auto industry established the model for mass production, **Figure 2-4**. As you learned in Chapter 1, manufacturing existed long before Henry Ford, but the sheer number of parts required for the production and assembly of automobiles required an entirely different approach to machining. Modern cars have between 800 and 1,200 machined parts in the engine alone. As the auto industry grew so did the need for manufacturers across the country to begin building parts for cars. Although areas like Detroit have suffered in recent years from competitive automakers and offshore manufacturing, there are still machine shops across the country building pieces and parts for automobiles.

OVKNHR/Shutterstock.com

Figure 2-4. Modern automotive assembly line.

2.2.2 Medical Technology

As the medical field has advanced and surgical techniques have become more technical, the need for high-precision components has increased. Current trends toward less invasive, outpatient procedures require devices to be much smaller and increasingly accurate. Many manufacturing jobs in this industry therefore require *micromachining*. Micromachining is the process of manufacturing extremely small parts, **Figure 2-5**.

The types of parts made for the medical industry are as diverse as they are critical. Medical machinists work with a wide variety of materials, including titanium, stainless steels, cobalt chrome, aluminum, and even plastics and polymers. The earliest medical machining primarily created tools such as forceps, clamps, or spreaders, but soon evolved into more advanced medical devices, such as screws and plates to mend broken

science photo/Shutterstock.com

Figure 2-5. Micromachining of printed circuit boards for computer components.

bones, **Figure 2-6.** In the last 50 years, medical advances have led to joint implants and artificial limbs. Medical machinists contribute to everything from hip and knee replacements to artificial heart valves and structural pinning for bone repair.

2.2.3 Aerospace Applications

The term *aerospace* refers to human efforts in design and engineering to achieve flight in the earth's atmosphere or in space. It is truly one of humankind's greatest achievements and adventures. Some of the most impressive advances in machining were spurred by aviation and space travel. For example, solid carbide or insertable cutters were developed to machine exotic aircraft alloys, such as titanium and molybdenum, **Figure 2-7.** Conversely, some of the greatest advances in flight could only have been achieved using precision machining and radical machining techniques. The first computer-controlled machines were developed to make helicopter rotors that were balanced with repeatable precision. Today's multi-axis machines were designed to machine the extremely complicated geometry of today's aerospace industry, **Figure 2-8.**

Critical components in aerospace applications range in size from as small as a grain of rice to as large as a car. It takes a wide variety of machines and skills to produce the critical pieces that fulfill the requirements of the aerospace industry.

Aerospace continues to push the limits of precision and human ability to machine exotic materials. The key to faster flight is increasing speed, or thrust, and decreasing weight, or mass. However, faster travel puts more stress on the materials. This added stress has pushed engineers to use exotic materials, such as titanium and high-strength aluminums. Many old machining techniques cannot be used for these tough materials. Instead, machinists must employ high-speed spindle feeds and develop new cutting tool technology. The result is that the aerospace industry can be one of the most challenging and exciting machining fields.

2.2.4 Defense

Manufacturing has always played a massive role in national defense. Between 1885 and 1914, the United States experienced tremendous manufacturing growth and prosperity. As World War I (WWI) started, this growth meant that the United States initially was the largest supplier of weapons and ammunition to European war efforts. The US capability to machine and

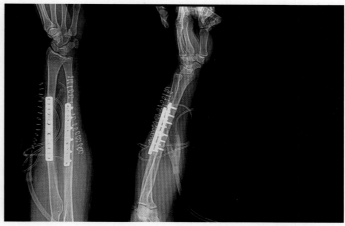

Praisaeng/Shutterstock.com

Figure 2-6. X-ray view of medical plate and screw placement for bone repair.

Anatoliy Lukich/Shutterstock.com

Figure 2-7. USAF F-22 Raptor in flight.

Castleski/Shutterstock.com

Figure 2-8. NASA space shuttle orbiting Earth.

manufacture parts for the defense sector led to stockpiling munitions as the United States entered WWI in 1917 and continued to support the Allies throughout the world.

After WWI ended in 1918, the United States began a major push to build aircraft of all shapes and sizes as well as naval vessels. Once again, machinists were asked to push the envelope of conventional knowledge and machine capacities. Production became increasingly precise and larger in size. Where machinists were once asked to make rifles and bullets, they were now making tanks, high-speed aircraft, and ships capable of becoming small cities. As engineering advanced, so did the requirements for skilled machinists.

By the beginning of World War II, American machinists were building battleships for the Allies in Europe. In 1947, President Harry S. Truman organized and created the Department of Defense (DOD) with the purpose of preparing for and protecting against future threats. Mass production techniques developed by Henry Ford and others were put in place to meet the demands of the new DOD.

Today, as always, there is ongoing demand in the defense industry (and beyond) for advances in technology and engineering capabilities. The defense industry is building helicopters, drones, nonlethal laser-guided weapons, and aircraft capable of withstanding speeds in excess of 1,200 miles per hour (mph). Defense machinists use exotic metals and alloys to build ships, submarines, and long-range missile systems that function in the harshest conditions, **Figure 2-9.**

Steve Allen/Shutterstock.com

Figure 2-9. The aircraft carrier HMS Queen Elizabeth docked in Portsmouth, England.

2.2.5 Computing and Technology

The effect of CNC machining on computing and technology industries is often overlooked, but the beginning of the modern computer age drove the manufacturing field to its current state of health. Everything from cell phone cases, plastic bodies for laptops, circuit boards, and even CNC machines themselves are all products of advanced CNC machining, **Figure 2-10.**

Think about the amazing cycle of building CNC machinery. As engineering advancements require faster, larger, and more accurate machines, the industry turns to its own machinists and designers to create the newest, most complex machines. Gene Haas, the founder of Haas Machines, started his own machine shop because he needed a better indexer for his jobs. Then Haas began building indexers for sale, and eventually he realized he could assemble a team to build his own CNC machines. The need to build a better tool led to building some of the best machinery in the manufacturing industry.

Similarly, as computing technology advances, innovative machining techniques are required to build new products. There is a story about Steve Jobs, the longtime CEO of Apple, bringing a rounded stone from the beach to Apple headquarters and declaring this was the feel he wanted for the Apple iPhone. The designers decided

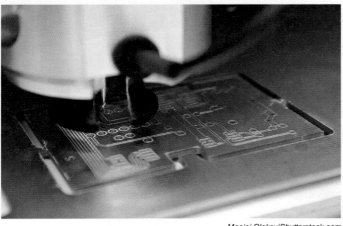

Maciej Oleksy/Shutterstock.com

Figure 2-10. Precision cutter being used to machine information technology components.

on a solid machined piece of aluminum to get the grip and feel for the iPhone body, **Figure 2-11**. Today, iPhone enclosures are still machined from aluminum pieces in CNC machines. Many internal components like boards and switches are machined on CNCs as well.

2.2.6 Woodworking

Machines for woodworking have been around for hundreds of years. Woodworkers are respected for their skilled use of hand tools to create decorative and intricate designs, but the woodworking industry is one of the most impacted by CNC technology. While there still is a high demand for handcrafted furniture, the advent of CNC routers and CNC lathes to create pieces of furniture has revolutionized furniture making. The mass production of repetitive pieces in drawers, cabinets, and doors has minimized cost and improved function. The ability to create complicated joints and joining features with high speed, accuracy, and repeatability has driven cost down significantly for the consumer, making sturdy, durable furniture affordable for all.

Today, routers are attached to gantry-style machines and programmed to move in three axes of movement, virtually replicating the CNC mill, **Figure 2-12**. Edge finishing and complicated design work can now be done in seconds instead of hours. For example, Louisville Slugger, one of the leading baseball bat makers, can create 5,000 bats per day—or over 1.8 million per year—using CNC lathes. Much like the rest of the manufacturing world, the woodworking industry has experienced tremendous growth as a result of CNC technology and machinery.

2.2.7 Sheet Metal Fabrication

Sheet metal fabrication is the process of cutting, forming, and bending pieces from flat sheet metal into a usable shape, **Figure 2-13**. This can be one of the most complicated processes in metal fabrication. Prints define the finished shape, but the part must be laid out flat, precision cut, and bent to match printed specifications. The bending process stretches the metal, depending on material thickness and type, and those deviations must be accounted for.

The sheet metal industry has been most creative in building and designing CNC machinery. For many years, handheld plasma cutters were the standard method for cutting sheet metal. Alternatively, machines called shears could slice the sheet metal along layout lines. Both methods created lots of metal waste with variable accuracy, depending on the operator. With the implementation of CNC technology, sheet metal

mama_mia/Shutterstock.com

Figure 2-11. Solid-bodied aluminum phone structure.

Azami Adiputera/Shutterstock.com

Figure 2-12. Cabinetmaker using a CNC router to cut cabinet doors.

zilber42/Shutterstock.com

Figure 2-13. A brake press operator bending sheet metal forms.

designers began using CNC lasers and water jets. A **CNC laser** is similar to a router or CNC mill, but it uses a laser cutting head to precision cut shapes to tolerances as close as ±0.002″. A **water jet** is similar to a laser, but it uses high-pressure water, up to 20,000 pounds per square inch (psi) to cut material up to 8″ thick. Both processes have faster speeds and minor material losses compared with earlier techniques, resulting in greater efficiency and cost reductions.

2.3 Careers in CNC Manufacturing

Within every industry influenced by CNC, there are additional career opportunities. Careers in CNC manufacturing cover a wide range of skills and abilities. CNC machine operators, technicians, programmers, designers, inspectors, engineers, managers, and even teachers are needed in the rapidly growing manufacturing field. As this field becomes more technical, your ability to compete will be directly related to maintaining your education and specialized training.

2.3.1 CNC Machine Operator

A **machine operator** is typically an entry-level employee responsible for the day-to-day operation of a CNC machine, **Figure 2-14**. Machine operators perform daily maintenance, including checking all fluid levels and initiating pre-operation and start-up procedures. Operators must know how to properly load material into the work area and have a working knowledge of any fixtures or tooling being used. They must also be able to evaluate cutting tools for wear or damage. The most important task for an operator is inspecting products as they are made. Operators are qualified to make adjustments to machines or tools to maintain the quality of the parts produced.

Goodheart-Willcox Publisher

Figure 2-14. A CNC mill operator performing a machining operation.

2.3.2 CNC Machine Setup Operator

The *machine setup operator* is responsible for ensuring the machine, tooling, and work holding are efficient and accurate to make parts as programmed. Machine setup operators connect the programmer's concept with the actual machine function. The setup operator installs cutting tools and sets their offsets, as well as establishing the origin of the work coordinate system for the parts being machined. Machine setup operators also install any fixturing or work holding into the machine tool and align it as necessary. Often the setup operator is an experienced machine operator with a working knowledge of the machines and the process of machining.

2.3.3 CNC Programmer

A *CNC programmer* writes the code, or instructions, to a CNC machine to machine finished parts, **Figure 2-15.** CNC programmers program CNC machines that turn raw materials like wood, plastics, and metals into usable components. This job involves reading blueprints and understanding the dimensions of the given component in order to properly program machines. In today's manufacturing environment, programming normally requires some knowledge of a computer-aided manufacturing (CAM) system, such as Mastercam, Esprit, Gibbs, or an equivalent system. A programmer is a highly skilled and experienced machinist. Programmers must read and understand machining requirements given in a print, evaluate the required tooling, and make determinations about fixturing and process order. These factors must be correct to machine an accurate part. Programmers must also consider time and cost to make programs as efficient and repeatable as possible.

FERNANDO BLANCO CALZADA/Shutterstock.com

Figure 2-15. A CAD/CAM operator preparing a file for CNC programming.

2.3.4 QC Inspector

Inspecting a finished product is an important step in modern manufacturing. *Quality control (QC) inspectors* work directly with manufacturing teams to make sure products meet form, fit, and function requirements. QC inspectors must possess knowledge of the machining process and an acute attention to detail. The inspector is the last line of defense in preventing the distribution of parts that are not to specification. In modern manufacturing, QC inspectors are also operators of highly advanced inspection machines, called coordinate measuring machines (CMM). CMMs operate similar to a CNC, but they are used to inspect parts to dimensions as close as 0.00005".

2.3.5 Engineer

Engineers are directly involved in the design and manufacturing of products and require an advanced degree. Engineers use math and science to develop new products or manufacturing processes. There are many subfields in engineering, but mechanical and manufacturing engineers are most common in the manufacturing industry. *Mechanical engineers* are responsible for machine design or modification, and they often work on new designs or improving existing equipment. *Manufacturing engineers*, or *tooling engineers*, often work to improve the process of making the finished products. Modifying machines or changing the placement of machines on the shop floor can often make a tremendous impact on the manufacturing process.

2.3.6 Manager

The manufacturing industry requires managers who can organize, plan, and supervise the manufacturing of finished goods. It often takes someone with good organizational skills to maximize labor, machines, and capital. The manager is the problem solver who evaluates all the resources available and supports the manufacturing team. Jobs at this level often require a bachelor's degree or higher, but many successful managers have direct experience from holding previous jobs as operators, setup personnel, or programmers on the machine shop floor.

2.3.7 Trainer

Like the rest of the manufacturing industry, the education sector is rapidly changing. Today more than ever, qualified, experienced CNC machinists are needed to train the next generation of production workers. Trainers work in a variety of roles, including instructors at schools, industry trainers, and professional trainers. The demand for instructors is at an industry high and appears to be growing. The best CNC machining instructors will come from inside the industry. Although some trainer positions do require a degree, often the requirements are industry experience and state licensing.

2.4 Applying for a Position

Now that you have examined several different manufacturing fields and career paths, you should be starting to understand the range of possibilities your future holds in machining. So, how do you get started? Just as there are many different fields and career paths, there are also many different ways to get into this industry. Before you decide *how* to get into the manufacturing industry and *what* you want to accomplish, you must first consider what your future employer wants to see in you and how to present yourself as a qualified applicant.

When you are ready to apply for employment, you will need to know the appropriate steps to take. Having a well-prepared résumé is an important first step. You will also need to organize and update your portfolio. Knowing how to write an acceptable letter of application is another goal. Finally, you will want to practice your interviewing techniques.

2.4.1 Your Résumé

A *résumé* is a brief outline of your education, work experience, and other qualifications for work. A well-written résumé can help you get an interview. You will need to include several sections on your résumé (for an example, see **Figure 2-16**). Make sure that your résumé is precise and error-free. When an employer requests you send a résumé via traditional mail, make sure you print your résumé on high-quality, neutral-colored paper (white, gray, or cream colors).

Along with the résumé, you need to develop a list of references. A *reference* is an individual who will provide important information about you to a prospective employer. A reference can be a teacher, school official, previous employer, or any other adult outside your family who knows you well.

You will need at least three references. Always get permission from each person to use his or her name as a reference before actually doing so. Your list of references, along with their titles, phone numbers, and addresses, should be kept private. Share this list only with an employer who has interviewed you and asks for your references.

You can also have your references write *letters of recommendation* for you. These give an employer a more in-depth look at your skills. Choose people who know you well. Make sure you choose references who are good writers, since they will be representing you. Ask as many people as possible—then you can choose the best letters to submit to employers.

2.4.2 Letter of Application

The letter of application is often the first contact you have with a potential employer. It can make a lasting impression. It should be neat and follow a standard form for business letters. If you are applying by traditional mail or submitting your letter in person, the paper you use should be ivory, white, or a neutral color and free of smudges and mistakes. Use a standard font to give the letter a professional look, be sure to check spelling and punctuation, and have several people read the letter and offer advice for improving it. You should include a copy of your résumé with your letter of application.

A sample letter of application appears in **Figure 2-17**. It provides a good example for responding to a job ad. The letter should be brief and to the point. It should include the following items:

- Title of the job you seek
- Where you heard about the job
- Your strengths, skills, and abilities for the job
- Reasons you should be considered for the job
- When you are available to begin work
- Request for an interview

Gina D. Pate

543 Itasca Street
Covington, TX 76055

(212) 555-1234
gpate@e-mail.com

Career Objective
To obtain an entry-level position in the CNC manufacturing field.

Work Experience
FRC Screw Products, Grand Prairie, TX   September 2017–present
Machine Operator
- Operated manual lathe.
- Performed part inspection and quality control.
- Lead member of safety team.

Builder's Supply, Arlington, TX   March 2017–September 2017
Receiver/Stocker
- Unload, store, and stock merchandise.
- Operated fork trucks and hand trucks.

Tony's Grocery Store, Burleson, TX   May 2016–March 2017
Cashier
- Operated checkout scanner and bagged groceries.
- Answered customer questions.
- General stocking and cleaning throughout store.

Education
Certificate Diploma in CNC Machining Technology   May 2018
Washington College of Technology
- GPA: 3.22/4.0
- Coursework included residential blueprint reading, G & M code programming, and CAM programming.
- Participated in NTMA student chapter.

Community Service
Habitat for Humanity, volunteer, summers of 2016, 2017, 2018
Presbyterian Night Shelter, volunteer, 2015–present

References
Available upon request.

Figure 2-16. Sample résumé.

Gina D. Pate
543 Itasca Street
Covington, TX 76055
(212) 555-1234
gpate@e-mail.com

April 23, 2020

Mr. James Williamson
Williamson Machining
4392 East 134th Street
Grand Prairie, TX 75052

Dear Mr. Williamson:

The entry-level CNC Machinist position you advertised on the Career Finder website is exactly the type of job I am seeking. After reviewing the job description and requirements, it was clear that my experience, skills, and interests are a perfect match for this opportunity. I would be available to begin working at Williamson Machining after providing reasonable notice to my current employer.

While obtaining my diploma in CNC Machining Technology from Washington College of Technology, I gained both the theoretical knowledge and the hands-on skills required for this position. In my current position as a machine operator at FRC Screw Products, I have developed strong customer service skills and gained a better understanding of the machining industry. I am eager to apply the skills I have learned and to continue gaining new skills.

Please find my résumé enclosed with this letter. I would greatly appreciate an opportunity to interview for this position. Please contact me at your convenience by phone or e-mail to schedule an interview. I look forward to hearing from you.

Sincerely,

Gina D. Pate

Gina D. Pate

enclosure

Figure 2-17. Sample job application letter.

2.4.3 Job Application Forms

A prospective employer may ask you to complete a job application form before having an interview. The job application form highlights information the employer needs to know about you, such as your education and prior work experience. Employers often use these forms to screen applicants for the skills needed on the job. You might complete a form in a personnel or employment office. Sometimes you may get the form by mail or online. When filling out an online application, it is extremely important to include key terms for which the employer may search. This will help you stand out from the many other applications the employer will receive.

The appearance of the application form can give an employer a first impression of you. Fill out the form accurately, completely, and neatly. How well you accomplish that can determine whether you get the job. When asked about salary, write *open* or *negotiable*. This means you are willing to consider offers.

Be sure to send or give the form to the correct person or department. This information often appears on the form. Tips for completing the job application appear in **Figure 2-18**.

When preparing your application online, be sure to save it in the appropriate format. If a preferred format is not given, it is best to save the application in document or pdf file format. This will enable the employer to find specific search terms in your document. Be sure to complete all the fields of the application. Many job-search sites have sample forms on which you can practice before attempting a real application.

2.4.4 The Job Interview

The interview gives you the opportunity to learn more about a company and to convince the employer that you are the best person for the position.

Tips for Completing a Job Application

✓ Follow the instructions for filling out the form. If you are asked to print, type, or use black ink, be sure to do so. Be as neat as possible.

✓ Complete every question in the form. If some questions do not apply to you, draw a dash or write "NA" (not applicable) so the employer knows you did not overlook it.

✓ You may wish to omit your Social Security number and write "will provide if hired."

✓ You can write "open" or "negotiable" for any question regarding salary requirements.

✓ For each former job, there may be a question asking your reason for leaving the job. Avoid writing any negative comments about yourself or a former employer.

Goodheart-Willcox Publisher

Figure 2-18. Tips for completing a job application.

The employer wants to know if you have the skills needed for the job. Adequate preparation is essential for making a lasting, positive impression. Here are some ways to prepare for an interview.

- **Research the employer and the job.** Know the mission of the employer and specifics about the job. Also, try to learn what the company looks for when hiring new employees.

- **Be prepared to answer questions.** Go over the list in **Figure 2-19** and prepare answers for each question.

- **List the questions you want answered.** For example, do you want to know if there is on-the-job training? Are there opportunities for advancement?

- **List the materials you plan to take.** This seems simple enough. However, if you wait to grab items at the last minute, you will likely forget something important.

- **Decide what to wear.** Dress appropriately, usually one step above what is worn by your future coworkers. For instance, casual clothing is acceptable for individuals who will do manual labor or wear a company uniform. If the job involves greeting the public in an office environment, a suit is more appropriate. Always appear neat and clean.

- **Practice the interview.** Have a friend or family member interview you in front of a mirror until you are happy with your responses.

- **Know where to go for the interview.** Verify the address of the interview location by checking the site beforehand, if possible. Plan to arrive ready for the interview at least 10 minutes early.

Common Interview Questions and Responses	
Question	**Response**
What can you tell me about yourself?	Briefly summarize your abilities as they relate to the job qualifications or your career goals. Do not provide a general life history.
Why do you want to work for this company?	Describe what you know about the company. Explain how your abilities match the company's needs.
Why do you think you would like this kind of work?	Relate the job requirements to your successful past experiences.
What are some of the projects you worked on in school?	Briefly summarize a project or coursework relevant to the job qualifications.
What other jobs have you had?	Focus on jobs with skills that relate to the jobs you are seeking.
Why did you leave your last job?	Be honest. However, avoid saying anything negative about your previous employer.
Have you ever been fired from a job? If so, why?	Answer honestly. If you have been fired, share what you learned from the experience. Avoid trying to blame others.
What are your major strengths and weaknesses?	Select a strength that relates to the job qualifications. Be honest when selecting a weakness, but give an example of how you've worked to improve on it.
Have you ever had a conflict with a coworker? How did you handle it?	Briefly describe the situation and how you handled it. Avoid placing all the blame on the other person. Explain what you learned from the experience.
What do you expect to be paid?	If possible, determine the salary range before the interview. Say that you are willing to discuss the salary or state a range you feel comfortable with.
What are your future plans?	Describe how the need to learn and grow is important to you. Confine your answer to the company with which you are interviewing.

Goodheart-Willcox Publisher

Figure 2-19. Common interview questions and sample responses. Take time to review possible interview questions, including additional questions related to your field, and prepare thoughtful answers.

Good preparation will make you feel more confident and comfortable during the interview. Be polite, friendly, and cheerful during the process. Use a firm handshake and maintain eye contact at all times. Avoid chewing gum and fidgeting. Answer all questions carefully and as completely as you can, and be honest about your abilities. Also be aware of questions you legally do not have to answer, such as those related to age, marital status, religion, or family background.

A prospective employer may ask you to take employee tests. Some employers administer tests to job candidates to measure their knowledge or skill level under stress. Since all employers support a drug-free workplace, most will likely require you to take a drug test if you are hired. You can ask those who have completed similar tests what to expect.

After the interview, send a letter to the employer within 24 hours thanking him or her for the interview. If you get a job offer, respond to it quickly. If you do not receive an offer after several interviews, evaluate your interview techniques and seek ways to improve them.

2.5 Succeeding in the Workplace

All potential employers want to know about your technical skills and education, but equally important attributes for an employer include *soft skills*. Soft skills are desirable qualities in an employee that are distinct from technical training or specific subject-matter expertise. Soft skills include communication skills, leadership skills, and other interpersonal skills.

After securing employment, adjusting to your new duties and responsibilities will occupy your first few weeks. Your supervisor and coworkers will help you learn the routine. An introduction to company policies and procedures as well as the particular safety rules that all employees must know is common for new employees.

Not only will your coworkers be watching what you do, they will also pay attention to "how you work." How to behave in the workplace is an important lesson all employees should learn. Making an effort to do your best will help you succeed.

 From the Shop

Success in the Shop
Technical training is a very important part of your career. You should always continue to grow your technical skills and advance your career through ongoing training. But, it's also just as important to be a good employee. Being a team player who's willing to do the extra work and help out your team will bring you added success. Every machine shop owner needs that person who is always willing to do what it takes to keep the business successful.

2.5.1 Health and Hygiene

As an employee, you are a representative of your company. Therefore, your employer expects you to be neat and clean on the job. Taking care of yourself gives the impression that you want people to view you as a professional.

Your daily grooming habits should consist of bathing or showering, using an antiperspirant, and putting on clean clothes. Regularly brushing your teeth and using mouthwash will promote healthy teeth and fresh breath. Keep your hair clean and styled in a way that will not be distracting.

Employers expect workers to dress appropriately. Many places of work have a dress code. If your workplace does not, use common sense and avoid extremes. Refrain from wearing garments that are revealing or have inappropriate pictures or phrases. Some employers have rules requiring that tattoos or piercings (apart from pierced ears) remain covered. Good appearance is especially important for employees who have frequent face-to-face contact with customers.

2.5.2 Work Habits

Employers want employees who are punctual, dependable, and responsible. They want their employees to be capable of taking initiative and working independently. Other desirable employee qualities include organization, accuracy, and efficiency.

A *punctual* employee is always prompt and on time. This means not only when the workday starts, but also when returning from breaks and lunches. Being dependable means that people can rely on you to perform your duties and meet your deadlines. If you are not feeling well, be sure to call in and let your employer know right away. If there are reasons you cannot be at work, discuss this with your employer and work out an alternative arrangement. Many people have lost jobs by not checking with their supervisor about time off.

Taking *initiative* means that you start activities on your own without being told. When you finish one task, you do not wait to hear what to do next. Individuals who take initiative need much less supervision. They have *self-motivation*, or an inner urge to perform well. Generally, this motivation will drive you to set goals and accomplish them. All these qualities together show that you are capable of working *independently*.

You are expected to be as accurate and error-free as possible in all that you do. This is why you were hired. Complete your work with precision and double-check it to assure accuracy. Your coworkers depend on the careful completion of your tasks.

2.5.3 Time Management

A good employee knows how to manage time wisely. This includes ability to prioritize assignments and complete them in a timely fashion. It also involves not wasting time. Time-wasting behaviors include visiting with coworkers, making personal phone calls, texting, sending e-mails, or doing other non-work activities during work hours.

While it is important to complete all your work thoroughly, you must also be able to gauge which assignments are most important. Avoid putting excessive efforts into minor assignments when crucial matters require your attention. Even though you are still accomplishing work, this is another way of wasting of time.

2.5.4 Attitude on the Job

Your attitude can often determine the success you have at your job. Your *attitude* is your outlook on life. It is reflected by how you react to the events and people around you. A smile and courteous behavior can make customers and fellow employees feel good about themselves and you. Clients and customers prefer to do business in friendly environments. Being friendly may take some effort on your part, but it does pay off.

Enthusiasm spreads easily from one person to another. Usually, enthusiasm means a person enjoys what he or she is doing. In a sales environment, enthusiasm increases sales. In an office, enthusiasm builds a team spirit for working together.

People who do a good job feel pride in their work. They feel a sense of accomplishment and a desire to achieve more. This attitude can inspire others as well.

2.5.5 Professional Behavior

You will be expected to behave professionally on the job. This includes showing respect for your boss and coworkers. Limit personal conversations and phone calls to break times or lunch. Act courteously; remember that others are focusing on their work. Interruptions can cause them to lose concentration.

Part of behaving professionally is responding appropriately to *constructive criticism*. Every employee, no matter how knowledgeable or experienced, can improve his or her performance. If you receive criticism from a supervisor or coworker, do not be offended. Instead, use the feedback to improve yourself. The more you improve, the more successful you will be in your work.

2.5.6 Decision-Making and Problem-Solving

Employers value workers who have the ability to make sound decisions. This process applies in the workplace as well as other aspects of life. The process will help you identify an issue and possible solutions, make a decision, implement the decision, and evaluate the results.

Having the ability to solve problems on the job shows an employer that you are able to handle more responsibility. Solving problems as a group can strengthen camaraderie and help employees feel more pride in their work.

The ability to make decisions and solve problems requires *critical-thinking skills*. These are higher-level skills that enable you to think beyond the obvious. You learn to interpret information and make judgments. Supervisors appreciate employees who can analyze problems and think of workable solutions.

2.5.7 Communication Skills

Communicating effectively with others is important for job success. Being a good communicator means that you can share information well with others. It also means you are a good listener.

Good communication is central to the smooth operation of any business. Communication is the process of exchanging ideas, thoughts, or information. Poor communication is costly to an employer, such as when time is lost because an order was entered incorrectly. Poor communication can result in lost customers, too.

The primary forms of communication are verbal and nonverbal. *Verbal communication* involves speaking, listening, and writing. *Nonverbal communication* is the sending and receiving of messages without the use of words. It involves *body language*, which includes the expression on your face and your body posture.

Listening is an important part of communication. If you do not understand what is being said, be sure to ask questions. Also give feedback to let others know you understand them and are interested in what they have to say. Leaning forward while a person is talking signals interest and keen listening. Slouching back in a chair and yawning give the opposite signal—that you are bored and uninterested.

Communication tools have advanced with the development of new technologies. To be an effective employee, you need to know how to communicate well with the common tools of your workplace. For example, when sending e-mail communications, remember to think through each message as you would before sending a letter by post. Messages are often sent quickly without thought of how the recipient may interpret them. The same is true of voicemail.

The development of good communication skills is an ongoing process. Attending communication workshops and practicing regularly can keep your skills sharp. You should periodically give yourself a communications checkup by asking your supervisor to suggest areas that need improvement.

2.5.8 Ethical Workplace Behavior

Ethical behavior on the job means conforming to accepted standards of fairness and good conduct. It is based on a person's sense of what is right to do. Individuals and society as a whole regard ethical behavior as highly important. Integrity, confidentiality, and honesty are crucial aspects of ethical workplace behavior. *Integrity* is firmly following your moral beliefs.

Unfortunately, employee theft is a major problem at some companies. Such theft can range from carrying office supplies home to stealing money or expensive equipment. Companies should have policies in place to address these concerns. In cases of criminal or serious behavior, people may lose their jobs. If proven, the charge of criminal behavior stays on the employee's record. Such an employee would have a difficult time finding another job.

2.5.9 Interpersonal Skills

Interpersonal skills involve interacting with others. Some workplace activities that involve these skills include training others, leading, negotiating, and working as a member of a team. Getting along well with others can require great effort on your part, but it is essential for accomplishing your employer's goals.

Teamwork

Employers seek employees who can effectively serve as good team members. Due to the nature of most work today, teamwork is necessary. A *team* is a small group of people working together for a common purpose. Cooperation often requires flexibility and a willingness to try new ways to get things done. If someone is uncooperative, it takes longer to accomplish the tasks. When people do not get along, strained relationships may occur, which get in the way of finishing the tasks.

A big advantage of a team is its ability to develop plans and complete work faster than individuals working alone. In contrast, a team usually takes longer to reach a decision than an individual worker does. Team members need some time before they become comfortable with one another and learn to function as a unit. You will be more desirable as an employee if you know how to be a team player.

Creative ideas often develop from building on another person's idea. Honesty and openness are essential. Also, trying to understand the ideas of others before trying to get others to understand your ideas is an effective skill to develop.

Leadership

All careers require leadership skills. *Leadership* is the ability to guide and motivate others to complete tasks or achieve goals. It involves communicating well with others, accepting responsibility, and making decisions with confidence. Those employees with leadership skills are most likely to be promoted to higher levels.

Leaders often seem to carry the most responsibility of a group. Other group members look to them for answers and direction. The most important role of leaders is to keep the team advancing toward its goal. Leaders do this by inspiring their groups and providing the motivation to keep everyone working together.

Good leaders encourage teamwork because a team that is working together well is more likely to reach its goals. They listen to the opinions of others and make sure all team members are included in projects. Leaders also want to set a good example by doing a fair share of the work. In these ways, leaders cultivate a sense of harmony in the group.

Negotiation

Often there are times when employees and employers must negotiate on a task or work-related issue. *Negotiation* is the process of agreeing on an issue that requires all parties to give and take. The goal is a "win-win" solution in which all parties get some or all of what they are seeking.

Negotiation begins with trying to understand the other party's or parties' interests. Possible solutions that meet their mutual concerns can be developed. Often the best solution becomes clear when all parties have ample time to explain what they are trying to accomplish.

2.5.10 Staying Safety Conscious

Safety on the job is everyone's responsibility. Many workplace accidents occur because of careless behavior. Poor attitudes can often cause unsafe behavior, too. Common causes of accidents include the following:

- Taking risks
- Showing off
- Forgetting safety details
- Disobeying company rules
- Daydreaming
- Losing your temper
- Falling asleep.

Practicing good safety habits is essential for preventing accidents and injuries on the job. Keep in mind that a healthy worker is more alert and less likely to make accident-prone mistakes. Knowing how to use machines and tools properly is the responsibility of both the employer and employees. Wearing protective clothing and correctly using safety equipment help keep workers safe. Your employer will emphasize the safety practices that employees must follow in your workplace.

The government agency that promotes safety in the workplace is the Occupational Safety and Health Administration (OSHA). You will be required to follow the specific OSHA regulations that apply to your workplace.

2.6 Technical Training

Once you have the soft skills you need and a career plan, how are you going to acquire technical skills? This will be an important decision based on how much time you can invest and the depth of the skills you want to obtain. While all of the following options are a good start, recognize that your education and skills will constantly need to improve as the industry changes. Take a look at some of the ways you can gain the technical skills and certifications you will need to succeed.

2.6.1 Apprenticeships

This may be one of the oldest forms of training. An *apprenticeship* refers to an instructional program of on-the-job training under the supervision of a journeyperson. The term *journeyperson* refers to a skilled laborer who has completed an apprenticeship. In practical terms, this means that an apprentice works directly with a skilled laborer and learns a trade under supervision. This process can take 2–5 years to complete and usually consists of some textbook or research-type curriculum.

Most apprentices are paid for their work. They are often assigned tasks machining parts or doing secondary operations. Companies invest in apprentices as potential long-term employees. Apprentices gain experience working side-by-side with a skilled machinist. Learning a trade from an established journeyperson with years of practical experience is a major benefit to the apprentice system. Unfortunately, apprenticeships are becoming much harder to find. Companies want to hire skilled workers or beginning operators, but many companies do not want to make the long-term commitment to building their own workforce.

2.6.2 On-the-Job Training

On-the-job training (OJT) has been the accepted method of training in manufacturing over the past few decades. Companies typically hire employees that have the soft skills they are looking for and rely on more advanced employees to train them on their products and machines. OJT is not a formalized training environment and most often has no curriculum or guidelines. OJT can be cost-efficient for a company in the short term because little to no time is invested initially in a new employee. Often, the employee will be assigned to a machine that is already running smoothly and given only basic instructions. But, because there are no benchmarks

 Safety Note

Staying safe is not just about you. Often, the person injured in an accident is a coworker. Unsafe conditions in a machine shop can cause materials or tools to be projected into someone else's work space. Make sure that you and your neighbor stay safe. Don't hesitate to speak up to a coworker and communicate any unsafe condition(s) you observe.

for advancement, an operator can have a difficult time advancing to the next level of training.

2.6.3 NIMS

NIMS formed in 1995 to develop a set of credentials for metalworking professionals. NIMS has since established a recognized set of standards developed by industry in operational areas including metalforming, machining, and industrial maintenance. The standards range from entry- to master-level and require hands-on performance tasks and an online theory test. Training programs use NIMS credentials as performance measures of attainment, often incorporating the credentials as completion requirements. The credentials may then form a basis for articulation among training programs. Industry also uses the credentials to recruit, hire, place, and promote individual workers.

NIMS also offers a competency-based apprenticeship program. The NIMS apprenticeship system represents a dramatic departure from traditional time-based programs and integrates NIMS standards and skill certifications in defining and measuring required competencies. The NIMS apprenticeship program was developed in partnership with the United States Department of Labor (DOL). More than 300 companies participated in the deliberations and design of the program. Apprentices earn a DOL industry-recognized national credential after completing the program, and they can expect higher earnings and more advancement potential as compared with non-apprentices.

2.6.4 Society of Manufacturing Engineers

The *Society of Manufacturing Engineers (SME)* has long provided training for the manufacturing industry. Typically, this training has been directed at manufacturing engineers or manufacturing technicians with extensive formal training and SME certifications. Most recently, SME has added Tooling-U, PRIME, and a wide range of other online courses that teachers can use in the classroom and machinists can use for professional development. SME and organizations like it recognize the manufacturing industry's need to further develop the current workforce and inspire entry-level machinists.

2.6.5 National Tooling and Machining Association

The *National Tooling and Machining Association (NTMA)* has also created an online machining education program. In 2011, NTMA launched NTMA-U. This online training course for machinists is open to anyone who is interested. NTMA has also been a partner, along with NIMS and other organizations, in growing and supporting the manufacturing industry. NTMA training is eligible for college credit through the University of Akron and is approved through the Federal Bureau of Apprenticeship Training.

2.6.6 Education

Traditional education has long supported the manufacturing industry in high schools, junior colleges, trade schools, and even four-year universities. Traditional education is still the best resource for training future machinists as well as improving the skills of the current workforce. Different types of schools provide different advantages. Some students want to pursue a four-year degree and so choose the university path. Other students opt for a shorter, more intense program at a technical or trade school. All paths can be beneficial, depending on the needs of the individual student.

In this chapter, you have explored the modern manufacturing environment, the scope of work in the manufacturing field, the skills needed to compete, and the training or certification required. As you can see, manufacturing is a vast field with tremendous opportunities just waiting for the next wave of machinists to move it forward.

Chapter Review

Summary

- CNC machines are specialized pieces of equipment that are operated by highly trained and skilled technicians.

- In 2017, there were 12.4 million workers in the manufacturing industry in the United States. Roughly 8.5% of all jobs were in manufacturing.

- CNC manufacturing supports industries such as automotive, medical technology, aerospace, defense, computing and technology, woodworking, and sheet metal fabrication.

- Career paths in CNC manufacturing can include machine operator, machine setup operator, programmer, QC inspection, manufacturing or mechanical engineer, manager, or trainer.

- Knowing how to prepare job application materials is essential for securing employment.

- The soft skills required to compete in the modern workforce include good health and hygiene, time management, positive attitude, professionalism, decision-making and problem-solving, communication, good ethics, teamwork, leadership, negotiation, and safety consciousness.

- Many technical skills are required in manufacturing. Machinists must be detail-oriented, willing to learn, and trained on multiple types of machines and operations.

- An apprenticeship refers to an instructional program of on-the-job training under the supervision of a journeyperson. An apprentice works directly with a skilled laborer to learn a trade.

- On-the-job training (OJT) is a long-accepted method of training in manufacturing. Companies typically hire employees that have the soft skills they are looking for and rely on more advanced employees to train them on specific products and machines.

- Organizations such as NIMS, SME, and NTMA all offer certifications at multiple levels for manufacturing and CNC machining.

Review Questions

Answer the following questions using the information provided in this chapter.

Know and Understand

1. With the advent of _____ machines, machinists no longer have to move the machine handles by hand.

   A. automatic lathe   C. CNC
   B. boring   D. milling

2. *True or False?* Today's CNC machines are highly specialized pieces of equipment that can be operated by unskilled workers.

3. *True or False?* The automobile industry is generally unconnected to machining and manufacturing.

4. As the medical field has advanced and surgical techniques have become more technical, the need for _____ components has increased.

   A. automatic   C. steel
   B. precision   D. titanium

5. *True or False?* Some of the greatest advances in machining have been brought about by our exploration of aviation and space travel.

6. _____ continues to push our limits of precision and our ability to machine exotic materials.

   A. Aerospace
   B. Lathing
   C. Milling
   D. Information technology

7. Manufacturing has always played a massive role in the _____ of our country.

   A. abilities
   B. defense
   C. farming
   D. lathing

8. *True or False?* The modern computer age was a key factor in the advancement of manufacturing improvements over the last 30 years.

9. While there still is a high demand for handcrafted furniture, the advent of CNC _____ and CNC lathes to create pieces of furniture has revolutionized furniture making.

A. lasers
B. operator
C. routers
D. apprenticeships

10. _____ is the process of cutting, forming, and bending pieces from flat sheet metal into a usable shape.

A. CNC water jetting
B. Routing
C. Setting up
D. Sheet metal fabrication

11. *True or False?* The machine setup operator is responsible for ensuring the machine, tooling, and work holding are efficient and accurate to make parts as programmed.

12. A quality control (QC) inspector will work directly with the _____ to make sure the product being produced meets the form, fit, and function of a customer's requirements.

A. engineer
B. manager
C. manufacturing team
D. setup technician

13. *True or False?* A résumé is an extensive explanation of your education, work experience, all your interests, and other qualifications for work.

14. All potential employers want to know about your technical skills and education, but equally important attributes for an employer include _____.

A. desired pay
B. religious beliefs
C. soft skills
D. work history

15. What is often the first contact you have with a potential employer?

A. Résumé
B. Letter of application
C. Interview
D. Phone call

16. What should you *not* do when preparing for an interview?

A. Research the job
B. Decide what to wear
C. Practice the interview
D. Decide on a salary

17. *True or False?* A big advantage of a team is its ability to develop plans and complete work faster than individuals working alone.

18. Many workplace accidents occur because of _____.

A. careful behavior
B. careless behavior
C. lack of supervision
D. poor management

19. *True or False?* The apprenticeship process can take 2–5 months to complete and never consists of textbook training.

20. NIMS has established a recognized set of standards developed by industry in operational areas including metalforming, _____, and industrial maintenance.

A. apprenticeships
B. leadership
C. machining
D. programming

Apply and Analyze

1. What are some of the significant improvements in machine safety over the last 30 years?

2. What impact did the advancements in aviation and defense have on the machining industry?

3. What are some of the soft skills required to be successful in the manufacturing industry?

4. After securing employment, what are some of the key attributes it will take to be successful with your new employer?

5. List some organizations that provide skills training and machining certifications.

Critical Thinking

1. Briefly outline the major historical factors responsible for many modern machining advances. How have those factors influenced modern machining?

2. Although there are multiple career paths in manufacturing, what are the some of the most important factors in your future success?

3. Briefly define "soft skills" and list some examples. Which soft skills do you think are the most important and why?

3

Machine Shop Safety

Chapter Outline

3.1 General Shop Safety
3.1.1 Attire and Appearance
3.1.2 Professional Shop Behavior
3.1.3 Cleanliness
3.1.4 Hazardous Materials
3.1.5 Chip Removal
3.1.6 Proper Use of Hand Tools

3.2 Fire Safety
3.2.1 Oily Rag Containment
3.2.2 Oil Spill Cleanup and Containment
3.2.3 Fire Extinguisher Classifications

3.3 Hazardous Materials Identification System

3.4 Safety Data Sheets
3.4.1 Record Keeping
3.4.2 Chemical Tagging

3.5 Safety Aids
3.5.1 Safety Guards
3.5.2 Lock Out/Tag Out (LOTO)
3.5.3 Personal Protective Equipment
3.5.4 Hazard Alert Labels

Learning Objectives

After reading this chapter, you should be able to:

- Define the five-second rule.
- Describe basic features of safe working environments.
- List hazards associated with clothing and jewelry.
- Summarize desirable shop behavior and standards of cleanliness.
- Explain possible hazards resulting from materials being machined.
- Describe the proper way to remove chips from machine workspace.
- Explain the hazards of using small hand tools around operating machines.
- Recognize common fire hazards that exist in machine shops.
- Identify oil hazards and describe oily rag containment procedures.
- List and describe the classes of fires.
- Discuss the hazardous material identification system (HMIS).
- Identify the components of a safety data sheet (SDS).
- Describe safe practices for tagging chemicals.
- List and describe common safety guards and safety aids.
- Describe the lock out/tag out (LOTO) procedure.
- Identify proper personal protective equipment (PPE) for your work environment.
- Select appropriate eye protection, ear protection, footwear, outerwear, and respirators.
- Recognize hazard alert labels.

Key Terms

alloy

chips

chuck key

class A fire

class B fire

class C fire

class D fire

five-second rule

hazard alert label

hazardous materials identification system (HMIS)

lock out/tag out (LOTO)

mill vise

personal protective equipment (PPE)

safety data sheet (SDS)

subtractive process

work envelope

3.1 General Shop Safety

Shop safety should be the primary concern for everyone in a machine shop. Accidents are often preventable as long as employees are careful and adhere to established safety procedures. One simple rule to learn is the *five-second rule*. Take five seconds before turning on or operating any piece of equipment to look around your environment and the equipment you are about to operate. In only a few seconds, you can most often see an impending danger. For instance:

- Always note anyone in the near vicinity of the machine. They may create a hazard or be impacted by any unsafe condition of the machine or environment.
- Make sure the machine is in the correct starting position.
- Check your personal protective equipment is in place and properly worn.
- No loose tools should be inside the working area of the machine.
- Make sure all safety guards are in place.
- All material needs to be properly secured before cutting.

It is not possible to list every condition that might be unsafe, but this chapter addresses some of the most common and preventable safety concerns:

- Dress and behave professionally and appropriately for the conditions in your shop.
- A safe environment should always be clean and free from debris or loose tools.
- Take care with hazardous materials and be attentive to proper treatment of machined chips.
- Handle the tools of the machine trade with the care and respect they deserve.

Generally, you should always abide by established safe practices for specific machines and materials. Know and understand all safety practices in your shop and make it a habit to stay safe.

3.1.1 Attire and Appearance

Clothing may not be the first thing you think about when operating equipment, but it is a leading cause of workplace injuries. Clothing should never be loose or sagging. Shirts should always stay tucked in. Long sleeves can be dangerous and must always be tucked in, as they can become entangled in moving parts of a machine and cause catastrophic injuries to hands and arms, **Figure 3-1**.

Jewelry, especially rings or dangling necklaces, are of special concern when working around rotating machine parts. A ring snagged on a turning spindle or tool can quickly remove a finger from the hand. Dangling or loose necklaces can also be caught by rotating parts and pull an operator into an unsafe area of a machine tool.

Loose hair is another concern. Any hair past shoulder length should be kept under a hat or tied up on top

TZIDO SUN/Shutterstock.com

Figure 3-1. A manual lathe machinist performing a secondary operation while wearing appropriate safety gear and work attire.

of the head to prevent it from becoming entangled in a machine. Remember that machines and tools are designed to cut metal with lots of power and force. If a machine tool grabs a loose item on your person, you are unlikely to stop it. Prevent dangerous situations before they happen by using the five-second rule to make sure you are safe and secure.

 From the Shop

Protect Yourself at All Times
Many years ago, I witnessed an operator with his shirt left untucked. Within just a few minutes the bottom of his shirt got tangled in a turning spindle. Within just a couple seconds he was standing in the middle of the machine shop with no shirt. He was lucky—it could have been much worse.

"Experience is a hard teacher because she gives
the test first, the lesson afterward."

—Vernon Law

3.1.2 Professional Shop Behavior

A machine shop is a place to work and learn. Take the machining environment seriously. Horseplay and fooling around should not occur in the shop. Any distraction can lead to permanent and devastating injuries and so should never be tolerated around machines. Any medication that may cause drowsiness should be avoided. If you are impaired in any way, do not endanger yourself or others by operating equipment.

3.1.3 Cleanliness

The most important factor in general machine shop safety is keeping the shop and work area clean. Keep metal chips or shavings in scrap or recycling bins to ensure they do not become a hazard later. Trip hazards, such as loose parts on the floor, pallets, and air lines, should be kept secure and out of any walkways.

Often when operating a machine, an operator has inspection tools, hand tools, or deburring tools near the machine. These tools should never be left in the *work envelope*, defined as any space within range of a machine's normal operation. It is often best practice to have a separate workbench or space to neatly and safely store any item needed to complete your assignment or job. See **Figure 3-2**.

3.1.4 Hazardous Materials

Be aware of the materials you are machining and any safety or environmental concerns they may pose. For example, magnesium is a metal that can be machined for a variety of different applications. It is used in aircraft components because it is the lightest structural metal. Magnesium dust and chips are flammable, especially in the presence of moisture or water. This is a rare occasion where water or some other machining coolant

Goodheart-Willcox Publisher

Figure 3-2. A clean, well-organized workbench to store inspection tools and hand tools.

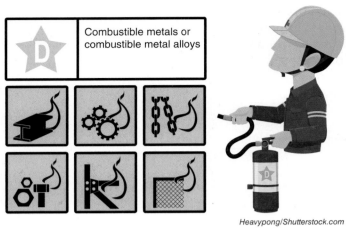

Combustible metals or combustible metal alloys

Heavypong/Shutterstock.com

Figure 3-3. Combustible metals warning label.

can actually cause a fire hazard. If magnesium ignites and water is added to fight the fire, the fire will spread and intensify. When machining exotic metals, be aware and have proper fire extinguishers in place, **Figure 3-3**.

Beyond safety hazards, some materials may have unexpected environmental impacts. For example, the use of lead in machine-grade steel used to be a common practice. Lead adds lubricity to the cutting edge and makes steel more machinable, or "softer." But, lead is a serious environmental hazard if not contained and separated properly before disposal.

These are just a couple of examples of material conditions you should be aware of before you start the machining process. Refer to **Figure 3-4** for additional examples of hazardous materials common in machine shops as well as permissible exposure limits set by the Occupational Safety and Health Administration (OSHA).

 Tech Tip

Potentially hazardous metals, like those listed in **Figure 3-4**, are added to base metals to create *alloys*, mixtures of two or more metals. For example, austenitic stainless steels are made from 16–26% chromium and up to 35% nickel. Check the safety data sheets for any materials being machined to determine exact material specifications.

Exposure Limits for Common Hazardous Materials			
Material	Fumes (mg/m³)	Dust (mg/m³)	Metal (mg/m³)
Cadmium	0.005	0.005	—
Chromium	0.05	0.05	1
Cobalt	0.1	0.1	0.1
Nickel	1	1	1

Goodheart-Willcox Publisher

Figure 3-4. Chart defining exposure limits of common metals, expressed in milligrams per cubic meter.

Some materials and processes can also produce hazardous fumes or dust. Be aware of these potential hazards and take time to do your research on any unfamiliar material. In the event that dangerous dust or fumes are produced, use an approved respirator or dust mask to prevent inhalation. Some abrasive processes, like grinding, cause usually safe materials to produce a dust or fume that requires respiratory protection. Fumes and dust can have residual, cumulative effects that cause damage over long exposure times. Start protecting yourself today so you can be healthy for a long time to come. Take care to know your materials and stay safe always.

3.1.5 Chip Removal

The process used in many machining operations is considered a *subtractive process*, or one that starts with a piece of stock and then subtracts material to leave the finished part. Think of it like a sculptor who starts with a large piece of stone and removes unwanted stone to reveal the piece of art inside. Subtractive processes can create *chips*, or smaller pieces of material cut from a larger workpiece. These chips can be dangerous to handle and can create safety concerns, **Figure 3-5**. As a result, machinists must consider the best way to handle chips in a safe and eco-friendly way.

David Tadevosian/Shutterstock.com

Figure 3-5. Typical chip formation after cutting solid material.

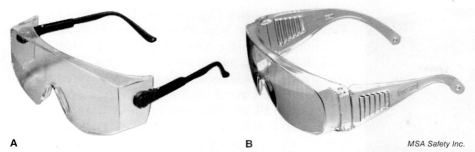

Figure 3-6. Safety glasses with wraparound side protection. A—Standard. B—Prescription.

First, how do machinists protect themselves from chips when machining? Eye protection is most important, **Figure 3-6**. In most shops or labs you cannot enter the facility without safety glasses. As parts are machined and chips are created, they can be thrown from the machine in any direction, sometimes over large distances. Safety glasses should completely cover the eye and prevent any chips from entering from the sides of the glasses. Safety glasses can be purchased with prescription lenses or be large enough to cover existing prescription frames.

Second, how do machinists handle loose chips on or around a machine? Chips are normally very sharp. The process of cutting the material leaves the leading edge with a literal knife edge, so extra caution should be taken when removing chips from machines. Gloves can be worn when removing or cleaning up chips, but gloves should *never* be worn around any turning tool or moving spindle. Make sure the machine is off and nothing is moving before trying to remove chips. A lathe can make chips that are long or stringy. In this case, a pair of pliers can be used to grab the chips and remove them.

Most often there will be small, loose chips on the milling machine table or the bed of a lathe. Use a small brush to brush these chips away, **Figure 3-7**. Never use compressed air to remove chips, as it unpredictably accelerates them and can blow them back in your direction, **Figure 3-8**.

Last, how do machinists dispose of or store loose chips created during machining? After machines and work areas are completely clean, chips should be stored for removal and recycling. Chips also need to be separated and stored with like materials. For example, place aluminum chips

Tech Tip

Stringy or long chips are typically created from an improper cutting edge or using improper feed rates. Monitor the output of any machine tool and adjust as needed for best results.

Goodheart-Willcox Publisher

Figure 3-7. Use a brush to remove loose chips on a machine table.

Goodheart-Willcox Publisher

Figure 3-8. Improper use of compressed air to remove loose debris. Compressed air poses a safety risk to the operator. Never use compressed air to remove chips.

Figure 3-9. Never leave a T-handled chuck key in a lathe chuck.

 **Safety Note**

Never take your hand off the key while it is in the chuck. If a chuck key is left in the lathe and the machine spindle begins rotation, the key may be ejected at a dangerously high speed with lots of force.

together, steel chips together, and brass chips together. This is the only way they can be recycled. Most machine shops place loose chips in large steel drums until they can be picked up for recycling. If these barrels are stored outside, ensure they are covered so that water does not get in and leak from the drums onto the ground. All machine shops should have procedures in place to handle scrap parts and chips. Take care to review and understand all relevant procedures.

3.1.6 Proper Use of Hand Tools

Small hand tools are commonly used in manufacturing. However, improper use of tools can create safety hazards. Additionally, some tools are sharp and dangerous in their own right. Never carry tools that are sharp or pointed in your pocket. Sharp tools should be laid out neatly on a bench or work surface with the sharp end facing away from you. Only use a hand tool for its intended purpose. Never use a hand tool around a machine in operation or close to rotating spindles or chucks.

There are many specialized tools designed for specific purposes related to machine operation or material workholding. One of these items is the *chuck key* for the lathe. See **Figure 3-9.** This tool is used to clamp material in the lathe chuck. This should *never* be left in the chuck, not even for a moment.

Another tool is the handle for a *mill vise*. A mill vise holds objects in place for milling by clamping a workpiece firmly between its jaws so that it stays completely still when the cutter rotates through the material. A mill vise handle is used to tighten a vise onto the material being machined. This handle should never be left inside the machine working envelope, **Figure 3-10.** Use the handle to tighten the vise, then remove the tool and place it on your workbench or in a designated storage device.

Figure 3-10. A mill vise handle left on the table creates a potential safety hazard. Remove all loose tools from the work area.

Hand tools are an essential part of machining, but use them safely and only as intended. CNC machines are enclosed and can be very safe when properly used. All machine tools currently made are required to be equipped with safety door latches and shatter-resistant glass.

3.2 Fire Safety

Fires are always a concern in a machining environment. Machining processes generally use heat and force to remove material. Some machining processes require the use of flammable oils as well. That combination can be a recipe for a fire. To minimize risk, take steps to prevent fires whenever possible and establish a plan for responding to emergencies. See **Figure 3-11**. Most shops should have a visible emergency evacuation map and marked locations for fire extinguishers. Know before a fire happens what the plan is and where the emergency evacuation point is for your work area. In an emergency situation, it is most important to ensure all personnel exit a building safely and call for help. Let the fire department fight the fire without having to worry about rescuing trapped workers.

3.2.1 Oily Rag Containment

Cutting fluids and oils have evolved over time. Machining fluids used to be very flammable and had long-term health risks. Today, machining oils are primarily water soluble, noncarcinogenic, and often vegetable oil-based. Despite these improvements, machinists still need to consider the storage of shop rags or clothing that may contain oil-based products. Any oily rags or wipes used to clean up oil spills need to be contained in a properly labeled and approved safety container. Cloth rags used to clean up metal shavings and oil may *spontaneously combust*, or ignite by rapid oxidation without external heat source, due to the interaction of oils with certain metals. Oily rag containers should be emptied daily and the contents disposed of properly in accordance with safety standards.

Emergency Action Plan Components	
Emergency Situation	**Action Plan**
Evacuation	Follow evacuation map.
Utility Outage	Contact appropriate utility company.
Gas Leak	Shut off main gas source and contact gas company.
Medical Emergency	Follow company procedures and phone emergency contacts.
Criminal Behavior	Follow company procedures and phone appropriate emergency contacts.
Hazardous Material Spill	Locate SDS and contact HazMat clean up.
Bomb Threat	Follow company procedures and contact proper authorities.
Fire/Explosion	Follow evacuation procedures and fire extinguisher guidelines.
Emergency Exit	Refer to detailed map of exits from all locations in the building.

Goodheart-Willcox Publisher

Figure 3-11. Chart of typical emergency action plan components. Familiarize yourself with your company's policies, relevant emergency contact information, and evacuation and exit maps.

Accuform

Figure 3-12. Removable barriers in a work zone alert personnel of a potential risk. Never cross a barrier without authorization.

3.2.2 Oil Spill Cleanup and Containment

Oil spill cleanup and containment is critical in preventing fall injuries and also has environmental implications. Once an oil-based product is on the floor, that floor will become extremely slippery and can cause a violent fall for anyone walking through. The first step in preventing such an injury is to block the area while clean-up supplies are obtained, **Figure 3-12.** Oil spills cannot be cleaned with water. Instead, absorbent sand must be applied, like cat litter, over the entire spill to begin absorbing the oil. Once the sand is applied, a heavy broom can be used to grind the sand into the oil to aid absorption. A thick oil cloth designed specifically to absorb a high volume of oil may also be used to clean up spills. Once the oil is absorbed, the sand or towel should be placed in a fire-resistant container with an appropriate lid. Remember that oil is not an environmentally safe product. It will soak through a floor and eventually into the ground below, causing possible groundwater contamination if left untreated.

 **Thinking Green**

Green Fire Safety
Fire safety is a concern in any machine shop, especially for those that use flammable, oil-based machining fluids. Recent advances in automated manufacturing have improved the manufacturing process but may leave combustible material and expensive equipment unattended. To protect against fires, some machine shops use on-machine fire suppression systems. These systems are set up to automatically detect fire and release a chemical fire extinguishing agent. Traditional fire extinguishers contain halon, a chemical that can cause ozone depletion when released into the atmosphere. New, potassium-based extinguishing agents are more environmentally friendly. They do not cause ozone depletion, and they are nontoxic and easier for shop workers to clean up.

3.2.3 Fire Extinguisher Classifications

Combustible materials are categorized into fire classes. Fire extinguishers are labeled to identify the fire classes they can safely extinguish. In the machine shop, fire extinguishers should be clearly labeled and used appropriately based on classification, **Figure 3-13.**

- *Class A fires* consist of normal combustible materials, such as wood, paper, and cloth. These types of fires require water and the cooling effect of a solution with high water content. This is the most common household fire, but class A fires are not as common in manufacturing. Never use a class A fire extinguisher on oil or metal fires.

- *Class B fires* result from flammable liquids or grease. Class B fires can be put out by removing oxygen or smothering the fire, so class B extinguishers use dry chemicals or carbon dioxide.

- *Class C fires* are electrical fires that require nonconducting chemicals and extinguishing agents. Never use a pressurized water extinguisher on electrical fires.

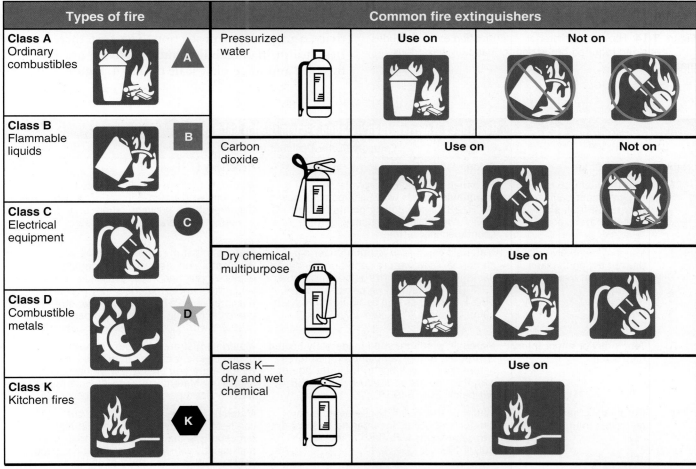

Goodheart-Willcox Publisher

Figure 3-13. Different types of fire extinguishers must be properly applied depending on fire type.

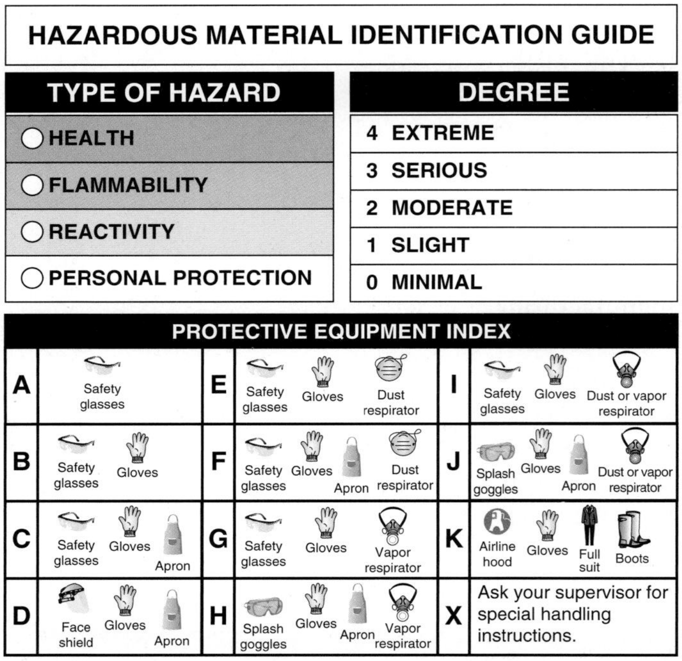

HAZARDOUS MATERIAL IDENTIFICATION GUIDE

TYPE OF HAZARD	DEGREE
○ HEALTH	4 EXTREME
○ FLAMMABILITY	3 SERIOUS
○ REACTIVITY	2 MODERATE
○ PERSONAL PROTECTION	1 SLIGHT
	0 MINIMAL

	PROTECTIVE EQUIPMENT INDEX		
A Safety glasses	**E** Safety glasses, Gloves, Dust respirator	**I** Safety glasses, Gloves, Dust or vapor respirator	
B Safety glasses, Gloves	**F** Safety glasses, Gloves, Apron, Dust respirator	**J** Splash goggles, Gloves, Apron, Dust or vapor respirator	
C Safety glasses, Gloves, Apron	**G** Safety glasses, Gloves, Vapor respirator	**K** Airline hood, Gloves, Full suit, Boots	
D Face shield, Gloves, Apron	**H** Splash goggles, Gloves, Apron, Vapor respirator	**X** Ask your supervisor for special handling instructions.	

Goodheart-Willcox Publisher

Figure 3-14. Hazardous material identification chart. Includes PPE levels and required protection for handling hazardous materials.

- *Class D fires* involve flammable metals, such as lithium or magnesium. Class D extinguishers contain specialized heat-absorbing dry powders. Do not use a class A extinguisher on these types of metal fires.

The most important step in fire safety is knowing where the exits and extinguishers are located before a fire occurs. Inspect extinguishers at least twice a year for service and usability.

3.3 Hazardous Materials Identification System

The *hazardous materials identification system (HMIS)* is a hazard rating system that uses labels, colors, and alphanumeric codes to communicate chemical hazards, **Figure 3-14**. The HMIS rating system was created by the American Coatings Association for compliance with OSHA. The system is meant to rate hazardous materials as well as any chemicals a machinist might encounter in four separate categories: health, flammability, reactivity, and personal protection. The first three categories, health, flammability, and reactivity, are indicated by color and rated on a scale of 0 to 4, **Figure 3-15**. The

	Hazardous Materials Identification System (HMIS) Ratings		
	Category (Color Designator)		
Rating	Health (Blue)	Flammability (Red)	Reactivity (Yellow)
4	Life-threatening, major, or permanent damage may result from single or repeated overexposures.	Flammable gases, or very volatile flammable liquids with flash points below 73°F (23°C) and boiling points below 100°F (38°C). Materials may ignite spontaneously with air.	Materials that are readily capable of explosive reactions with water, detonation or explosive decomposition, polymerization, or self-reaction at normal temperature and pressure.
3	Major injury likely unless prompt action is taken and medical treatment is given.	Materials capable of ignition under almost all normal temperature conditions. Includes flammable liquids with flash points below 73°F (23°C) and boiling points above 100°F (38°C), as well as liquids with flash points between 73°F and 100°F.	Materials that may form explosive mixtures with water and are capable of detonation or explosive reaction in the presence of a strong initiating source. Materials may polymerize, decompose, self-react, or undergo other chemical change at normal temperature and pressure with moderate risk of explosion.
2	Temporary or minor injury may occur.	Materials which must be moderately heated or exposed to high ambient temperatures before ignition will occur. Includes liquids with flash points at or above 100°F (38°C) but below 200°F (93°C).	Materials that are unstable and may undergo violent chemical changes at normal temperature and pressure with low risk for explosion. Materials may react violently with water or form peroxides upon exposure to air.
1	Irritation or minor reversible injury possible.	Materials that must be preheated before ignition will occur. Includes liquids, solids and semi-solids with flash points above 200°F (93°C).	Materials that are normally stable but can become unstable (self-react) at high temperatures and pressures. Materials may react nonviolently with water or undergo hazardous polymerization in the absence of inhibitors.
0	No significant risk to health.	Materials that will not burn.	Materials that are normally stable, even under fire conditions, and will not react with water, polymerize, decompose, condense, or self-react. Nonexplosives.

Goodheart-Willcox Publisher

Figure 3-15. Explanation of threat levels used in the HMIS. Includes health, flammability, and reactivity categories.

HMIS Personal Protection Ratings

Code Letter	Equipment	Symbol(s)
A	Safety glasses	
B	Safety glasses and gloves	
C	Safety glasses, gloves, and apron	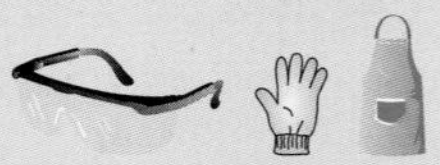
D	Face shield, gloves, and apron	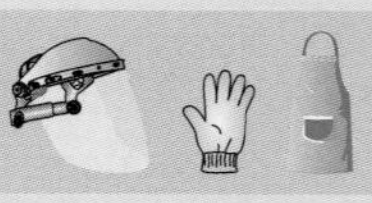
E	Safety glasses, gloves, and dust respirator	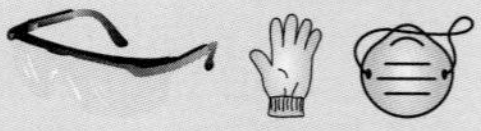
F	Safety glasses, gloves, apron, and dust respirator	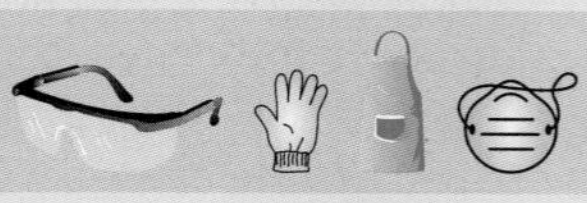
G	Safety glasses and vapor respirator	
H	Splash goggles, gloves, apron, and vapor respirator	
I	Safety glasses, gloves, and dust or vapor respirator	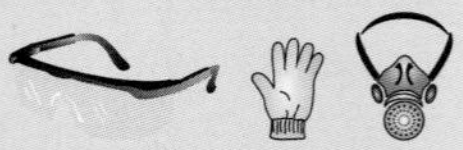
J	Splash goggles, gloves, apron, and dust or vapor respirator	
K	Airline hood or mask, gloves, full suit, and boots	

Goodheart-Willcox Publisher

Figure 3-16. Letter code designators for HMIS personal protection ratings. Lists personal protective equipment (PPE) required to handle any given material.

fourth category, personal protection, receives a letter designation corresponding to personal protective equipment (PPE) required. See **Figure 3-16.**

3.4 Safety Data Sheets

A *safety data sheet (SDS)* is a detailed informational document prepared by the manufacturer of a hazardous chemical. It should describe the physical and chemical properties of the product. An SDS contains useful information such as procedures for spills and leaks, toxicity, flash point, storage guidelines, and exposure control, **Figure 3-17.**

3.4.1 Record Keeping

Employers are required by law to provide workers with access to information contained in an SDS. Most often, these records are kept in a highly

Safety Data Sheet (SDS) Sections	
Section	**Description**
Section 1. Identification	Includes a product identifier, contact information for the manufacturer or distributor, recommended use, and restrictions on use.
Section 2. Hazard(s) identification	Lists all hazards regarding the chemical and required label elements.
Section 3. Composition/information on ingredients	Lists information on chemical ingredients and trade secret claims.
Section 4. First-aid measures	Describes potential symptoms or effects and required treatment.
Section 5. Fire-fighting measures	Describes potential fire hazards and lists suitable extinguishing techniques and equipment.
Section 6. Accidental release measures	Lists emergency procedures, required protective equipment, and proper methods of containment and cleanup.
Section 7. Handling and storage	Lists precautions for safe handling and storage, including incompatibilities.
Section 8. Exposure controls/personal protection	Specifies exposure limits, including OSHA's Permissible Exposure Limits (PELs), ACGIH Threshold Limit Values (TLVs), and any other exposure limits recommended by the manufacturer, importer, or employer. Also includes recommended engineering controls and personal protective equipment (PPE).
Section 9. Physical and chemical properties	Specifies the chemical's characteristics.
Section 10. Stability and reactivity	Lists the stability of the chemical and the possibility for hazardous reactions.
Section 11. Toxicological information	Describes the toxicity of the chemical, including routes of exposure, symptoms of toxicity, acute and chronic effects of toxicity, and numerical measures of toxicity.
Section 12. Ecological information	Lists potential ecological impacts of the chemical.
Section 13. Disposal considerations	Explains how to dispose of the chemical safely.
Section 14. Transport information	Describes transportation requirements and safety considerations.
Section 15. Regulatory information	Lists regulations pertaining to the chemical.
Section 16. Other information	Lists the date of preparation or last revision of the SDS.

Adapted from OSHA

Figure 3-17. Safety data sheet (SDS) section descriptions.

Goodheart-Willcox Publisher

Figure 3-18. A "Right to Know" station that contains material safety data sheets for common access.

visible binder in a "Right to Know" station, **Figure 3-18.** It is every worker's right and obligation to know and understand all information concerning any chemical with which they may have contact.

3.4.2 Chemical Tagging

Normally, it is not a good practice to remove chemicals from their original storage containers. Original containers feature pertinent hazard warnings and chemical descriptions. In the rare case when a smaller container is required to hold fluids, ensure that the new container is clearly marked with the name of the contents. Once you have finished with the small container, return the contents to the original storage container.

 Safety Note

Never mix chemicals or use a contaminated container. The reaction of dissimilar chemicals can be unpredictable and dangerous.

3.5 Safety Aids

Thinking about safety before any job or operation can help keep you safe. In addition, there are numerous safety aids designed specifically to contribute to a safe work environment. Safety aids are present in all industries and can include personal protective equipment (PPE), machine guards, lock out/tag out, or hazard warning signs.

3.5.1 Safety Guards

Machines contain numerous driven parts, including many components that can cause severe injuries. Although moving components can be dangerous, machines come equipped with guards and covers to prevent physical injury, **Figure 3-19**. Machine guards are designed to prevent injuries and should only be removed for maintenance, **Figure 3-20**. In the event that a guard is removed, the machine must be locked out and tagged out to prevent operation.

3.5.2 Lock Out/Tag Out (LOTO)

Lock out/tag out (LOTO) is a system to secure machines and prevent accidental operation during maintenance or other unsafe conditions. LOTO procedures are especially important for large machines, where it is possible for someone to power on a machine without realizing that the machine is under maintenance or otherwise unsafe to operate.

Every shop should come equipped with a LOTO station, **Figure 3-21**. This kit should include an electrical clamp, padlocks, cord covers, and red machine tags. Any time maintenance is required, the power needs to be shut off at its source and a clamp or plug cover

Goodheart-Willcox Publisher

Figure 3-19. Two examples of guards placed on a horizontal bandsaw to prevent the operator from interfering with blade movement.

Goodheart-Willcox Publisher

Figure 3-20. The gear drive end of a manual lathe with guards removed. Observe that removing a machine guard poses a risk to operators from pinch points or moving components. Only remove safeguards when necessary for maintenance, and replace them promptly after work is complete.

Master Lock

Figure 3-21. The equipment in this kit is used for performing lock out/tag out (LOTO) procedures.

A

B

Master Lock

Figure 3-22. These LOTO devices are used to secure machinery as required. A—Cord ends. B—Quick disconnect boxes.

put in place to prevent anyone from restoring power to the equipment, **Figure 3-22.** The lock is secured by the operator performing maintenance, and they keep the key until the machine is safe to operate. A tag is also placed on the on/off switch of the machine to notify all other workers that the machine is locked out for maintenance.

3.5.3 Personal Protective Equipment

Personal protective equipment (PPE) should be determined by the job being performed. PPE is any equipment worn to minimize exposure to hazards that cause workplace injuries or illnesses. Safety glasses, gloves, aprons, earplugs, steel-toed shoes, and respirators are examples of PPE. Your HMIS and SDS should be used to determine any additional PPE that might be required.

Eye Protection

Different levels of eye protection are available, depending on your job and shop conditions, **Figure 3-23.** Safety glasses are required whenever you enter a manufacturing facility. Safety glasses should be rated at Z87.1-2015 or higher by the American National Safety Institute (ANSI), indicating impact resistance and full protection. The rating should be imprinted on the frame of the glasses for reference. Make sure the glasses

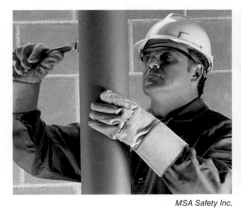

MSA Safety Inc.

MSA Safety Inc.

Reggie Lavoie/Shutterstock.com

Figure 3-23. Eye protection is required at all times in a machine shop. Different tasks require different types of protection.

you select provide full eye coverage and prevent debris from entering at the sides. Prescription safety glasses are available if necessary. In the event you are wearing prescription glasses, side shields should be installed for full protection.

Earplugs

Earplugs are required in any machining environment where employees are exposed to over 85 decibels of sound over an 8-hour period. Earplugs come in several styles, including expanding foam, fitted rubber, or ear-over full coverage. Earplugs must be properly fitted and worn at all times in high-noise-level areas.

Footwear

Required footwear varies with the machining environment, but at a minimum all footwear should have certain features. Oil-resistant, nonslip soles are required in any environment where oils or lubricants are present. Shoes should be nonathletic with leather uppers that come up above the ankle. This provides support in case of a slip and protection to the ankle. Workers who regularly handle materials weighing more than 35 pounds might need steel-toed boots. Any footwear should meet minimum safety standards and be comfortable for multiple hours of standing or walking.

Outerwear

The appropriate type of outerwear creates a barrier between you and any hazard, including chemical and physical hazards. Your employer will advise you of any specific outerwear that may be required to safely perform the job. Most often in machine shops, outerwear protects you from flying metal chips that may be hot or have sharp edges. Polyester fabric can burn through with a metal chip, so cotton or denim is generally preferred. Always make sure you have no loose articles of clothing—unbuttoned shirt sleeves or untucked shirttails can become entangled in moving machine parts and create a hazard. A shop apron or jacket can be worn to provide additional protection.

Respirators

Respirators are not common equipment in machining, but they may be required with some materials or machining applications. Grinding is a common operation that requires breathing protection, as it creates dust from both the grinding stone and the material removed. Many materials are more hazardous when in fine particulate form than in normal chip form.

Respirators can be as simple as a filter mask or as sophisticated as rebreather units with full face shields. Refer to relevant safety data sheets for specific risks involved with chemicals and materials, and protect yourself appropriately.

3.5.4 Hazard Alert Labels

Hazard alert label is a generic term referring to any additional label or warning that might be placed around a shop or classroom lab to identify potential risks. See **Figure 3-24**. Pay special attention to these warnings, as they identify area-specific safety concerns.

Always bear in mind that all warnings and labels are designed to provide the safest possible working environment. You are responsible for your own safety. Remember to take that extra five seconds to observe your surroundings and look for labels or warnings that might present a risk to your short- and long-term health.

Goodheart-Willcox Publisher

Figure 3-24. Warning signs or hazard alert labels alert you to risks and should raise awareness of potential hazards.

Chapter Review

Summary

- The five-second rule dictates that, before turning on any piece of equipment, the operator should stop and take five seconds to ensure the safety of their work area.

- To maintain a safe work environment, dress and behave professionally, keep your area clean, take care with hazardous materials, be attentive to proper treatment of machined chips, handle tools carefully, and always abide by established safe practices.

- Loose-fitting clothing and dangling jewelry risk potential entanglement around a turning spindle or tool.

- Sharp metal shavings, hazardous fumes, and fire danger are all potential risks of machining metal.

- Machined shavings or chips can be extremely sharp or hot and can pose potential safety hazards.

- Machined chips should be brushed away from the working area.

- Small hand tools should not be used around any moving equipment and should be secured away from the working envelope of any machine.

- Oily rags, hot metal, electrical equipment, and some materials like magnesium can cause fire hazards in a machine shop.

- Oil and oily rags are flammable. They should remain contained in sealed metal bins to prevent ignition.

- When using a fire extinguisher, be sure it is labeled for use on the appropriate class of fire.

- The hazardous material identification system (HMIS) rates hazards in three categories, Health, Flammability, and Reactivity, on a scale of 0 to 4. A fourth category, personal protection, assigns a letter code indicating required PPE for specific hazards.

- There are 16 different categories included in a safety data sheet (SDS) that classify and describe material handling procedures.

- All chemicals should be contained and marked for easy identification, including HMIS information.

- A safety guard is any machine attachment that protects the operator while the equipment is in use. Safety warning signs or hazard alert labels may also be present to alert the operator to potential risks of injury.

- Lock out/tag out (LOTO) procedures are used to isolate sources of power from a piece of equipment while maintenance or repair is occurring.

Review Questions

Answer the following questions using the information provided in this chapter.

Know and Understand

1. *True or False?* Accidents are most often preventable as long as workers are careful and adhere to safe practices.

2. Keeping _____ in scrap or recyclable bins will ensure they do not become a hazard later.
   A. oils
   B. rags
   C. metal chips
   D. strings

3. Some materials and processes can produce _____ fumes or dust.
   A. environmental
   B. hazardous
   C. healthy
   D. stringy

4. *True or False?* Horseplay is acceptable in a machine shop or lab as long as you have free time.

5. The machining processes can create _____, or smaller pieces of material that were cut off the larger piece.
   A. chunks
   B. hazardous waste
   C. metal chips
   D. oils

6. *True or False?* Safety glasses should be rated at Z72 or higher from the American National Safety Institute.

7. *True or False?* Gloves should *never* be worn around any tool or spindle that is turning.

8. Use a _____ to remove machined chips.
   A. brush
   B. compressed air hose
   C. hand
   D. rag

9. Some tools can be sharp and dangerous themselves, and improper use of these tools can also create many _____.

    A. good parts   C. safe operations

    B. metal chips   D. safety hazards

10. *True or False?* Never take your hand off the chuck key while it is in the chuck.

11. Cleaning up an oil spill and containing its affected area are critical in preventing _____ and _____.

    A. a mess; cleanup   C. fire; cleanup

    B. work; overtime   D. slips; falls

12. *True or False?* Class B fires are electrical fires.

13. *True or False?* It takes three things to sustain a fire: heat, space, and wood.

14. The hazardous materials identification system (HMIS) is an alphanumeric hazard rating system that uses _____ and _____ to communicate chemical hazards.

    A. colors; hand signals

    B. labels; colors

    C. numbers; warning signs

    D. signs; safety data sheets

15. Safety data sheets (SDS) are detailed informational documents prepared by the manufacturer of a _____.

    A. hazardous chemical   C. metal bar

    B. machine   D. textile

16. *True or False?* SDS records are often kept in a highly visible binder at a "Right to Know" station.

17. *True or False?* Moving components are not dangerous.

18. _____ procedures secure machines and prevent accidental operation while maintenance or any unsafe condition is ongoing.

    A. Lockdown/secure

    B. Lock out/power down

    C. Lock out/tag out

    D. Look out/watch out

19. Make sure the safety glasses you select provide full _____ and prevent entrance of debris from the sides.

    A. eye coverage   C. head protection

    B. face coverage   D. shade

20. _____ are generically any additional label or warning that might be placed around a shop or classroom lab to identify potential risks.

    A. American National Safety Institutes

    B. Hazard alert labels

    C. Hazardous material information sheets

    D. Safety data sheets

Apply and Analyze

1. What are some of the ways you can be responsible for your own safety while machining?

2. Why is it important to understand the classes of fire extinguishers and where to locate them?

3. What are some examples of personal protective equipment (PPE) and when should they be worn?

4. Why are lock out/tag out (LOTO) procedures a required safety practice when performing machine maintenance or repairs?

5. Who creates the best chance of keeping you safe at work and why?

6. Briefly describe the four categories included in the hazardous material identification system (HMIS).

7. What is a safety data sheet (SDS)? What are some key pieces of information included in an SDS?

8. What is the difference between a safety guard and a hazard alert?

Critical Thinking

1. What are some tangible ways that you can improve safety at home and at work today?

2. What are the economic impacts of working unsafely? How will this affect both your employer and you if you are hurt at work?

3. There are multiple organizations dedicated to safe workplaces. How do national standards and regulating bodies help improve safety?

4

Machining Math

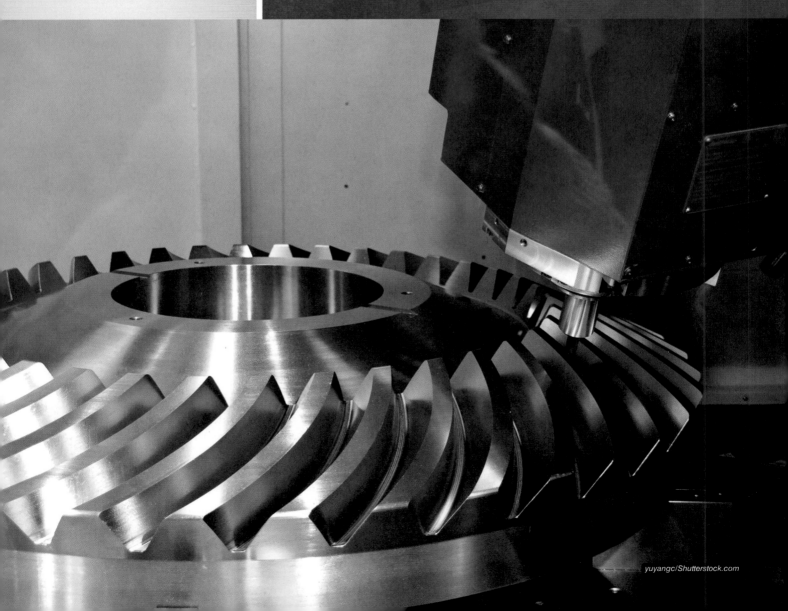

Chapter Outline

4.1 Introduction to Machining Math

4.2 Fractions

    4.2.1 Adding Fractions

    4.2.2 Subtracting Fractions

    4.2.3 Converting Fractions to Decimals

    4.2.4 Use of Thousandths Accuracy

4.3 Geometric Shapes

    4.3.1 Circles

    4.3.2 Polygons

    4.3.3 Lines

4.4 Angles

    4.4.1 Supplementary Angles

    4.4.2 Complementary Angles

4.5 Trigonometry

4.6 Cartesian Coordinate System

    4.6.1 Number Lines

    4.6.2 Four-Quadrant Plotting

    4.6.3 Three-Dimensional Plotting

4.7 Calculating Speeds and Feeds

    4.7.1 Mill

    4.7.2 Lathe

Learning Objectives

After reading this chapter, you should be able to:

- Practice performing basic mathematical operations.
- Demonstrate how to use a simple scientific calculator.
- Identify the parts of a fraction.
- Practice adding and subtracting simple fractions.
- Convert fractions to decimals.
- Explain the use of thousandths accuracy.
- Identify basic geometric shapes.
- Name four types of triangles.
- Use the Pythagorean theorem to solve problems involving triangles.
- Explain the use of right triangles in machining math.
- Define line and tangent line.
- Define supplementary and complementary angles.
- Name the parts of a right triangle.
- Use basic trigonometry to solve for sides and angles in a right triangle.
- Describe the Cartesian coordinate system.
- Understand 2-dimensional and 3-dimensional plotting in a CNC machine.
- Practice calculating feeds and speeds.

Key Terms

acute angle	geometric shape	quadrilateral
angle	hypotenuse	radius
bolt circle	improper fraction	rectangle
Cartesian coordinate system	isosceles triangle	rhombus
chip load	line	right angle
circle	line segment	right triangle
common denominator	mixed number	scalene triangle
complementary angles	numerator	sine
cosine	obtuse angle	square
denominator	origin	supplementary angles
diameter	parallelogram	surface footage
equilateral triangle	polygon	tangent
flutes	Pythagorean theorem	tangent line
	quadrant	trigonometry

Goodheart-Willcox Publisher

Figure 4-1. A simple scientific calculator, such as this TI-30Xa, is a useful tool for completing mathematical calculations.

5/8″

3/8″

Goodheart-Willcox Publisher

Figure 4-2. A fractional print designation of a round part with an extended hub.

4.1 Introduction to Machining Math

Math is essential in the machining and manufacturing trades. The math covered in this chapter is limited to the basics needed for blueprint conversions and creating CNC programs. This text assumes access to a simple scientific calculator, such as a TI-30, **Figure 4-1**. The calculator used and formulas covered in this text are tools, just like a file or wrench, to support your work in the shop. Any calculator with fractions and sine, cosine, and tangent functions is sufficient to perform machining calculations.

4.2 Fractions

Many prints define dimensions in fractional format. Although fractions usually indicate a level of accuracy for inspection, machinists cannot enter fractions into a CNC program. Instead, machinists must be able to convert fractions to decimals as well as add and subtract fractions. A fraction has two parts, a *numerator* and *denominator*. The **numerator** is the number on top of the fraction. It indicates the number of parts. The **denominator** is the number on the bottom of the fraction, and it indicates the size of the parts. In the fraction 1/4″, the 1 is the numerator and the 4 is the denominator.

$$\frac{1}{4} \begin{array}{l} \leftarrow \text{Numerator} \\ \leftarrow \text{Denominator} \end{array}$$

4.2.1 Adding Fractions

Machinists often need to add fractions. In the machining environment, you typically only work with fractional denominators of 2, 4, 8, 16, 32, and 64, so that is all this text covers. These are the same denominators used in tape measures or rulers.

There are two scenarios for adding fractions. In some cases, the fractions to be added have a common (or same) denominator. The rest of the time, the fractions have different denominators. Consider fractions with a common denominator first. This is the easier scenario. In **Figure 4-2**, what is the overall length of the part?

This is a simple fractional addition problem.

$$\frac{5}{8} + \frac{3}{8} =$$

Because this fraction has a common denominator, the denominator is the same in the answer.

$$\frac{5}{8} + \frac{3}{8} = \frac{}{8}$$

Now the numerators can be added together. The sum becomes the numerator of the answer as follows.

$$\frac{5}{8} + \frac{3}{8} = \frac{8}{8}$$

The resulting fraction can then be reduced to 1, meaning the part is 1″ long.

Consider the problem again with a calculator. Scientific calculators have a simple button to calculate fractions, **Figure 4-3**. To enter fractions in the calculator, type in the numerator, press the fraction button, then type in the denominator. For example, to enter the fraction 5/8, press 5, fraction button, and then 8. In this text, green buttons illustrate exactly which buttons on the calculator to press and in which order:

To solve 5/8 + 3/8 with your calculator, press the following keys:

$$\boxed{5}\ \boxed{a^{b/c}}\ \boxed{8}\ \boxed{+}\ \boxed{3}\ \boxed{a^{b/c}}\ \boxed{8}\ \boxed{=}$$

The calculator displays the answer, 1, already reduced.

The second scenario for adding fractions involves fractions with different denominators. This scenario is more common than the first, so it is important to practice adding unlike fractions. Consider a different part and calculate the overall part length. See **Figure 4-4**.

Set up a fractional math problem to express total part length:

$$\frac{1}{2} + \frac{1}{4} + \frac{3}{4} =$$

Notice that the denominators are not the same, so they cannot be carried over to the answer as in the previous problem. Instead you must find a **common denominator**, or make all the denominators in the expression the same. A simple way to accomplish this is to multiply the "unlike" denominator, 2, by the other denominator, 4, to arrive at a common denominator. The common denominator in this case is 8.

$$\frac{1}{2^{\times 4}} + \frac{1}{4^{\times 2}} + \frac{3}{4^{\times 2}} = \frac{}{8} + \frac{}{8} + \frac{}{8} =$$

You have successfully multiplied the denominator by a number to increase it and make it common. Now, to maintain the equality of the expression, you must also multiply each numerator by the same number as the denominator.

$$\frac{1^{\times 4}}{2^{\times 4}} + \frac{1^{\times 2}}{4^{\times 2}} + \frac{3^{\times 2}}{4^{\times 2}} = \frac{4}{8} + \frac{2}{8} + \frac{6}{8} =$$

After finding a common denominator, simply add the numerators and reduce as in the first example.

$$\frac{1^{\times 4}}{2^{\times 4}} + \frac{1^{\times 2}}{4^{\times 2}} + \frac{3^{\times 2}}{4^{\times 2}} = \frac{4}{8} + \frac{2}{8} + \frac{6}{8} = \frac{12}{8} = 1\frac{1}{2}$$

To perform this operation on a calculator, follow these same steps. A big advantage of using a calculator is that you do not need to worry about common denominators. Here are the buttons in order:

$$\boxed{1}\ \boxed{a^{b/c}}\ \boxed{2}\ \boxed{+}\ \boxed{1}\ \boxed{a^{b/c}}\ \boxed{4}\ \boxed{+}\ \boxed{3}\ \boxed{a^{b/c}}\ \boxed{4}\ \boxed{=}$$

Once again the calculator returns the correct answer, 1 1/2, with no need to worry about denominators or reducing.

Goodheart-Willcox Publisher

Figure 4-3. The key used to input fractions on a scientific calculator.

Goodheart-Willcox Publisher

Figure 4-4. A fractional print reference with noncommon denominators.

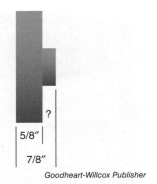

Goodheart-Willcox Publisher

Figure 4-5. A fractional print reference that notes the overall part length, 7/8″, and one feature length of 5/8″ for the left-hand feature.

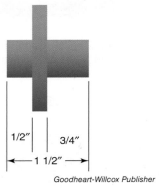

Goodheart-Willcox Publisher

Figure 4-6. A fractional print reference that includes overall part length and two feature lengths. The missing feature dimension must be calculated.

4.2.2 Subtracting Fractions

Subtracting fractions is very similar to adding fractions. You still need a common denominator to calculate the answer. Subtracting fractions can be useful in several situations, such as when interpreting prints. A print might dimension the overall length of a part, with one or two dimensions, but not define the last dimension. You can start with the overall length and subtract the known values to find the unknown dimension.

Consider a part print that defines overall length and one dimension, **Figure 4-5.** What is the dimension of the undefined feature? When setting up the subtraction problem, place the longest dimension first to avoid calculating a negative answer:

$$\frac{7}{8} - \frac{5}{8} =$$

The common denominator carries directly to the opposite side of the equal sign.

$$\frac{7}{8} - \frac{5}{8} = \frac{}{8}$$

Now, simply subtract the numerators.

$$\frac{7}{8} - \frac{5}{8} = \boxed{\frac{2}{8}} = \frac{1}{4}$$

The answer is 2/8, which can be reduced to 1/4. The problem on the calculator uses the same function keys as the addition problem:

In the next example, **Figure 4-6,** you have a print that gives the overall length of a part and two step dimensions, but the center feature is not dimensioned. Set up the problem with the largest number or overall length first in your equation to prevent a negative answer. Also note that the denominators are not common.

$$1\frac{1}{2} - \frac{1}{2} - \frac{3}{4} =$$

This problem includes a ***mixed number***, or a whole number mixed with a fraction, which must also be simplified. Simplifying a mixed number is a simple process that should be completed before trying to solve for the unknown value. To simplify, multiply the whole number by the denominator, and then add the numerator to that number. The result becomes the numerator, while the denominator stays the same.

$$1\frac{1}{2}$$

Multiply Add   Same

Numerator: $(1 \times 2) + 1 = 3$
Denominator: 2

$$1\frac{1}{2} = \frac{3}{2}$$

The new ***improper fraction***, or fraction whose numerator is larger than its denominator, is 3/2. Place that value back in the original problem, then find the common denominator. Be sure to multiply each numerator by the same value as the denominator to maintain equality.

$$\frac{3}{2} - \frac{1}{2} - \frac{3}{4} = \frac{12}{8} - \frac{4}{8} - \frac{6}{8}$$

Now subtract 4 and 6 from 12 and simplify the result.

$$\frac{12}{8} - \frac{4}{8} - \frac{6}{8} = \frac{2}{8} = \frac{1}{4}$$

The missing dimension is 1/4″.

Confirm this result with your calculator and practice inputting a mixed number. The "trick" is inputting the whole number first, followed by the fraction key, and then proceeding as usual with the rest of the fractions:

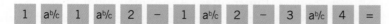

This may seem like a lot of key strokes, but this function becomes easy with practice.

4.2.3 Converting Fractions to Decimals

Now that you have an understanding of adding and subtracting fractions, practice converting fractions to decimals. Fractions cannot be used in CNC programs, so you must always convert them. The conversion from fractions to decimals is also important in measurements because most measuring tools operate in decimal mode.

The process of converting fractions to decimals is a straightforward division problem. The denominator is divided into the numerator, and the result is a decimal. Take a look at a sample problem:

$$\frac{3}{4} = 4\,\overline{\smash{\big)}\,3} \text{ or } 3 \div 4$$
$$\frac{3}{4} = .750$$

The process is even easier using a calculator:

$$\boxed{3}\ \boxed{\div}\ \boxed{4}\ \boxed{=}$$

4.2.4 Use of Thousandths Accuracy

Notice that the answer to the previous problem is listed as .750. Mathematically, it would be just as correct to have expressed the answer as .75, or even .7500. The trailing zero after a decimal does not hold any value. Similarly, any leading zero on the leftmost side of the decimal has no value.

$$00.75 = 0.75 = .750 = 0.750000 = 0000.7500000$$

All these numbers are equal. However, machinists have a specific way of communicating. To prevent confusion, machinists always "talk" or "work" to 3 places to the right of the decimal. This is called the *thousandths* place:

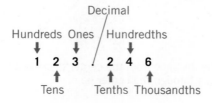

For example, a print dimension of .750 is read as "seven hundred fifty thousandths" (.750). If you were measuring a drill bit that was .5 in diameter, you would say "five hundred thousandths" (.500). This system provides an efficient and effective method of communication.

4.3 Geometric Shapes

Geometric shapes include 2-dimensional objects, such as circles, arcs, triangles, polygons, and quadrilaterals; and 3-dimensional objects, such as spheres, pyramids, cubes, and polyhedrons, **Figure 4-7.** Two-dimensional geometric shapes are flat. Imagine drawing a shape on a piece of paper. You can draw left and right or up and down, but only in 2 dimensions. A 3-dimensional shape has depth. Consider the difference between a circle and a sphere. A circle is 2-dimensional, and spheres are 3-dimensional. You should be familiar with basic geometric shapes and how they apply to CNC machining and calculations.

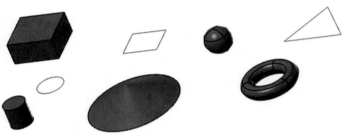

Goodheart-Willcox Publisher

Figure 4-7. Assorted 2-dimensional and 3-dimensional geometric shapes.

4.3.1 Circles

Circles are 2-dimensional objects often encountered in machining, such as drilling holes or cutting round shafts. A *circle* is defined as a closed plane curve that is an equal distance at all points from its center point. In simpler terms, picture a closed loop with a center point. If the distance from that center point to every point on the outside is exactly the same, the closed loop is a circle, **Figure 4-8.**

The distance from the center point to the edge of a cicle is defined as the *radius*. The radius is frequently used in creating CNC programs and calculating locations. The distance across the circle through the center point is the *diameter*. Hole sizes are often defined on a print in terms of diameter, whereas partial circles or corner blends are identified by radius. Pay close attention to printed specifications and verify the dimension as a radius or a diameter, **Figure 4-9.**

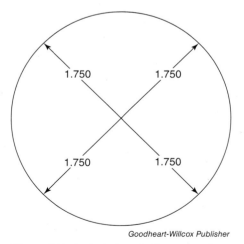

Goodheart-Willcox Publisher

Figure 4-8. A circle is a 2-dimensional shape defined by a center point that is equidistant from all points on a closed plane curve.

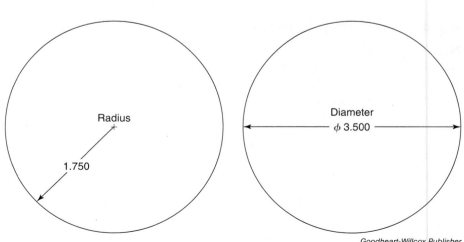

Goodheart-Willcox Publisher

Figure 4-9. A radius is the distance from the center to circle edge. The diameter is the distance from edge to edge through the center point.

4.3.2 Polygons

A *polygon* is a 2-dimensional shape with straight sides. Examples of polygons include triangles, rectangles, pentagons, and hexagons, **Figure 4-10**. Many of the machining projects in later chapters make use of polygons. Often, machining involves cutting rectangular blocks or multi-sided objects. The prefix *poly-* means many, and indeed polygons can have an unlimited number of sides.

Triangles

Triangles are particularly important polygons in machining applications. A triangle is made up of three sides that form three angles. There are several types of triangles, which are defined by the relationship between their sides and angles.

- A *scalene triangle*, **Figure 4-11**, has no equal sides or equal angles.

- An *isosceles triangle*, **Figure 4-12**, has two equal sides and two equal angles.

- An *equilateral triangle*, **Figure 4-13**, has all equal sides and all equal angles.

- A *right triangle*, **Figure 4-14**, has one 90° angle. A 90° angle is defined as two lines that are exactly perpendicular to each other. On a print, these triangles are designated by a box in the 90° corner of the triangle. The right triangle has special mathematical properties that make it advantageous for use in machining.

All triangles share the mathematical property that the sum of all three angles must equal 180°.

$$180° = \text{Angle 1} + \text{Angle 2} + \text{Angle 3}$$

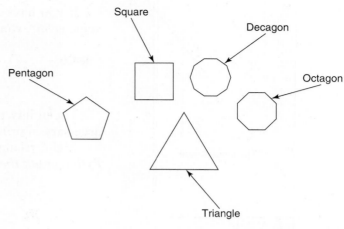

Goodheart-Willcox Publisher

Figure 4-10. Several examples of polygonal geometric shapes.

Scalene Triangle

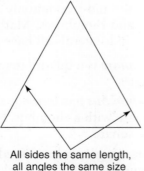

Goodheart-Willcox Publisher

Figure 4-11. A scalene triangle has no equal sides or angles.

Isosceles Triangle

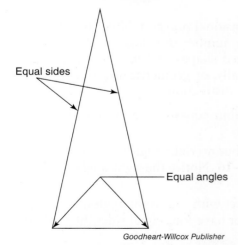

Goodheart-Willcox Publisher

Figure 4-12. An isosceles triangle has two equal sides and two equal angles.

Equilateral Triangle

All sides the same length, all angles the same size

Goodheart-Willcox Publisher

Figure 4-13. An equilateral triangle has all equal sides and all equal angles.

Right Triangle

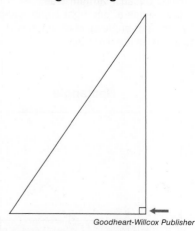

Goodheart-Willcox Publisher

Figure 4-14. A right triangle, denoted by a right angle symbol.

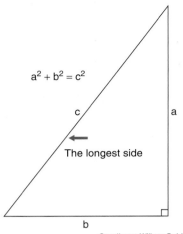

Goodheart-Willcox Publisher

Figure 4-15. Right triangle where sides can be calculated using the Pythagorean theorem.

If you have a triangle where you know one angle is 45° and another angle is 55°, you can calculate the third angle:

$$180° – \text{Angle 1} – \text{Angle 2} = \text{Angle 3}$$
$$180° – 45° – 55° = \text{Angle 3}$$
$$80° = \text{Angle 3}$$

The ability to calculate angles without all the angles defined is often necessary in print reading.

Right triangles also have a unique characteristic defined by the *Pythagorean theorem*:

$$a^2 + b^2 = c^2$$

The Pythagorean theorem states that, in a right triangle, the square of one side plus the square of the second side is equal to the square of the longest side, **Figure 4-15.** Note that the longest side in a right triangle is always across from the right angle and is called the *hypotenuse*.

Consider the following example: If side *a* is 2″ and side *b* is 1.5″, what is the length of side *c*?

$$a^2 + b^2 = c^2$$
$$(2″)^2 + (1.5″)^2 = c^2$$
$$4″ + 2.25″ = c^2$$
$$c = \sqrt{6.25} = 2.5″$$

To solve this problem on your calculator, press the following keys:

The last input key takes the square root. This last step changes c^2 to the desired solution, *c* or 2.5″. Using standard algebraic procedures, you can calculate any side of a right triangle if you know the other 2 sides. Additional mathematical methods for working with right triangles and finding all their sides and angles are covered in Section 4.5, Trigonometry.

Quadrilaterals

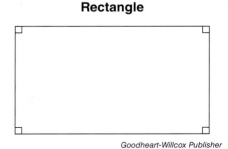

Goodheart-Willcox Publisher

Figure 4-16. Squares and rhombuses have four equal sides. A square has four right angles, while a rhombus has no right angles. Because all four sides are equal, the presence of one right angle, indicated by the right angle symbol, implies that all four angles measure 90°.

Rectangle

Goodheart-Willcox Publisher

Figure 4-17. A rectangle with four right angles and two sets of equal sides.

A *quadrilateral* is a 4-sided, 2-dimensional polygon. Note that the prefix *quad-* means four and indicates the number of sides. Quadrilaterals are among the most commonly machined shapes and include squares, rectangles, and rhombuses. Mathematically, or geometrically, these three different quadrilaterals all have specific distinctions.

- A *square* is a quadrilateral with four equal sides and four right (90°) angles.
- A *rhombus* has four equal sides but no right angles. Compare a square with a rhombus, **Figure 4-16.** Notice the right angle symbol in the square.
- A *rectangle* is also a 4-sided object with four right angles, **Figure 4-17,** but rectangles do not have four equal sides. Instead, they have two pairs of equal sides.

Observe that opposite sides are equal and parallel in all three quadrilaterals. This means squares, rhombuses, and rectangles are *parallelograms*.

4.3.3 Lines

Lines are an important geometric feature in machining. With the exception of machining circles, most CNC machining commands are along a line. A *line* is a continuous, straight, 1-dimensional geometric element with no end. This means a line goes into infinity. Since it is impossible to machine an infinitely long part, when the term "line" is used in machining, it actually refers to a *line segment*. A line segment is a line with a definitive beginning and end. Line segments, like lines, do not curve or waver. A curved line is actually an arc, or partial circle.

A line that touches an arc or circle at exactly one point is referred to as a *tangent line*. If a line crosses an arc or touches an arc at more than one point, it is not tangent, but intersecting, **Figure 4-18**.

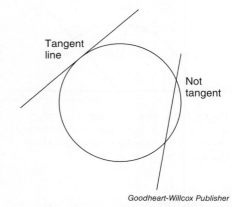

Goodheart-Willcox Publisher

Figure 4-18. A circle displaying a tangent and nontangent (intersecting) line.

4.4 Angles

Most machining requires a good working knowledge of angles. An *angle* measures the rotational distance between two intersecting lines or line segments from their point of intersection (or vertex) and is usually given in degrees, **Figure 4-19**. Always remember two mathematical truths about angles: Circles contain 360 degrees and lines by definition have 180 degrees. Because that is true, when one line intersects another line, it has a calculable relationship, **Figure 4-20**.

There are three main categories of angles:

- Recall that a *right angle* measures exactly 90° between two intersecting lines. When two lines intersect and form even one right angle, **Figure 4-21**, all other angles of intersection are also 90°. Observe that the sum of those angles is 360°.

- An *acute angle* measures less than 90°. Anything less than a right angle is an acute angle, **Figure 4-22**.

- An *obtuse angle* measures more than 90°, **Figure 4-23**. Notice that the second line intersects the first at an angle greater than a right angle.

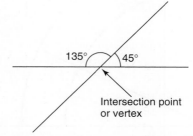

Goodheart-Willcox Publisher

Figure 4-19. Two lines intersecting at known angles.

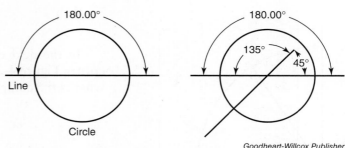

Goodheart-Willcox Publisher

Figure 4-20. Lines create angles with known relationships when they pass through a circle's center point.

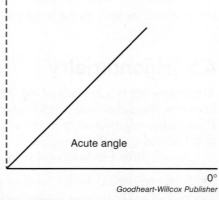

Goodheart-Willcox Publisher

Figure 4-21. Two lines intersecting at 90° angles.

Goodheart-Willcox Publisher

Figure 4-22. An acute angle, or one that is less than 90°.

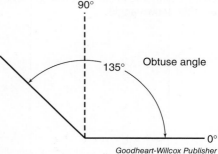

Goodheart-Willcox Publisher

Figure 4-23. An obtuse angle, or one that is more than 90°.

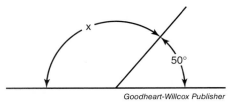

Figure 4-24. When one supplementary angle is known, the other supplementary angle can be calculated by subtracting the known angle from 180°.

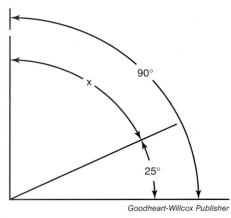

Figure 4-25. An unknown angle can be calculated by subtracting its known complementary angle from 90°.

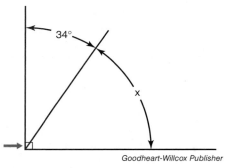

Figure 4-26. The right angle symbol indicates that the total angle measures 90°. An unknown complementary angle is easily calculated given a right angle and another known angle.

4.4.1 Supplementary Angles

Two angles are *supplementary angles* if they sum to 180°. Recall that a straight line is 180°, so if one line intersects another at 50°, you can calculate the second angle using supplementary angles, **Figure 4-24.**

$$50° + x = 180°$$
$$x = 180° - 50°$$
$$x = 130°$$

The prints used in a normal machining environment do not define every feature. In fact, it is incorrect to overidentify an object in a print. Often, one angle is defined, and you must calculate the second or supplementary angle.

 Tech Tip

Although you may need to calculate other data, the print defines what is inspected and to what tolerance. Chapter 5 covers this in more detail.

4.4.2 Complementary Angles

Complementary angles sum to 90°. Problems involving complementary angles normally start with a 90° or right angle that has been intersected by a line. You can find the missing angle that completes the 90° angle using your knowledge of complementary angles, **Figure 4-25.**

In this example, the given angle is 25°. What is the complementary angle, x?

$$90° = 25° + x$$
$$90° - 25° = x$$
$$x = 65°$$

In the next example, notice the right angle symbol in the lower left corner, **Figure 4-26.** This indicates a 90° angle. The known angle is 34°, so what is the complementary angle, x?

$$90° = 34° + x$$
$$90° - 34° = x$$
$$x = 56°$$

As you move forward with print reading, you will see prints that have right angles or right triangles. Often, the print may define only one angle. You must know how to calcualte any unknown angle using the principles given above as well as basic trigonometry described in the next section.

4.5 Trigonometry

Trigonometry is a branch of mathematics that deals with the relationships between the sides and angles of triangles and with functions of angles. Essentially, trigonometry is the mathematics of triangles and angles. This text focuses on right triangles.

Recall three fundamental principles for working with right triangles:

- Right triangles always have one angle of 90°.
- The sum of all three angles in a triangle is 180°.
- The Pythagorean theorem states that $a^2 + b^2 = c^2$.

Start by familiarizing yourself with the nomenclature of right triangles, **Figure 4-27**. Immediately, you should identify this as a right triangle because it has the right angle symbol. The hypotenuse, or longest side, is directly across from the right angle and designated by the letter H. The other two sides are named based on their relation to a given angle or for the angle for which you are trying to solve. They are called the *opposite side* and *adjacent side*, indicated by the letters O and A, respectively, **Figure 4-28**.

The Pythagorean theorem summarizes the relationship between sides of a right triangle, which is directly related to the angles they create. The mathematical relationships between sides and angles are given by three trigonometric functions:

- *Sine* of a given angle is the ratio of the opposite side to the hypotenuse.
- *Cosine* of a given angle is the ratio of the adjacent side to the hypotenuse.
- *Tangent* is equal to the ratio of the sides opposite and adjacent to an angle.

"SOHCAHTOA" is a helpful mnemonic for remembering the definitions of sine, cosine, and tangent.

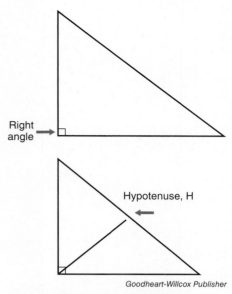

Figure 4-27. The right angle symbol denotes a right triangle. The side directly across from the 90° angle is known as the hypotenuse.

S = sine
O = opposite
H = hypotenuse
C = cosine
A = adjacent
T = tangent

$$(S)\ Sine = \frac{Opposite\ (O)}{Hypotenuse\ (H)}$$

$$(C)\ Cosine = \frac{Adjacent\ (A)}{Hypotenuse\ (H)}$$

$$(T)\ Tangent = \frac{Opposite\ (O)}{Adjacent\ (A)}$$

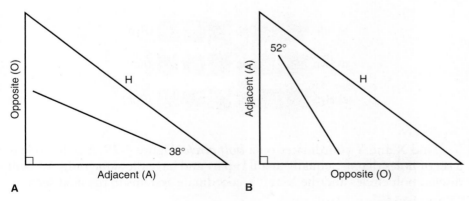

Goodheart-Willcox Publisher

Figure 4-28. The opposite and adjacent sides of a right triangle are determined relative to the known angle. A—This right triangle has one angle defined as 38°. The side directly opposite that angle is known as the *opposite side*. The side of the triangle that is next to or touching the 38° angle is called the *adjacent side*. B—This is the same triangle as in part A, but the known angle is now 52°, so the opposite and adjacent sides change positions.

Trigonometric functions can be used to determine all the sides and angles of a right triangle. Sine, cosine, and tangent can be calculated manually using charts or with a scientific calculator. Although charts are helpful tools, the probability of error rises with manual calculations. The quickest and most accurate calculations are made with a calculator. First, identify what sides and angles you know and what you are trying to find. Then, choose the appropriate formula and calculator inputs, shown below.

Find an angle when you know any two sides:

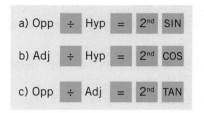

Find a side when you know one angle and one side:

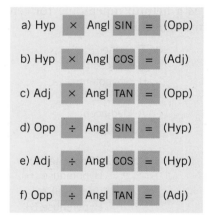

Find a side when you know two sides:

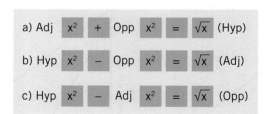

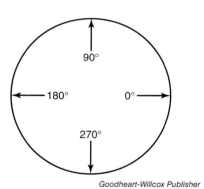

Goodheart-Willcox Publisher

Figure 4-29. To find the coordinates of a bolt circle, the 0° position is at 3 o'clock.

Find X and Y coordinates on a ***bolt circle***, **Figure 4-29**. A bolt circle is a set of holes that are equally angled apart and have the same radii. We will discuss bolt circles and the X and Y coordinate systems in the next section.

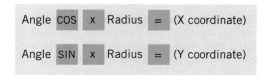

4.6 Cartesian Coordinate System

The structure used to plot and calculate machine positions and movement is known as the Cartesian coordinate system. This system is named for its developer, French scientist and mathematician René Descartes. The *Cartesian coordinate system* specifies each point uniquely in a plane with a pair of alphanumeric coordinates. You have used this system if you have ever bought an item from a vending machine like the one in **Figure 4-30**. You select the row of the item you want and then the number in that row. Specifying two points of intersection tells the machine where to go to find your item. The Cartesian coordinate system is also at work in locating a particular street address, **Figure 4-31**. To find 1244 Maple Street, you would travel up to Maple Street and then move along Maple until you reached 1244. These are two specific points on a single plane that intersect.

4.6.1 Number Lines

To understand how Cartesian coordinate grids work, look at a single-axis number line. You already learned about lines and line segments, but what if you need to find a specific point on a line, say 5? The starting point is known as the *origin*, or zero point. Establishing a common origin is essential for replicating movements. From the origin, you need to move 5, but is that right or left? On a number line, left is a negative (–) number and right is a positive (+) number, **Figure 4-32**.

Goodheart-Willcox Publisher

Figure 4-30. The rows and columns of a vending machine are an everyday example of the Cartesian coordinate system.

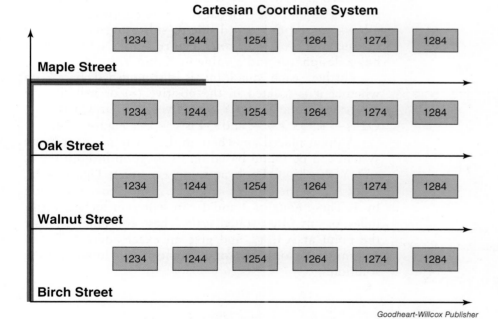

Goodheart-Willcox Publisher

Figure 4-31. A street map or address is a coordinate location system.

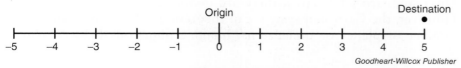

Goodheart-Willcox Publisher

Figure 4-32. A number line gives distance and direction of travel.

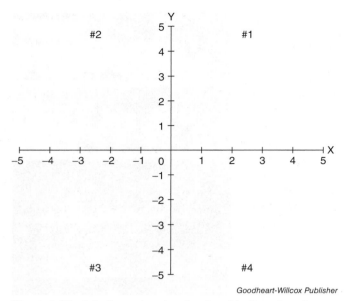

Goodheart-Willcox Publisher

Figure 4-33. A 2-dimensional or 2-axis grid allows locations to be determined in four quadrants. In machining, these two axes are designated as the X axis and Y axis.

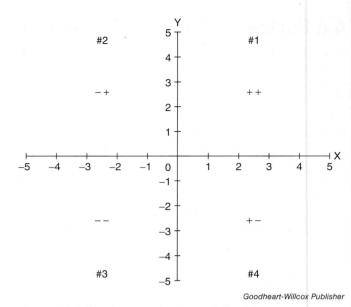

Goodheart-Willcox Publisher

Figure 4-34. The four quadrants are defined in terms of whether they are positive or negative along each of the two axes.

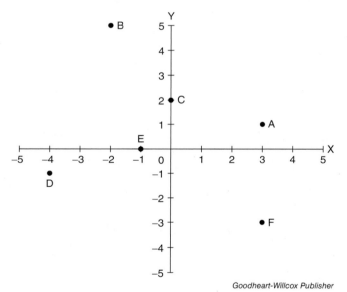

Goodheart-Willcox Publisher

Figure 4-35. Positions plotted in the four quadrants of an X and Y coordinate system.

(Solution: A—X 3, Y 1; B—X –2, Y 5; C—X 0, Y 2; D—X –4, Y –1; E—X –1, Y 0; F—X 3, Y –3.)

4.6.2 Four-Quadrant Plotting

The 2-axis coordinate system expands the number line into two directions of travel, or two axes known as the X axis and the Y axis. A 2-axis coordinate system allows up-and-down and side-to-side movement in four distinct areas, or *quadrants*, **Figure 4-33.** Each quadrant has a designator and a value in X and Y.

Further, each quadrant is distinguished based on whether it is located in the positive (+) or negative (–) direction on the X and Y axes. Everything in Q1 is + +, Q2 is – +, Q3 is – –, and Q4 is + –. See **Figure 4-34.**

A point placed anywhere in the Cartesian coordinate system can be found based on its location in both the X and Y direction. Look at some examples, **Figure 4-35.** Point A is to the right of origin by 3 and above the origin by 1. These units of 3 and 1 can represent inches, yards, miles, or any other stated scale. These coordinates locate the point at X3, Y1 and give an exact address. Try to determine the locations for all points before looking at the solution.

4.6.3 Three-Dimensional Plotting

So far you have been dealing with 2-dimensional coordinate systems. On a standard CNC lathe, there are two axes, X and Z, **Figure 4-36.** However, the Cartesian system can theoretically be used in an infinite number of planes or dimensions. Imagine numerous lines crossing a

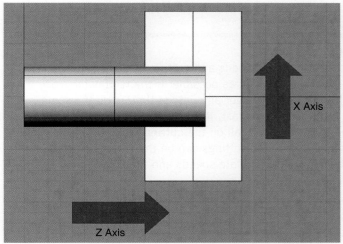

Goodheart-Willcox Publisher

Figure 4-36. A typical lathe 2-axis configuration.

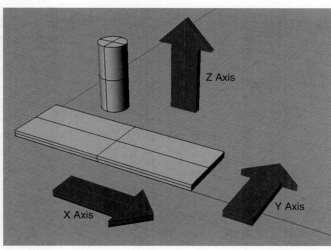

Goodheart-Willcox Publisher

Figure 4-37. A typical mill 3-axis configuration.

single intersection point and all those lines being designated as different axes. Any point anywhere can be plotted.

A CNC mill has three axes of movement, X, Y, and Z, **Figure 4-37.** The Z axis represents the tool going up and down. This additional axis allows machinists to move the cutting tool to the left and right (X), forward and back (Y), and up and down (Z). The origin remains fixed, as in a 2-dimensional machine, and anything above the origin is Z positive while anything below is Z negative.

An example of a machine with more than three axes is a 5-axis machining center where the table is allowed to rotate, **Figure 4-38.**

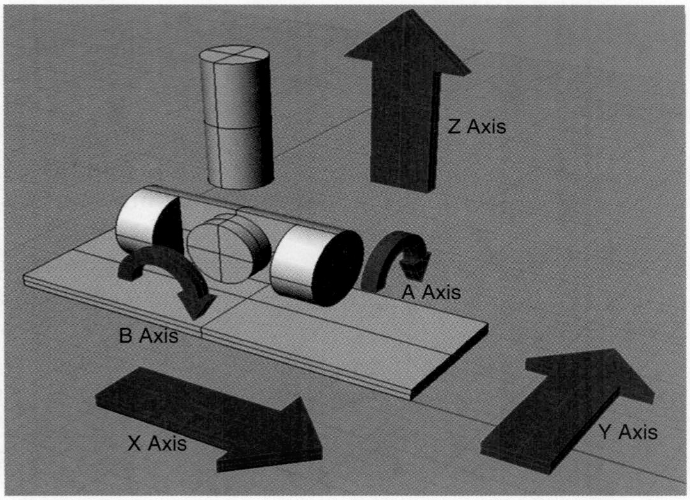

4.7 Calculating Speeds and Feeds

When cutting material on a CNC machine, the relationship between how fast the spindle turns and how fast the cutting tool moves across the material is vitally important. Machine shops are trying to get parts cut as quickly as possible. In addition, tooling reacts differently at different speeds and feeds. For example, if you turn an end mill very slowly and feed it into the material rapidly, the tool may suffer a catastrophic failure and shatter. Conversely, if you run an end mill at high speed but feed it slowly, it takes excessive time to cut the material and can dull the cutting edges of the tool. Calculating speeds and feeds in the modern machining environment is one of the most overlooked and miscalculated aspects of machining. Fortunately, there is a relationship between the material, the cutting tool, and the machine that can be easily calculated. Every good CNC programmer should be able to accurately calculate a tool's proper speed and feed.

Goodheart-Willcox Publisher

Figure 4-38. A 5-axis mill configuration, showing X, Y, and Z axes. The A and B axes are rotational axes.

From the Shop

Feeds and Speeds

Calculating the correct spindle speeds and cutting feeds is the best way to make your tools last as long as possible. If the feed is too slow, it means the tool is cutting for longer amounts of time to do the same amount of work. If the feed is too fast, it means the tool does not have enough time to cut and can fail on the cutting edge. Often, the best resource for speed and feed settings can be the tool distributor or a tooling engineer. Always calculate speeds and feeds to remain as efficient as possible.

4.7.1 Mill

The first factor in calculating speeds and feeds is the cutting tool. The modern machining environment primarily uses tools made from solid carbide, so that material is assumed in this text. Consult manufacturer specifications when working with other cutter materials. Milling calculations also depend on the diameter of the cutter and how many cutting edges it has.

The next important factor is the type of material being cut. Stainless steels do not cut as easily as aluminum, for example. Different materials have different rates of material removal or surface footage. *Surface footage* is the number of linear feet a location on a rotating component travels in one minute. This number can vary based on the tooling manufacturer, so it is best to get the surface footage range directly from the manufacturer.

Practice calculating the speed of two different end mills with two different types of materials. Consider a 1/2″ end mill cutting aluminum at 700 surface feet per minute (sfm) and a 3/8″ end mill cutting 304 stainless steel at 450 sfm. The formula for calculating speed, measured in revolutions per minute (rpm), is given by the following equation:

$$\frac{3.82 \times \text{sfm}}{\text{Diameter of Tool}} = \text{rpm}$$

Calculate the speed for the 1/2″ (.500″) end mill in aluminum.

$$\frac{3.82 \times 700}{.500} = 5{,}348 \text{ rpm}$$

The speed for this operation is 5,348 rpm. Remember that number for use in calculating the feed. Now calculate the speed for the 3/8″ end mill in stainless steel.

$$\frac{3.82 \times 450}{.375} = 4{,}584 \text{ rpm}$$

Now that you know speed, you can calculate feed rate. The formula for feed rate, measured in inches per minute (ipm), is given by the following equation:

Feed Rate (ipm) = rpm × Chip Load × Number of Cutting Edges (Flutes)

Observe that feed rate depends on chip load and the number of cutting edges as well as speed. *Chip load* refers to the actual thickness of the chip

Tech Tip

3.82 is a constant used in speed calculations. This value is the result of dividing the number of inches per foot by pi: $\frac{12}{3.14} = 3.82$.

being cut or the depth of each cutting edge as it passes through the material. Chip load varies with each cutter, but it should be between .005″ and .010″. Any end mill, whether solid carbide or inserted, will have multiple *flutes*, or cutting edges. Tools with more flutes can feed faster, but they are also weaker.

Return to the previous example and calculate feed rate for each cutter. The 1/2″ end mill in aluminum has a speed of 5,348 rpm and 3 flutes. Calculate feed rate as follows:

$$\text{Feed Rate (ipm)} = \text{rpm} \times \text{Chip Load} \times \text{Number of Cutting Edges (Flutes)}$$
$$= 5{,}348 \text{ rpm} \times .007 \times 3$$
$$= 112.3 \text{ ipm}$$

Now calculate the feed rate for the 3/8″ end mill in stainless steel with 2 flutes.

$$\text{Feed Rate (ipm)} = \text{rpm} \times \text{Chip Load} \times \text{Number of Cutting Edges (Flutes)}$$
$$= 4{,}584 \text{ rpm} \times .007 \times 2$$
$$= 64.2 \text{ ipm}$$

The feed rate for this operation is 64.2 ipm at a speed of 4,584 rpm.

4.7.2 Lathe

Calculating cutting speed for a lathe is the same as for a mill, except the diameter of the part being cut is used in place of tool diameter:

$$\frac{3.82 \times \text{sfm}}{\text{Cut Diameter}} = \text{rpm}$$

As above, 3.82 is a constant and the surface footage is determined by the material. Divide by the size of the part you are cutting to determine speed in rpm. Consider cutting a 1.500″ diameter bar of aluminum with a 750 sfm rating:

$$\frac{3.82 \times 750}{1.5} = 1{,}910 \text{ rpm}$$

Machining a bar on a lathe often requires multiple cuts to achieve the finished size. In such cases, speed should be changed with every cut. This problem is addressed in greater detail in Chapter 10.

To calculate the feed rate for a lathe, there are two options. Feed rate can be given in inches per minute (ipm) as above, but lathes can also be programmed in inches per revolution (ipr). Look at the ipm method first:

$$\text{rpm} \times \text{Chip Load} = \text{ipm}$$
$$1{,}910 \times .007 = 13.37$$

The most common method, and natural choice for CNC lathes, is to program in ipr. The previous formulas for mills used a chip load of .007 to control the thickness of the chip. In a lathe, this is already achieved by a rotating bar that controls tool advancement. The tool is advanced a certain distance every time the bar makes one rotation. In this case, you could program in ipr and use .007 with the result that every time the bar rotates the tool advances .007. You can simplify the process and reduce chances for error by using the desired value for ipr in your program.

 **Tech Tip**

Note that the number of flutes is always 1 because lathes cut with a single cutter edge.

Chapter Review

Summary

- There are three parts to a fraction: the numerator, denominator, and (if applicable) whole number.
- To add or subtract a fraction, you must first have a common denominator.
- To convert a fraction to a decimal, divide the numerator by the denominator.
- In the machining environment, it is standard practice to express decimals to 3 places, or in thousandths.
- Geometric shapes, such as circles, polygons, and lines, are commonly used in machining.
- Triangles are three-sided polygons that can be classified into four types: isosceles, scalene, equilateral, and right. The sum of all three angles in any triangle is 180°.
- A right triangle contains exactly one 90° angle. The longest side is known as the hypotenuse, and the other two sides are referred to as the opposite and adjacent sides.
- The Pythagorean theorem states that in a right triangle, $a^2 + b^2 = c^2$, where a and b represent the opposite and adjacent sides and c represents the hypotenuse, or longest side.
- Quadrilaterals are four-sided polygons. Common quadrilaterals in machining applications include squares, rhombuses, and rectangles, which are also classified as parallelograms.
- A line is a continuous, straight, 1-dimensional geometric element with no end; a line segment is a line with a definitive beginning and end; and a line that touches an arc or circle at exactly one point is referred to as a tangent line.
- Two angles are supplementary if they sum to 180°, and two angles that sum to 90° are complementary.
- Trigonometry is a branch of mathematics that deals with the relationships between the sides and angles of triangles and with functions of angles. Common trigonometric functions include sine, cosine, and tangent.

- The Cartesian coordinate system specifies the location of a point with coordinates that represent the distance of the point from the origin on each axis.
- A 2-axis coordinate system allows up-and-down and side-to-side movement in four distinct areas, or quadrants. A 3-axis coordinate system adds a vertical plane bisecting the origin.
- CNC machinists must be able to calculate correct speeds and feeds based on machine, cutting tool, and material being cut.

Review Questions

Answer the following questions using the information provided in this chapter.

Know and Understand

1. Many _____ define dimensions in fractional format to maintain a high level of accuracy.

   A. calculators
   B. inspectors
   C. prints
   D. supervisors

2. Machinists are often required to work with _____ that have different denominators, so it is important to be able to add and subtract them as needed.

   A. angles
   B. fractions
   C. numbers
   D. triangles

3. *True or False?* A mixed number is a whole number mixed with a fraction.

4. To prevent confusion, machinists always "talk" or "work" to 3 places to the right of the decimal. This is called the _____ place.

   A. hundredths
   B. tenths
   C. thousands
   D. thousandths

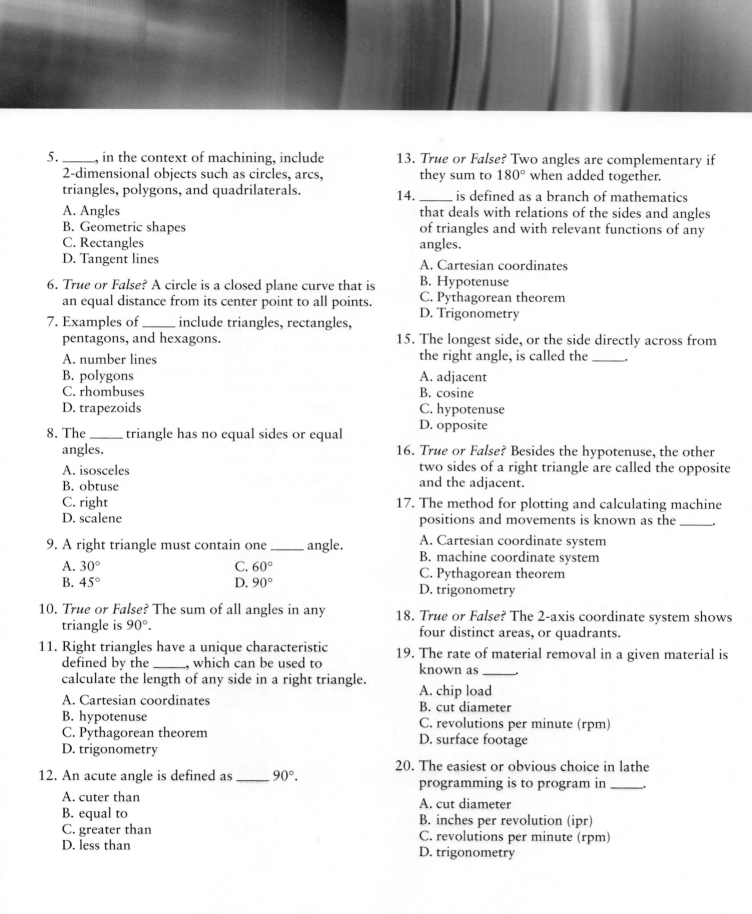

5. _____, in the context of machining, include 2-dimensional objects such as circles, arcs, triangles, polygons, and quadrilaterals.

   A. Angles
   B. Geometric shapes
   C. Rectangles
   D. Tangent lines

6. *True or False?* A circle is a closed plane curve that is an equal distance from its center point to all points.

7. Examples of _____ include triangles, rectangles, pentagons, and hexagons.

   A. number lines
   B. polygons
   C. rhombuses
   D. trapezoids

8. The _____ triangle has no equal sides or equal angles.

   A. isosceles
   B. obtuse
   C. right
   D. scalene

9. A right triangle must contain one _____ angle.

   A. 30°   C. 60°
   B. 45°   D. 90°

10. *True or False?* The sum of all angles in any triangle is 90°.

11. Right triangles have a unique characteristic defined by the _____, which can be used to calculate the length of any side in a right triangle.

   A. Cartesian coordinates
   B. hypotenuse
   C. Pythagorean theorem
   D. trigonometry

12. An acute angle is defined as _____ 90°.

   A. cuter than
   B. equal to
   C. greater than
   D. less than

13. *True or False?* Two angles are complementary if they sum to 180° when added together.

14. _____ is defined as a branch of mathematics that deals with relations of the sides and angles of triangles and with relevant functions of any angles.

   A. Cartesian coordinates
   B. Hypotenuse
   C. Pythagorean theorem
   D. Trigonometry

15. The longest side, or the side directly across from the right angle, is called the _____.

   A. adjacent
   B. cosine
   C. hypotenuse
   D. opposite

16. *True or False?* Besides the hypotenuse, the other two sides of a right triangle are called the opposite and the adjacent.

17. The method for plotting and calculating machine positions and movements is known as the _____.

   A. Cartesian coordinate system
   B. machine coordinate system
   C. Pythagorean theorem
   D. trigonometry

18. *True or False?* The 2-axis coordinate system shows four distinct areas, or quadrants.

19. The rate of material removal in a given material is known as _____.

   A. chip load
   B. cut diameter
   C. revolutions per minute (rpm)
   D. surface footage

20. The easiest or obvious choice in lathe programming is to program in _____.

   A. cut diameter
   B. inches per revolution (ipr)
   C. revolutions per minute (rpm)
   D. trigonometry

For questions 21–27, match each term with the correct geometric shape.

    A. Isosceles triangle
    B. Right triangle
    C. Square
    D. Scalene triangle
    E. Equilateral triangle
    F. Rectangle
    G. Rhombus

21.

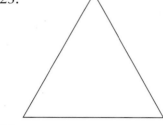

22.

23.

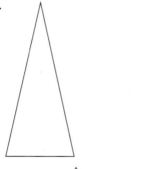

24.

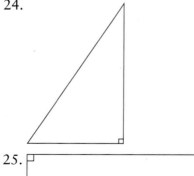

25.

26.

27.

Apply and Analyze

1. Solve for 3/8 + 1/4 and reduce as needed.

2. Solve for 5/8 − 1/2 and reduce as needed.

3. Convert 3/32 to its decimal equivalent. Round to the nearest thousandths.

4. Convert 1 3/16 to its decimal equivalent. Round to the nearest thousandths.

5. List and briefly describe the four types of triangles discussed in the chapter.

6. Use the Pythagorean theorem to solve for the unknown side of this right triangle.

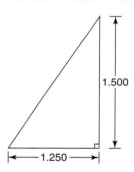

1.500

1.250

7. Use the Pythagorean theorem to solve for the unknown side of this right triangle.

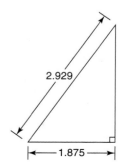

2.929

1.875

8. Label all the sides of this right triangle.

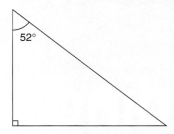

9. Use a calculator to solve for the missing angle of this right triangle.

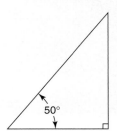

10. Use a calculator to solve for the missing side of this right triangle. Report answer to the nearest thousandths.

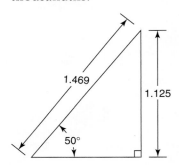

11. Use a calculator to solve for the two missing angles of this right triangle. Round answers to the nearest whole number.

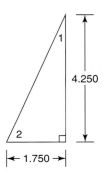

12. Using a 5/8″ diameter end mill, what is the correct speed, in rpm, for cutting aluminum at 750 surface feet per minute (sfm)?

13. When cutting mild steel in a milling machine with a calculated speed of 3,726 rpm, what is the feed rate in inches per minute (ipm)? Assume .005″ chip load with a 3-flute end mill.

Critical Thinking

1. Take a look around your home, school, or workplace and identify various geometric shapes. Can you measure and calculate any irregularities in these shapes? How do these irregularities affect the construction of these shapes?

2. Consider how you use the Cartesian coordinate system every day, in simple tasks such as giving directions, buying from a vending machine, or organizing a task. Can you write out directions from your work or school to home, using only distance and axis direction?

3. Depending on metal composition and cutting tools used, surface footage calculations can vary. How does this impact your machining time? Why does this make it so important to analyze material and cutting tools?

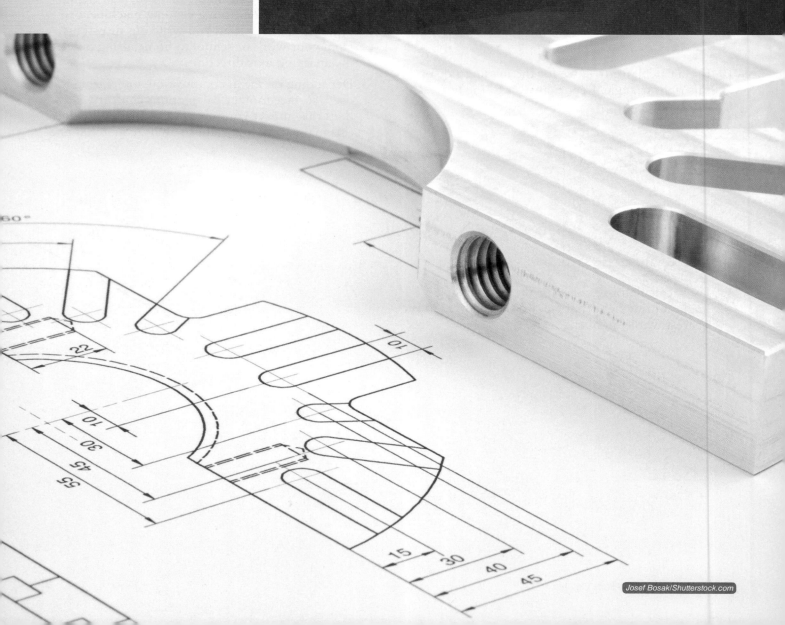

5 Print Reading

Chapter Outline

5.1 Introduction to Print Reading

5.2 Drawings and Prints
5.2.1 Notes
5.2.2 Visualizing Shapes and Drawing Views
5.2.3 Sheet Sizes

5.3 Title Blocks
5.3.1 Part Number
5.3.2 Drawing Number
5.3.3 Material Description
5.3.4 Tolerance Block

5.4 Auxiliary and Sectional Views
5.4.1 Cutting Plane Lines
5.4.2 Section Lines
5.4.3 Half Sections
5.4.4 Broken-Out Sections

5.5 Line Usage
5.5.1 Visible Lines
5.5.2 Extension Lines
5.5.3 Dimension Lines
5.5.4 Leader Lines
5.5.5 Hidden Lines
5.5.6 Centerlines
5.5.7 Symmetry Lines

5.6 Dimensions
5.6.1 Fractional Dimensions
5.6.2 Decimal Dimensions
5.6.3 Reference Dimensions
5.6.4 Tabular Dimensions
5.6.5 Basic Dimensions
5.6.6 Polar Coordinates

5.7 Tolerances
5.7.1 Limiting Tolerance
5.7.2 Bilateral Tolerance
5.7.3 Unilateral Tolerance
5.7.4 Unspecified Tolerance

5.8 Holes
5.8.1 Countersink
5.8.2 Counterbore and Spotface

5.9 Arcs
5.9.1 Fillets
5.9.2 Radius

5.10 Threads

5.11 Geometric Dimensioning and Tolerancing

Learning Objectives

After reading this chapter, you should be able to:

- Explain the basic use of a print.
- Identify notes on a print.
- Identify drawing views and projections.
- Interpret parts of a title block.
- Explain auxiliary and sectional views.
- Describe the use of lines, both hidden and visible.
- Explain the use of dimensions, both fractional and decimal.
- Identify various dimensions.
- Demonstrate proper use of polar coordinates.
- Identify correct use of tolerances.
- Identify multiple types of tolerances.
- Describe various features of holes, arcs, and radii.
- Identify proper thread print identification.
- Identify basic geometric dimensioning and tolerancing (GD&T) symbols.

Key Words

alphabet of lines
arcs
auxiliary view
bilateral tolerance
broken-out section
centerline
counterbore
countersink
cutting plane line
dimension
dimension line
drawing number
extension lines
fillets

geometric dimensioning and tolerancing (GD&T)
go/no-go gages
half section
hidden line
International Organization for Standardization (ISO)
leader line
limiting tolerance
limits
multiview drawing
note
orthographic projection

part number
pitch
polar coordinate
print
section lines
sectional view
spotface
symmetry line
title block
tolerance
tolerance block
unilateral tolerance
unspecified tolerance
visible lines

5.1 Introduction to Print Reading

Print reading is a fundamental skill for a CNC machinist. Machinists must be able to understand and translate a print before a part can be made or a project can be started. A **print** is the plan that transfers information from the designer to the builder. The designer, often an engineer or architect, has a grand vision of the finished project, but the actual building is completed by a carpenter, machinist, welder, or some other skilled craftsman. A print lets the builder know exactly what the designer envisioned.

5.2 Drawings and Prints

Today's prints are modernized blueprints. **Blueprint** refers to an old process for copying prints. In the mid-1860s, a process was developed in which light-sensitive chemical was used to coat paper to be copied. When the paper was exposed to light, its contents transferred onto another surface in a blue tint, **Figure 5-1**. The arduous process of making multiple drawings by hand was over. The designer could draw the plan once and copy it as a blueprint as many times as was needed.

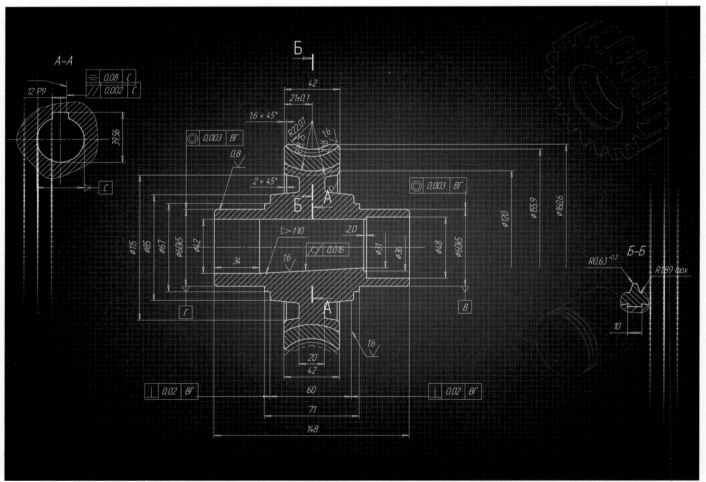

Figure 5-1. A chemically treated paper exposed to light creates the "blue" print.

The process of blueprinting lasted until the 1940s, when it was replaced by a more modern and less expensive method known as the cyanotype method. The cyanotype method allowed copies to be made onto white paper. These were often called white prints because the blue background was now gone. However, the term *blueprint* is still commonly used by machinists, even though it is not technically accurate anymore.

Modern printing, copying, and plotting have long made blueprinting obsolete. In the last 20 years, even white prints are becoming obsolete. Today's modern means of information transfer is digital, particularly through the use of computer-aided drafting (CAD) programs. To put this in perspective, consider an example: a modern car has about 30,000 parts. That car can be built in a CAD program, piece by piece, and assembled and tested virtually, without a single physical prototype or test piece. Every piece and part can be tested for form, fit, and function before it ever goes to production. When a part needs to be manufactured, that single piece can be isolated and sent to the manufacturer, just as designed and with only a few keystrokes. If that single piece needs some modification, it can be done in a master file and retested, with all the assembly being updated. Modern cars, aircraft, and ships are all being designed, assembled, tested, and manufactured without a single plan ever printed. Various parts of a print encode and transfer information from designer to builder.

Thinking Green

Electronic CAD Drawings

Digital or electronic prints have become more common for companies in construction and machining. This greener practice of drawings can be used to increase jobsite efficiency, cut costs, and significantly reduce paper consumption. When there are miscalculations or errors in prints, more paper consumption is required for corrections and reprints. Digital prints can be distributed and accessed through either PDF or CAD files. In addition, digital prints allow designers to make changes, calculations, and status updates to their drawings while eliminating extra time for printing and additional paper use. Other users then have access to the latest prints immediately. CAD programs can also produce 3D sketches of prints to provide a more accurate view of the design.

5.2.1 Notes

A typical 2-dimensional print, **Figure 5-2**, describes the plan for a part with specifically sized and positioned features. Often the print contains additional *notes* beyond graphical information about part shape. Notes provide supplementary information needed for manufacturing or post-manufacturing processes, **Figure 5-3**.

Notes can include special instructions for painting or plating, specialized inspection, or a wealth of other information. Examine notes carefully prior to machining to understand the full requirements of the finished part. Notes can easily be overlooked, resulting in projects being scrapped because of missing detail.

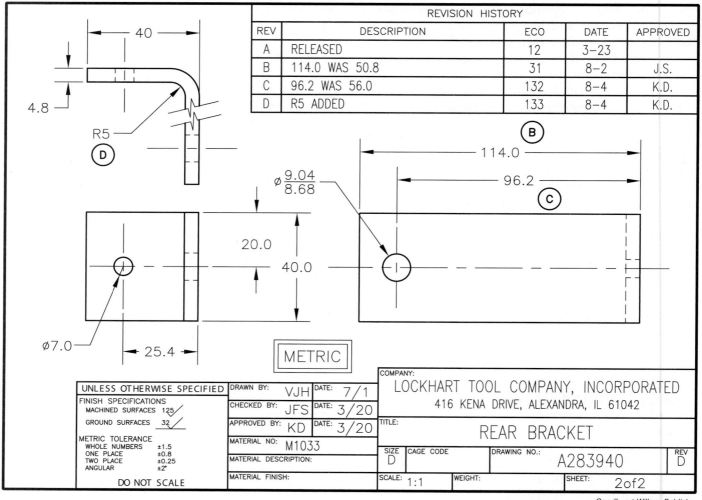

Figure 5-2. A 2-dimensional multiview print.

Goodheart-Willcox Publisher

NOTES: 1. ALL DIAMETERS CONCENTRIC TO ⌀ "A" WITHIN .0005
2. ±.005 ON FRACTIONAL DIMENSIONS
3. BREAK SHARP EDGES EXCEPT AS NOTED
4. REMOVE ALL BURRS
5. DELIVER WITH RUSTPROOFING LPS #1 OR EQUIV.

Goodheart-Willcox Publisher

Figure 5-3. Notes include additional instructions, such as treating sharp edges and removing burrs. In this example, a note is also added about post-machining treatments, which might include rustproofing.

5.2.2 Visualizing Shapes and Drawing Views

Identifying what you are "seeing" on a print and how that information moves around a drawing is the first step in understanding the piece being produced. The most common way to describe a 3-dimensional part in a 2-dimensional drawing is through the use of a *multiview drawing.*

A multiview drawing includes multiple views of a part. Generally, the number of views included on a multiview drawing is the minimum number of views needed to fully describe the part.

The views are created using a process called *orthographic projection*. The views created using orthographic projection are offset by 90° from one another. To visualize this, imagine a part placed in a glass box. Each view produced by orthographic projection would be what you see when looking through one side of the glass box. The most common way to orient a part in orthographic projection is *third-angle projection*. Third-angle projection establishes the rotation of the part in the print.

For example, consider the simplified part shown in **Figure 5-4**. This part would be difficult to define in a 2-dimensional print without additional views. In orthographic projection, the front view is defined first, then subsequent views are described based on that front view. Because this is a third-angle projected print, the faces of the part can be laid out in a standard format, **Figure 5-5**.

Defined in multiple views, the part in **Figure 5-4** appears as **Figure 5-6**. Take note of feature locations, such as the angle, slots, and partial slot. These features remain in the same alignment through all views and are only completely visible when shown in multiple views.

5.2.3 Sheet Sizes

Prints come in a variety of sizes, **Figure 5-7**. If you have a larger part, or even an assembly that requires more views or more space, the physical size of the print can vary. Standard print sizes and other features are defined by the *International Organization for Standardization (ISO)*, an international standard-setting body that promotes global industrial standards. You may also encounter standards set by the American Society of Mechanical Engineers (ASME), an American association that sets and promotes standards for the design and construction of industrial mechanical equipment.

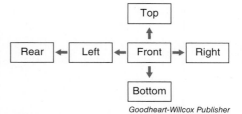

Goodheart-Willcox Publisher

Figure 5-4. An example of a 3-dimensional part that requires multiple views to define all features. In this third-angle projection print, six views are named starting from the thin top face, which is designated as the top view. From there, the right, left, front, back, and bottom views are established.

Goodheart-Willcox Publisher

Figure 5-5. Two-dimensional view of third-angle projection part rotation in a print.

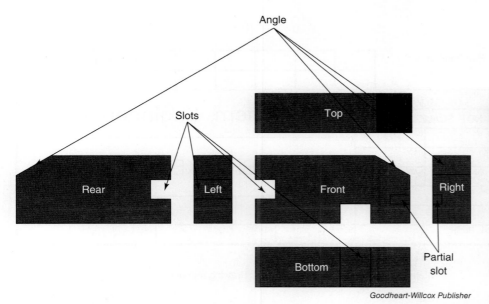

Goodheart-Willcox Publisher

Figure 5-6. Orthographic projection of a part through six possible viewing directions.

Standard Print Sizes		
Letter Size	**ISO Standard Size**	**Engineering Standard**
A	A4	8.5 × 11
B	A3	11 × 17
C	A2	17 × 22
D	A1	22 × 34
E	A0	34 × 44
F	—	28 × 40

Goodheart-Willcox Publisher

Figure 5-7. Standard print sizes as defined by ISO and ASME.

5.3 Title Blocks

A basic *title block* contains identifying information about the part and the print, **Figure 5-8**. It also contains supplementary information, including tolerances, scale, materials, and even designer. The title block is normally located in the bottom-left corner of the print layout.

5.3.1 Part Number

All parts should be given a unique *part number* to identify and distinguish them from very similar parts, or families of parts. Imagine the number of similar parts that might be used in an aircraft. Without unique part numbers to identify the parts being produced, confusion could arise in ordering and manufacturing.

5.3.2 Drawing Number

The *drawing number* is a print identifier that can be associated with a particular part number or appear as a stand-alone engineering number. Check the drawing number against the order to ensure the correct revision and drawing are being applied.

5.3.3 Material Description

The material description might be the most important—yet overlooked—aspect of a print. Materials can appear identical but be significantly different. For example, aluminums all look alike, but 6061 grade and 2024 grade have very different properties. Steels look alike, but 1018 and 4140 steels are different structurally. Using the wrong material to make a part can have catastrophic results. Double check the material specified against the material provided to make sure you have the right material for machining.

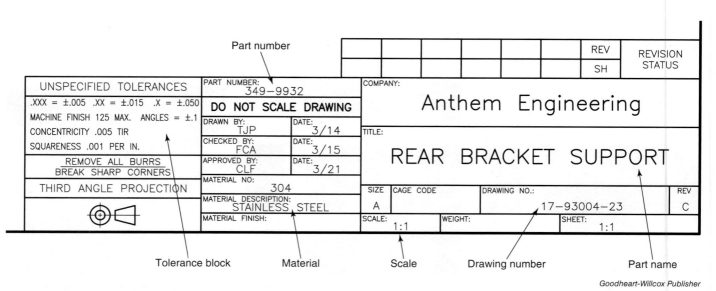

Goodheart-Willcox Publisher

Figure 5-8. Components of a print title block can include part name, part number, material, scale, and tolerance block.

5.3.4 Tolerance Block

The *tolerance block* sets limits for machining dimensions. Most features are dimensioned with specific limits for size, but not all tolerances are completely defined. On a print, a specific feature often has a dimension with no limits specified. For example, a chamfer that is defined as 45° × .125 with no tolerance limits does not have to be "exactly" that size. Every feature needs some tolerance because there are no perfect parts. Refer to the tolerance block in **Figure 5-8** and note that angles are ±1°. This means a 45° angle must fall between 44° and 46°. Also notice that .125 is defined by the 3-place decimal tolerances .XXX of ± .005. This means the dimension can be between .120 and .130.

5.4 Auxiliary and Sectional Views

Beyond normal 2-dimensional views, prints often show additional views to fully define all the features of a part. Parts may need to be cut or sectioned to show internal or hidden features that cannot be shown in normal views. These views are called *sectional views*. Views may also be required for parts oriented at an angle other than 90°. These projected orientations are referred to as *auxiliary views*.

5.4.1 Cutting Plane Lines

A part may need to be sectioned or rotated and shown in an additional view. In such cases, the print must define where a cut occurs and in which direction it is shown in additional views. A cut on a print is indicated with a *cutting plane line*. The print in **Figure 5-9** includes a cutting plane line.

In **Figure 5-9**, this part needs to be sectioned to show internal hole features. The dashed line with the letter "A" on both ends shows where the part is cut and defines Section A–A. The direction of the arrows at the end of the cutting plane line shows the direction of view for the segment. **Figure 5-10** shows the cut side from the direction of view noted. Also observe that the sectioned view has crosshatched lines through it to indicate that the part is not solid from that view.

5.4.2 Section Lines

In **Figure 5-10**, Section A–A, a set of diagonal or crosshatched lines appears. These lines indicate a feature that is not on a normal external side of a part and are known as *section lines*. The arrows in the cutting plane lines indicate direction of view for a given section. The part is then rotated in third-angle projection and references the front view.

5.4.3 Half Sections

Hidden or internal features are especially difficult to display in prints. As in the previous example, sometimes section lines are used to show

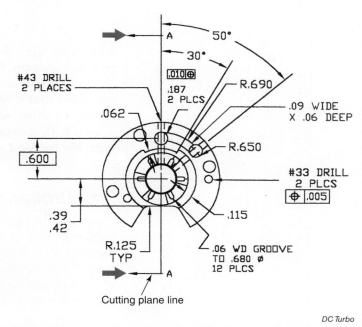

DC Turbo

Figure 5-9. The line marked A–A indicates a cutting line with arrows noting direction of view.

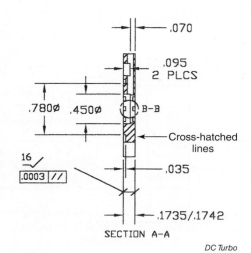

DC Turbo

Figure 5-10. A right-side view of the A–A cut part. The crosshatched lines denote a cut surface.

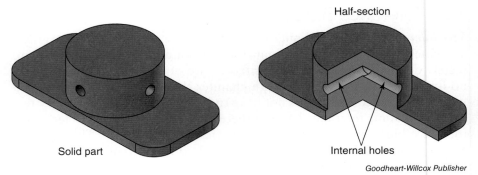

Half-section

Solid part

Internal holes

Goodheart-Willcox Publisher

Figure 5-11. A solid part with internal features requiring a half-section view to define.

hidden features. Alternatively, the part can be split by halves or quarters in a **half section**. Consider a solid model drawing, **Figure 5-11**. You can see in the half-section drawing on the right that there are two internal holes that intersect in the center of the part. That feature would be harder to show without cutting the part. The front view of the part is drawn with crosshatching to show the section that is cut away from the part, **Figure 5-12**.

5.4.4 Broken-Out Sections

Occasionally a print cuts away only a small portion of the part to show a hidden feature or give more detail. Consider a drawing with four identical features on a single plane, **Figure 5-13**. There is no need to identify each feature independently, and it is not correct to do so. The limited cutaway of a small section is known as a **broken-out section**. In this case, a partial section is enough to show the detail of the hidden features.

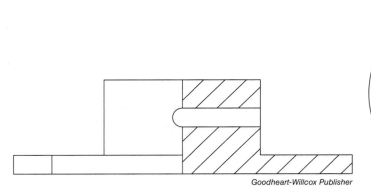

Goodheart-Willcox Publisher

Figure 5-12. A print representation of the solid part in Figure 5-11. The crosshatched lines indicate the part has been cut away to show internal features.

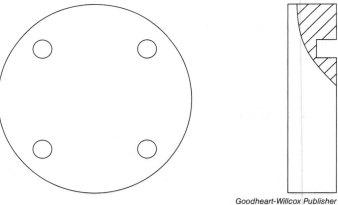

Goodheart-Willcox Publisher

Figure 5-13. A broken-out view to remove a section of part and display internal features.

5.5 Line Usage

Typical prints make use of many different types of lines. Lines define visual objects, hidden objects, and even dimensions. Lines can have different thicknesses or be dashed in certain configurations to determine intent. Lines on prints have specific definitions and intentions defined by the ASME Y14.2, Line Conventions and Lettering standard. This standard is referred to as the *alphabet of lines*.

5.5.1 Visible Lines

Visible lines are also referred to as *object lines*. Visible lines define the part shape or surface of an object. They show the actual edges or outlines of the object that are visible in any given view. Visible lines are solid, thick lines that should stand out on a print, **Figure 5-14**.

5.5.2 Extension Lines

When creating or interpreting dimensions on a print, there are a series of lines, numbers, and arrows that specify any given dimension. *Extension lines* extend from the edges of a feature and contain the dimension required, **Figure 5-14**. Extension lines define the exact limits of a feature.

5.5.3 Dimension Lines

Dimension lines appear between extension lines and indicate the extent of a part dimension. Dimension lines usually have an arrowhead on each end that can be solid or open and point to extension lines. Although dimension lines appear as two separate lines with the dimension in the middle, they are actually broken solid lines, **Figure 5-14**.

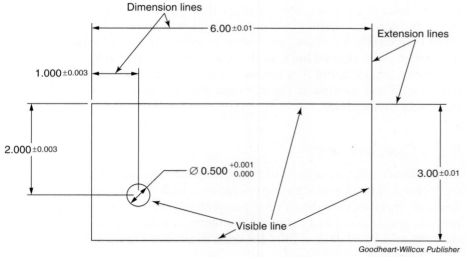

Goodheart-Willcox Publisher

Figure 5-14. Multiple lines, including visible lines, extension lines, and dimension lines, are used to indicate various features and part information.

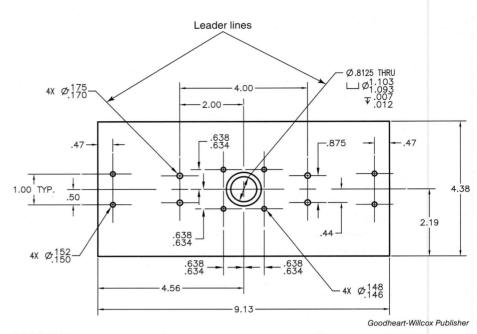

Figure 5-15. Leader lines are used for clarity to extend away from a part and add a note or dimension.

5.5.4 Leader Lines

A *leader line* is used to point to a specific location or note on a print. A leader line can be used for a special notation or reference, or it can even have a dimension if required. These lines are thin and continuous with an arrowhead on one end and a vertically centered note on the opposite end. See **Figure 5-15**.

5.5.5 Hidden Lines

Hidden lines are used to show a feature that may be visible in another view but is hidden or invisible in the current view, **Figure 5-16**. Hidden lines often help clarify a feature position in an alternate view. Hidden lines are medium-weight, dashed lines. In the example in **Figure 5-16**, a circle is visible in the top view, but it is unclear exactly what the circle is without the left view. When looking at the left view, the feature is not visible without the use of hidden lines.

5.5.6 Centerlines

Centerlines designate centers of holes, diameters, radii, or other symmetrical features. In many machining operations, such as drilling or boring, the designer specifies coordinates for the center point of a round feature. To note the center point of a round or symmetrical feature, prints use alternating long and short dashed lines for centerlines. See **Figure 5-17**.

5.5.7 Symmetry Lines

Symmetry is the quality of having the same features facing each other across a line or around an axis. In machining, parts are often symmetric around an axis. When creating a print for a symmetric part, it is not necessary to list all dimensions for both sides of the part. Instead, simply

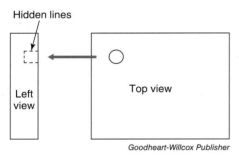

Figure 5-16. Hidden lines show features that cannot be seen in another view.

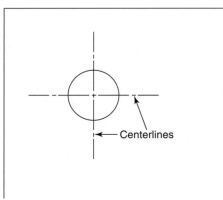

Figure 5-17. Centerlines indicate the center position of a feature. They are most commonly used in round or circular features.

dimension one set of features and show a *symmetry line*. In **Figure 5-18**, objects on the left half are the same, or mirrored, on the right half. By using a symmetry line, you need only define one set of objects.

5.6 Dimensions

A *dimension* defines the measurable extent of an object. Dimensions indicate an object's length, width, depth, height, or even girth and can be defined by any unit of measurement, including all US customary and metric units. Carefully examine a print and evaluate units of measurement to identify and understand the part you are machining. Most prints in US manufacturing are either in inches or millimeters. This text examines dimensions only in inches. Although for many years the US machining industry has attempted to convert to a global standard of metric dimensioning and part definition, most prints in US machine shops are still in inches or easily converted to inch increments through CAD or CAM software.

5.6.1 Fractional Dimensions

Dimensions can give insight into *design intent*, or how to interpret a designer's idea. For example, a dimension can be defined in a fractional format known as a *fractional dimension*, **Figure 5-19**. A dimension called out with a fraction is usually not as critical as other dimensions. High and low tolerances for fractional dimensions should also be defined by fractional limits.

 Tech Tip

Refer to the tolerances in the title block to determine the tolerance on fractional dimensions.

5.6.2 Decimal Dimensions

A *decimal dimension* is the most common form of dimensioning. As the name implies, decimal dimensions use numeric decimals to define dimensions of a part. These dimensions can be stated with as few as one decimal place or as many as required, but decimal dimensions do not normally have more than four decimal places. The numbers after the decimal also denote the accuracy level of a dimension. A decimal dimension with one number after the decimal is not as critical as a dimension with three numbers after the decimal place.

The print in **Figure 5-20** includes one-, two-, and three-place decimal dimensions. Note that the tolerance listed in the title block indicates different tolerances for appropriate decimal dimensions.

5.6.3 Reference Dimensions

A *reference dimension* is a dimension on a print or drawing for information or reference only. Reference dimensions are often used to clarify a dimension that might be defined in other views or a stack of other dimensions if added together. Reference dimensions should not be used to manufacture

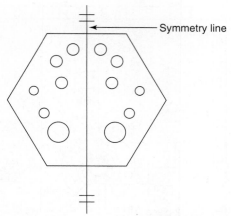

Goodheart-Willcox Publisher

Figure 5-18. Symmetry lines define a mirror-imaged feature and can eliminate the need for duplicate dimension definitions. In this case, the part is symmetric around the Y axis.

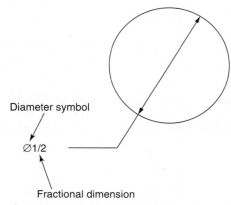

Goodheart-Willcox Publisher

Figure 5-19. A fractional dimension preceded by a symbol for diameter. This example shows a hole size of 1/2″. This is a common way of dimensioning a drilled hole using a 1/2″ drill.

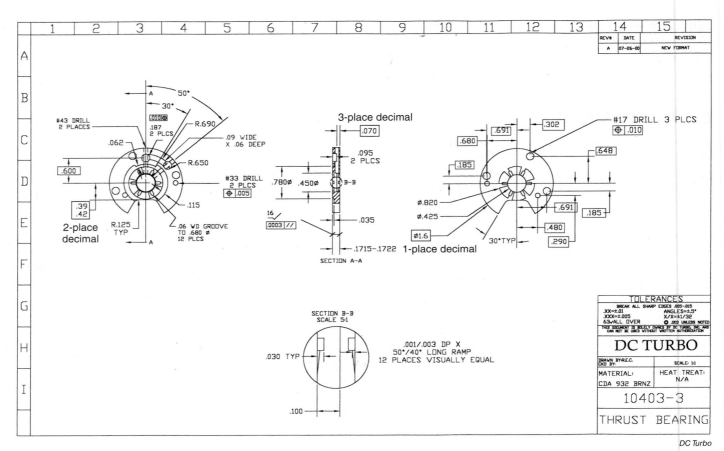

Figure 5-20. Decimal dimensions expressed with a one-, two-, or three-place decimal.

parts or considered in inspecting finished parts. These dimensions appear in parentheses to note they are for reference only. See **Figure 5-21**.

5.6.4 Tabular Dimensions

A *tabular dimension* can be used when multiple part numbers or families can be made from the same basic part print. For example, a print for a 2″ × 2″ square with a hole in the center could be used to make five different part numbers with five different hole sizes. Parts with 1/4″, 3/8″, 1/2″, 5/8″, and 3/4″ holes can all be defined with one print by noting the part number and corresponding hole size in a table on the print.

A tabular dimension can also simplify dimensions on a crowded print. In **Figure 5-22**, a tabular dimension is used to define some hole sizes and radii. Placing those dimensions inside the drawing could create a print that is hard to read and interpret. Instead, dimensions are listed in the table and labeled A, B, C, and so on, corresponding to labels on the print.

5.6.5 Basic Dimensions

Basic dimensions define the theoretical exact location, form, or orientation of a feature, regardless of the rest of the part. Basic dimensions appear in a rectangular box and are not toleranced. See **Figure 5-23**. The section on geometric dimensioning and tolerancing (GD&T) describes how the form and orientation of a part can be dictated by a series of other features,

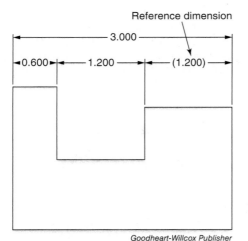

Reference dimension

Goodheart-Willcox Publisher

Figure 5-21. A reference dimension is shown to add clarity but has no tolerance and is not subject to inspection. In this example, the (1.200) dimension is a reference dimension. Practically, what this means is that the .600, 1.200, and 3.000 dimensions must be in tolerance, while the (1.200) is only calculated.

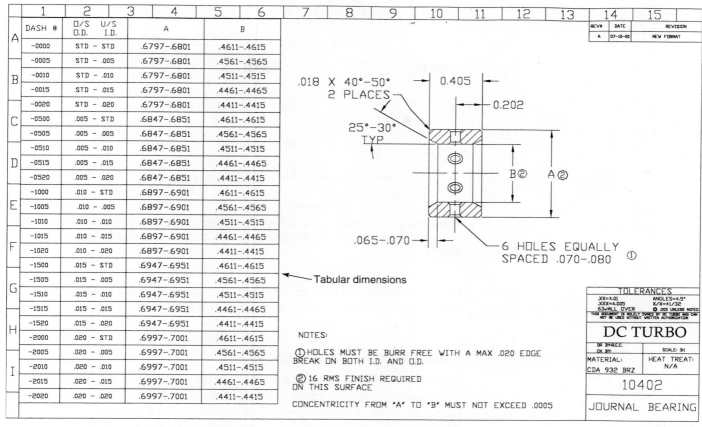

Figure 5-22. A tabular dimension table allows for families of parts to be defined on a single print.

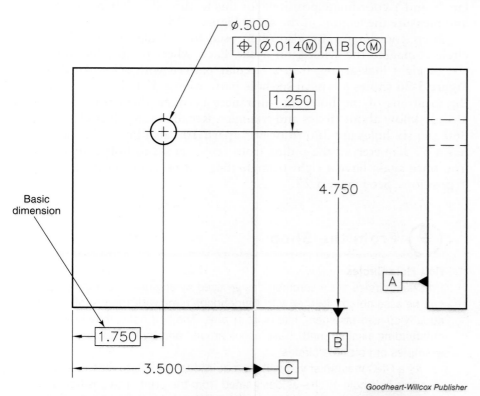

Goodheart-Willcox Publisher

Figure 5-23. Basic dimensions are placed in a box and note the theoretical exact position of a feature.

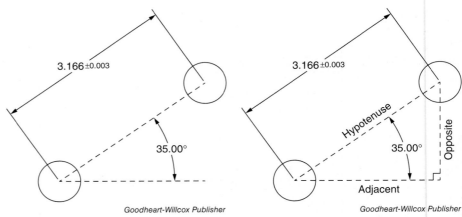

Goodheart-Willcox Publisher

Figure 5-24. Polar coordinates define the distance and angular direction of a feature.

Goodheart-Willcox Publisher

Figure 5-25. A right triangle trigonometry calculation can solve for the distance between two features.

known as datums. These datums are reference points for basic dimensions, which are most often encountered in GD&T.

5.6.6 Polar Coordinates

A *polar coordinate* is a dimension that references a distance and an angle. Polar coordinates are often necessary to control the location of a feature that starts from a known point and moves across a distance in an angular direction. **Figure 5-24** shows an example of a dimension from one hole center to the second hole center defined as a distance and angle. To calculate X and Y coordinate positions for this feature, sketch in a right triangle and measure the lengths of the sides, **Figure 5-25**.

Polar coordinates are also often used for calculations involving bolt circles. Imagine the wheels on a car. Each wheel usually has four, five, six, or eight lugs arranged in a circular pattern with a known diameter. **Figure 5-26** shows a typical six-hole pattern on a 5″ bolt circle. Calculate the locations of the holes and distances between them using what you already know about circles and triangles. Recall that a full circle measures 360°, so six holes are 360°/6 or 60° apart. You also know this bolt circle has a 5″ diameter, so the radius from center point to hole center is 2.5″. You once again have a right triangle that can be used to calculate X and Y positions. See **Figure 5-27**.

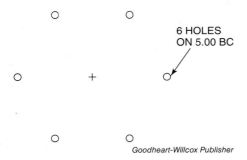

6 HOLES
ON 5.00 BC

Goodheart-Willcox Publisher

Figure 5-26. A typical bolt hole pattern with a common center point.

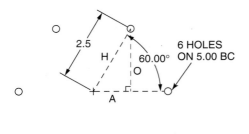

6 HOLES
ON 5.00 BC

Goodheart-Willcox Publisher

Figure 5-27. A right triangle trigonometry calculation can solve for the location of each hole in a bolt hole pattern.

From the Shop

Bolt Hole Circles

Bolt hole circles are a common design used by engineers. Components can be assembled together with tremendous force with the use of equally spaced fasteners, like bolts or pins. The bolt hole circle provides outstanding axial strength. Take a look around and try to find some other examples of bolt hole circles.

As a CNC machinist you will be required to be able to machine and measure bolt hole circles as designated from the print. The position and machining of these holes will be critical in the final assembly of assembled components.

5.7 Tolerances

In machining, there is no perfect part. Every feature must have some *tolerance*, or an amount of acceptable deviation from a desired size. A designer or engineer must consider all components that are being manufactured and determine the maximum allowable deviation. This is referred to as the *form*, *fit*, and *function* of a machined component.

Consider a simple operation like machining a 1/2″ hole into a plate to receive a 1/2″ dowel pin. Several factors must be considered prior to the hole being machined. First, consider pin size and allowable deviation in each pin. The standard for dowel pins is defined by ANSI ASME B18.8.2. By this standard, a 1/2″ dowel pin must be .5001″ to .5003″ in diameter. Next, decide how the pin should fit, based on planned use. If the pin is going to be press fit into place to ensure it does not move, the diameter of the hole needs to be .4995″ to .5000″. If the pin is meant to slide into the hole, then the hole needs to be .5005″ to .5010″ in diameter, depending on how easily you want the pin to slide into place. All these factors go into machining one simple hole. Imagine all the specifications required in prints for entire assemblies and machines.

Tolerances can be stated in multiple ways depending on the piece being machined and the design intent of a part. Normally, machinists try to machine to the middle, or *mean*, of a tolerance to allow for fluctuation in size while still creating acceptable parts.

5.7.1 Limiting Tolerance

Limits are the definitive maximum and minimum sizes of a part. When using a *limiting tolerance*, upper and lower dimensions must be given to define part tolerance. **Figure 5-28A** shows a feature defined with a dimension and a limiting tolerance. This feature is basically designed to be 1″ in size, but it has a minimum size of .997″ and maximum size of 1.003″, or a .003″ variation from the designed size. Note that this tolerance defines the entire feature, and nowhere along that feature can the part vary from the .997″–1.003″ tolerance.

5.7.2 Bilateral Tolerance

Bilateral tolerance is also called *plus-minus (+/–) tolerance*. This type of tolerancing indicates a tolerance in both directions. A bilateral tolerance can be equal or unequal. **Figure 5-28B** shows a bilateral equal tolerance, meaning the dimension has an equal limit in both directions. The designed size of this feature is 1.437″, but it is allowed to vary .003″ in either direction. This feature may be as large as 1.440″ or as small as 1.434″ when actually machined. **Figure 5-28C** displays an unequal bilateral tolerance. There is a dimension of 2.150″ with an upper limit (+) of .001″ and a lower limit (–) of .005″. This feature can range from 2.151″ to 2.145″. Notice that all three of these tolerances have a .006″ range, but all are defined differently.

5.7.3 Unilateral Tolerance

Tolerances in both positive and negative directions are not allowable in every case. *Unilateral tolerance* describes a dimension with tolerance in a single direction. Unilateral tolerances indicate that the designer wants a feature to be a certain size, but if the machining is off, it has to be off in

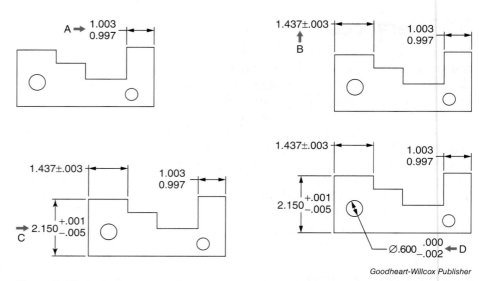

Goodheart-Willcox Publisher

Figure 5-28. A bilateral, unilateral, or limiting tolerance can be used to define part feature limits.

one direction only. For example, **Figure 5-28D** shows a hole dimensioned with unilateral tolerance, displaying a desired size with a tolerance in one direction only. This hole can vary from .600″ to .598″ in diameter.

5.7.4 Unspecified Tolerance

On many engineering drawings or prints, there are dimensions for features that appear to have no tolerance limits. These are known as ***unspecified tolerances***. Think of unspecified tolerances as "default" settings that are used when no other tolerance is given. Along with basic information about the print, unspecified tolerances are included in the title block, **Figure 5-29**. Although these dimensions may appear not to have tolerances, they actually do. Remember that it is *never* acceptable to have a dimension without a tolerance.

5.8 Holes

Holes are common features in machining that need to be fully defined and dimensioned on any print. A hole is defined in the top view of a print with a simple circle and often represented in subsequent views with hidden lines. Holes with additional features can have more than one circle on a print. The dimensions of a hole influence the machining process used to create that hole. For example, a hole dimensioned at 1/2″ can be drilled with a 1/2″ diameter drill. A hole dimensioned as .502″/.498″ needs to be drilled and reamed. A reamer is a fixed-diameter tool used after drilling to machine a straight hole with a controlled size, **Figure 5-30**. If a hole is dimensioned at .500″/.501″, the basic machining process would be to bore the hole out with a boring head. Additionally, holes can be *counterbored*, *spotfaced*, or *countersunk*.

TOLERANCES	
.XX=±.01	ANGLES=±.5°
.XXX=±.005	X/X=±1/32
63 ALL OVER	⭕ .002 UNLESS NOTED

CPK Precision Machine	
DR BY:R.E.C. CK BY:	SCALE: 3:1
MATERIAL: CDA 932 BRZ	HEAT TREAT: N/A
10402	
JOURNAL BEARING	

Goodheart-Willcox Publisher

Figure 5-29. Unspecified tolerances are controlled by the tolerance block in the title block section of a print. In this example, the tolerance block states .X = ± .032″, .XX = ± .015″, and .XXX = ± .005″. If a print has a dimension with 1 decimal, such as 2.1″, then the tolerance is ± .032″. If the dimension has 3 decimal places, as in 1.250″, then the tolerance is ± .005″. Although these tolerances are "unspecified," they are actually defined within the limits of the print.

Sugrit Jiranarak/Shutterstock.com

Figure 5-30. A solid body reamer used for close hole tolerance machining.

5.8.1 Countersink

A *countersink* is a tapered angle, or chamfer, placed on the outermost edge of a hole. A countersink can be a simple edge break to prevent sharp edges or a specific depth and angle to allow flush-mounted hardware, such as a screw. **Figure 5-31** shows the solid view of a countersink and corresponding 2-dimensional top and front views. In prints, the symbol for a countersink appears with the definition of the countersink dimension, **Figure 5-32**. A countersink specification stipulates the diameter of the hole, the diameter of the countersink, and the angle of the countersink.

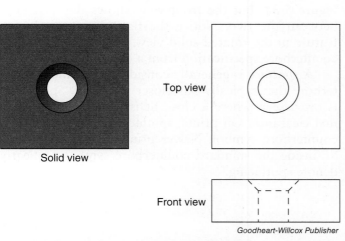

Solid view

Top view

Front view

Goodheart-Willcox Publisher

Figure 5-31. A countersink hole feature shown in a top and front view.

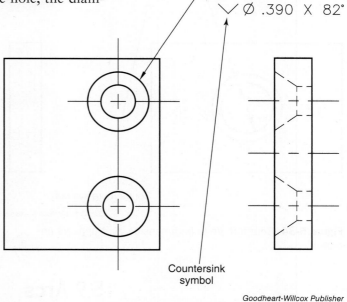

2X Ø .191
∨ Ø .390 X 82°

Countersink symbol

Goodheart-Willcox Publisher

Figure 5-32. A countersink hole dimension with hole size, countersink size, and angle.

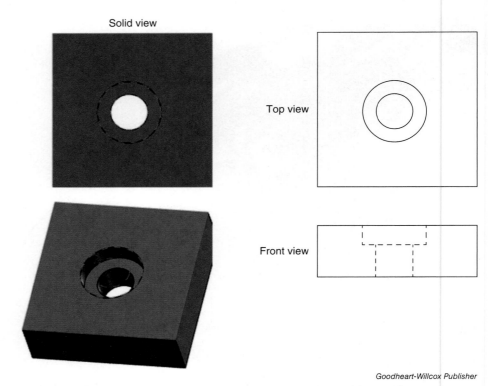

Goodheart-Willcox Publisher

Figure 5-33. A counterbored hole feature in a solid model with top and front print view.

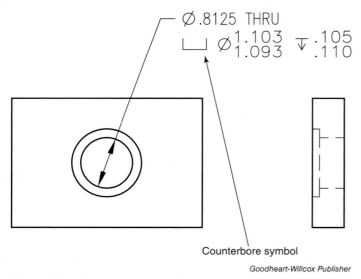

Goodheart-Willcox Publisher

Figure 5-34. Counterbore hole dimension as it appears on a print.

5.8.2 Counterbore and Spotface

A *counterbore* is a flat-bottomed hole that might be used to mount hardware, such as a socket-head cap screw, flush with or below surface level. A counterbored hole looks identical to a countersunk hole in the view in which the feature appears as a circle. The top view in **Figure 5-33** looks identical to the top view in **Figure 5-31**, but the front view shows the difference between the two. Notice the flat-bottom machining feature in the rotated solid view. **Figure 5-34** shows a counterbore specification from a drawing.

A *spotface* is generally considered a shallow counterbore that typically has closer tolerance. Spotfacing is often seen where a close fit is required with minimal clearance. On prints, spotfacing is indicated by a counterbore symbol. Newer prints include the letters *SF* inside the standard counterbore symbol to clearly denote spotfacing.

5.9 Arcs

In dimensioning or in general machining, *arcs* can include full circles, partial circles, or even small blends between lines. These small arcs, often called *fillets*, are very common in machining because they help eliminate sharp edges or corners in favor of small round edges. The placement and distinction of an arc is usually dimensioned by its center point and radius.

5.9.1 Fillets

Fillets are small arcs that blend two intersecting lines. Fillets are often dimensioned for size only, not placement. Remember, even if these fillets are not specifically toleranced, they must still conform to unspecified tolerances in the title block. An example of how a fillet might appear on a print is given in **Figure 5-35**.

5.9.2 Radius

Recall that *radius* refers to the distance from the center point of an arc to its edge. Generally, a radius is defined by center point and dimension, or size. Carefully interpret a print to determine whether an arc is defined as a radius, diameter, or full circle. Normally, closed arcs, or 360° arcs, are defined by diameter, and less-than-360° arcs are defined by radius. **Figure 5-36** shows an example of a fully defined, or fully dimensioned, radius.

5.10 Threads

There are a multitude of ways to define a thread on a print, depending on form and application. A thread is the raised helical groove around a screw, nut, or bolt. Threads can be internal threads, external threads, tapered threads, or specialty threads, such as buttress or Acme. Be aware of customer requirements, specific tolerances for all forms and classes of threads, and applicable standards that define thread forms and fits.

Threads are defined by two numbers, *size* and *pitch*. In American Standard thread, the standard for inch-based threaded fasteners and washers, *pitch* refers to the number of peaks, or total turns, in a 1″ span and is measured in threads per inch, **Figure 5-37**.

Most often on prints, threads are illustrated as a solid line feature accompanied by a hidden, or broken line. **Figure 5-38** shows a typical print configuration of both internal and external threads. A leader line is drawn to the feature with an annotation of thread size and pitch.

Measuring exact thread specifications is simplified with *go/no-go gages*. Go/no-go gages may be internal or external and are manufactured at the high and low limits of a thread specification. The "go" side of the gage should thread onto existing threads, and the "no-go" side should

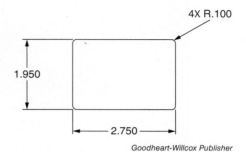

Goodheart-Willcox Publisher

Figure 5-35. Notice the dimension for this fillet is labeled with a size (.100″) and a note of 4 places. This note signifies that a fillet is required on all 4 corners.

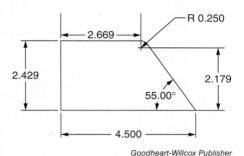

Goodheart-Willcox Publisher

Figure 5-36. A .250″ radius center location dimensioned.

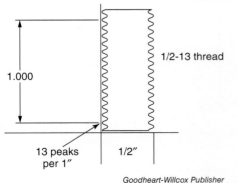

Goodheart-Willcox Publisher

Figure 5-37. The pitch of a thread is defined by the number of peaks in a 1″ distance.

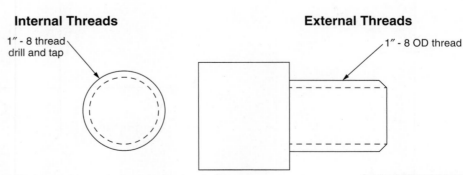

Goodheart-Willcox Publisher

Figure 5-38. Both internal and external threads can be drawn with an object line and a hidden line, representing the major and minor diameters.

not. Verify thread size and classification to determine the correct gage. Go/no-go gages are certified by manufacturer for accuracy.

Tech Tip

Micrometers may also be used to measure external threads in combination with a system of measuring over a series of three wires to calculate correct thread dimensions. The use of these tools requires more extensive discussion and the tables can be looked up online for future reference.

5.11 Geometric Dimensioning and Tolerancing

Geometric dimensioning and tolerancing (GD&T) is a more comprehensive system to define nominal, or theoretically perfect, geometry and its allowable tolerance. With increasing use of solid models and electronic prints, GD&T has become the accepted standard for dimensioning prints.

To understand how "normal" tolerancing differs from GD&T, consider an example of a normally dimensioned part, **Figure 5-39**. This seems like a fairly straightforward print for a .875″ diameter hole located 1″ from the left edge of the part and 1.375″ down from the top. However, with the tolerancing given, this hole can range in size from .870″ to .880″, and the location can move from .995″ to 1.005″ from the left and from 1.370″ to 1.380″ in both the X and Y directions. The hole size and location can vary enough to make future assembly impossible unless the designer limits the dimension tolerances to a very small amount to ensure assembly fit and function.

All these variables are controlled through the use of GD&T, where the part is dimensioned to a theoretical perfection. In this case, as the hole gets smaller, the X and Y locations are more controlled, and, as the hole gets larger, more tolerance is allowed in both X and Y directions.

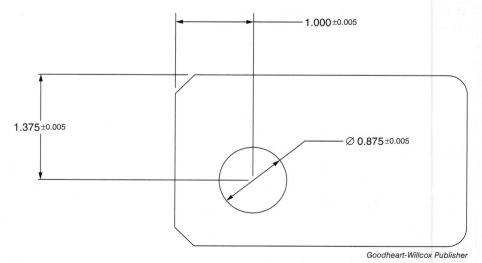

Goodheart-Willcox Publisher

Figure 5-39. A typical drawing of a hole location only controlled by bilateral tolerances.

Understanding and applying geometric dimensioning and tolerancing is an entire subject unto itself. Begin to familiarize yourself with the symbols in **Figure 5-40** and understand their definitions as they apply to parts and prints. A simple tolerance does not always fully define the desired form, fit, and function of a part. GD&T helps eliminate excessive dimensioning and poor-fit assemblies.

Geometric Characteristics			
Type of Tolerance	**Geometric Characteristic**	**Symbol**	**Definition**
Form	Straightness	—	A flat surface controlled by only one line
	Flatness	▱	A flat surface as measured over an entire surface feature
	Circularity	○	Amount of roundness controlled around a single circle
	Cylindricity	⌭	Amount of roundness over the length of a diameter
Profile	Profile of a line	⌒	Amount of deviation of a nonflat surface over one line
	Profile of a surface	⌓	Amount of deviation over an entire surface feature
Orientation	Perpendicularity	⊥	Two lines or surfaces meeting at exactly 90°
	Angularity	∠	The amount of angular distance between two lines or surfaces
	Parallelism	//	The equal distance of two lines or surfaces to each other
Location	Symmetry	≡	The deviation from half an object's features to its other half through a given centerline
	Position	⊕	The exact location of a feature
	Concentricity	◎	The distance between the centerpoints of two arcs
Runout	Circular runout	↗	The deviation from the center point of one arc to another in a single location
	Total runout	⌰	The deviation from the center point of one arc to another over an entire surface

Goodheart-Willcox Publisher

Figure 5-40. Symbols and explanations for various geometric characteristics used in GD&T. Note that the symmetry and concentricity symbols were removed from the 2018 ASME standard, but you may still see these symbols on older prints.

Chapter Review

Summary

- A print is the plan that transfers information from the designer to the builder.
- Notes provide supplementary information needed for manufacturing or post-manufacturing processes.
- Multiview drawings created using orthographic projections allow for the interpretation of all sides and features of a finished part print.
- Auxiliary and sectional views allow for the visualization of part features that could be hidden from normal orthographic projections. They may also add detail in complicated features.
- Typical prints make use of many different types of lines. Lines define visual objects, hidden objects, and even dimensions. Lines can have different thicknesses or be dashed in certain configurations to determine designer intent.
- A dimension defines the measurable extent of an object. Dimensions indicate an object's length, width, depth, height, or even girth and can be defined by any unit of measurement.
- Polar coordinates can be used to define distance and angular direction of a feature from a reference point.
- Tolerances define the acceptable deviation of a dimension from the designed specification. Tolerances can be unilateral, bilateral, or limiting.
- Holes are common features in machining. Holes may be countersunk, counterbored, or spotfaced. In all cases, holes must be fully defined and dimensioned on any print.
- Arcs include full and partial circles as well as small blends between lines called fillets. The placement and distinction of an arc is usually dimensioned by its center point and radius.
- A thread is the raised helical groove around a screw, nut, or bolt. Threads can be internal, external, tapered, or specialty-type threads. A print must define size, threads per inch, class, and location of all threads.
- Geometric dimensioning and tolerancing (GD&T) allows for the definition of the form and shape of a part relative to other features on the part.

Review Questions

Answer the following questions using the information provided in this chapter.

Know and Understand

1. _____ are used to transfer information from designers to builders.

   A. Boards   C. Prints

   B. Papers   D. Tolerances

2. Computer-aided drafting (CAD) programs allow print information to be delivered _____.

   A. automatically   C. manually

   B. digitally   D. weekly

3. *True or False?* The most common way to present multiple views of a part is referred to as orthographic projection.

4. The _____ is an international standard-setting body that promotes global industrial standards.

   A. American Society of Mechanical Engineers

   B. International Organization for Standardization

   C. Manufacturing Standards Academy

   D. World Standards Organization

5. The basic _____ contains information that identifies the part and the print.

   A. auxiliary view   C. reference dimension

   B. information block   D. title block

6. *True or False?* A reference dimension provides specific limits for machining dimensions.

7. A part may need to be cut to show internal or hidden features that cannot be shown in normal views. These views are called _____.

   A. alternate views

   B. informational views

   C. orthographic projections

   D. sectioned views

8. _____ are also referred to as object lines.

   A. Hidden lines   C. Visible lines

   B. Section lines   D. Yellow lines

9. A(n) _____ is used to point to a specific location or note on the print.

   A. dimension line   C. leader line
   B. extension line   D. visible line

10. _____ designate the centers of holes, diameters, radii, or other symmetrical features.

    A. Centerlines   C. Leader lines
    B. Dimension lines   D. Phantom lines

11. *True or False?* A dimension can be defined as a measurable extent of an object.

12. A polar coordinate is a dimension that references a _____ and _____.

    A. direction; distance
    B. distance; angle
    C. distance; coordinate
    D. view; angle

13. *True or False?* Machined features do not require tolerances, or any amount of acceptable deviation from a desired size.

14. On many engineering drawings or prints, there are dimensions that appear to have no tolerance limits. These are known as _____.

    A. bilateral tolerances
    B. unified dimensions
    C. unilateral tolerances
    D. unspecified tolerances

15. A _____ is a tapered angle, or chamfer, placed on the outermost edge of a hole.

    A. counterbore   C. relief
    B. countersink   D. thread

16. *True or False?* Threads are defined by two numbers: size and threads per inch.

17. _____ is a comprehensive system to define nominal geometry and its allowable tolerance.

    A. Geometric dimensioning and tolerancing
    B. Geometric direction and timing
    C. Government dimensioning and tolerancing
    D. Government direction and tolerancing

Apply and Analyze

1. How do prints transfer information from a concept to a manufactured part?

2. Why is it important to have standards for print drawing and information exchange?

3. Describe when you might use a sectioned view or an auxiliary view.

4. List six different line types used in a print.

5. Dimensions can be given in either fractional or decimal designation. Is there a difference in levels of accuracy or design intent in these 2-dimensional designations? If so, explain the difference.

6. Explain the difference between a bilateral and unilateral tolerance as applied to a part dimension.

7. What type of fastener might require a countersunk hole for proper installation? What type of fastener might require a counterbored hole?

8. Why is it necessary to annotate both the size and pitch of a thread designation on a print?

9. How does geometric dimensioning and tolerancing (GD&T) differ from standard dimensioning?

10. Why is GD&T considered a more comprehensive form of dimensioning and more acceptable when using electronic modeling?

Critical Thinking

1. Print dimensions and tolerances influence the processes by which parts are machined. What are some key factors in print reading to consider before a machining process can be performed?

2. Reading prints and interpreting them correctly is a skill required of every machinist. What are some of the potential errors associated with incorrect print reading?

3. What is the impact of dimensioning and tolerances in multipart assemblies? Consider form, fit, and function of the entire assembly as those factors are impacted by the individual pieces assembled together.

6

Semiprecision and Precision Measurements

Chapter Outline

6.1 Introduction to Precision and Semiprecision Measurements

6.2 Steel Rules

6.3 Protractors

6.4 Calipers

6.5 Micrometers

6.5.1 Depth Micrometer

6.5.2 Inside Micrometer

6.5.3 Bore Micrometer

6.6 Comparison Gages

6.6.1 Thread Gages

6.6.2 Feeler Gages

6.6.3 Fillet Gages

6.6.4 Screw Pitch Gages

6.6.5 Pin Gages

6.7 Height Gages

6.7.1 Vernier Height Gages

6.7.2 Height Gages with Indicators

6.7.3 Digital Height Gages

6.8 Gage Blocks

6.9 Sine Bars

6.10 Telescoping Gages

6.11 Indicators

6.11.1 Dial Indicator

6.11.2 Drop Indicator

6.11.3 Coaxial Indicators

6.12 Advanced Gaging and Inspection Tools

6.12.1 Coordinate Measuring Machine (CMM)

6.12.2 Optical Comparator

6.12.3 Air Gaging

6.12.4 Electronic Gaging

Learning Objectives

After reading this chapter, you should be able to:

- Define precision and semiprecision measurement.
- Perform measurement with calipers.
- Identify measurements with micrometer.
- Identify components of a micrometer.
- Explain basic use of a height gage.
- Identify appropriate gage blocks for measurement.
- Perform simple angle setup with sine bar.
- Interpret multiple indicator readings.
- Identify correct application of a drop indicator.
- Explain the application of a coaxial indicator.
- Identify multiple go and no-go gages.
- Discuss the application of a coordinate measuring machine.
- Describe the application of an optical comparator.

Key Words

air gaging	feeler gage	protractor
bore micrometer	fillet gage	screw pitch gage
calipers	gage blocks	semiprecision measurement
coaxial	go/no-go gage	
coaxial indicator	grades	sine bar
comparison gage	height gage	steel rule
coordinate measuring machine (CMM)	indicator	telescoping gage
	inside micrometer	thread gage
depth micrometer	micrometer	vernier height gage
dial indicator	optical comparator	vernier scale
drop indicator	precision measurement	wringing
electronic gaging	pin gage	

6.1 Introduction to Precision and Semiprecision Measurements

Conventional wisdom in machining says, "Anyone can make a part, but very few can actually make one the right size." The last few chapters have covered math and prints to support actual measurement, a skill that most new machine operators are responsible for. Machine shops must produce finished parts that meet all print requirements. Machinists must verify that all parts meet dimensioning and tolerancing requirements, as well as other specifications, by measuring features on a part. Prints give guidance on the machining process and information on how to inspect parts, including which measuring tools to use to verify their dimensions.

Measurements may be described as *precision* or *semiprecision*, depending on the accuracy of the inspection tools used. Generally, *precision measurement* is defined as any measurement smaller than .005″ (five-thousandths of an inch) or less than 1° of accuracy. *Semiprecision measurement* is any measurement greater than .005″ or 1° of accuracy. Normal parts in a machining environment can have accuracies as close as .0001″ or as distant as 1/4″. How we measure these items is vitally important. For example, a part that is dimensioned at ±.005″ cannot be measured with a tape measure. Alternatively, a part with a tolerance of ±1/8″ does not require the use of precision micrometers or a coordinate measuring machine (CMM).

Each measuring tool has an accuracy based on the smallest unit of measurement the tool measures. For example, if the calipers' smallest delineation is .001″, its precision is ±.001″. It cannot measure closer than that. In general, you cannot measure a dimension on a part with a tool that is the same degree of accuracy as the dimension. All the tolerance of the part is taken up by the accuracy of the tool. The rule of thumb is to measure with a tool four times more accurate than the dimension.

Throughout this chapter, only US customary measurements and tools are described. The vast majority of US machine shops use only inch-based prints and tools. For those rare instances where conversion may be required from metric to US customary, consult an appropriate conversion chart or make use of electronic modeling software to convert printed specifications.

6.2 Steel Rules

Common rulers are often made of wood and are not designed to take semiprecision measurements in a manufacturing environment. *Steel rules* are precision-made rulers, usually laser engraved, and manufactured especially for the machining environment. They are available in several basic types and graduations, including fractional inch, decimal inch, and metric. Tape measures are a common steel rule often used during sawing operations when preparing material for machining, **Figure 6-1**. Steel rules can also be rigid, laser-engraved, flat pieces of steel, usually 6″ or 12″ long, that can be carried easily in the pocket, **Figure 6-2**.

Apex Tool Group, 2017

Figure 6-1. A typical tape measure used for extended measurements over long distances.

Manrit/Shutterstock.com

Figure 6-2. A machinist's pocket rule. The bottom scale is in inches and the top scale is in centimeters.

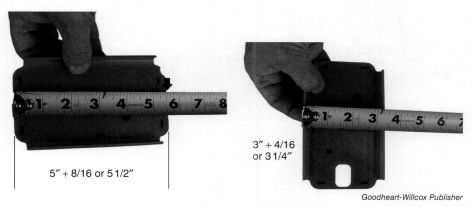

Goodheart-Willcox Publisher

3″ + 4/16
or 3 1/4″

5″ + 8/16 or 5 1/2″

Figure 6-3. A tape measure measurement of a phone case.

Standard rules are marked with several divisions. In **Figure 6-3**, the large bold numbers indicate inches from the end of the rule. The 16 divisions between each inch delineation each represent 1/16″. When measuring a piece, make sure you reduce the fraction as needed.

6.3 Protractors

Protractors used in manufacturing are not the same as the plastic protractors used in school, but they are similarly used to measure angles between two surfaces, **Figure 6-4**. A protractor can be a semiprecision or precision measuring device, depending on its level of accuracy. The protractor in **Figure 6-4** is a semiprecision protractor. It uses a movable leg to locate one edge of an angle against a fixed head. On the fixed head is a scale that denotes the measured angle. Semiprecision protractors do not measure closer than 1° of accuracy. The protractor in **Figure 6-5** is a precision protractor measuring the hexagonal head of a bolt. This protractor has a magnified scale that is accurate to 1/10th of 1°.

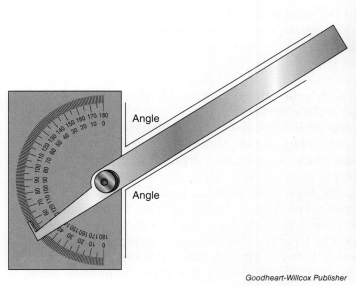

Angle

Angle

Goodheart-Willcox Publisher

Figure 6-4. A steel protractor used to make semiprecision angular measurements.

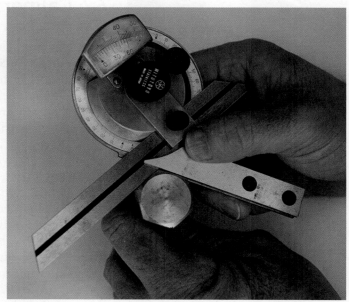

Goodheart-Willcox Publisher

Figure 6-5. A precision protractor with a magnified scale reader.

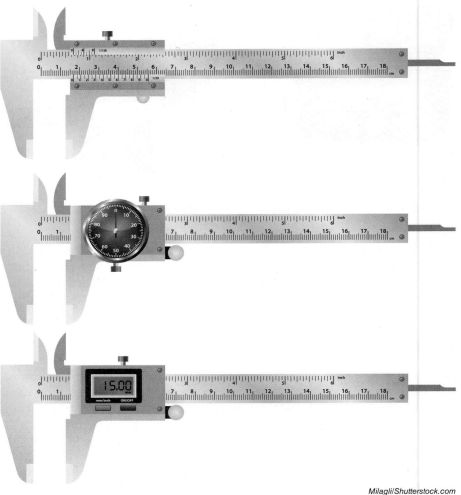

Figure 6-6. Examples of vernier, dial, and digital calipers.

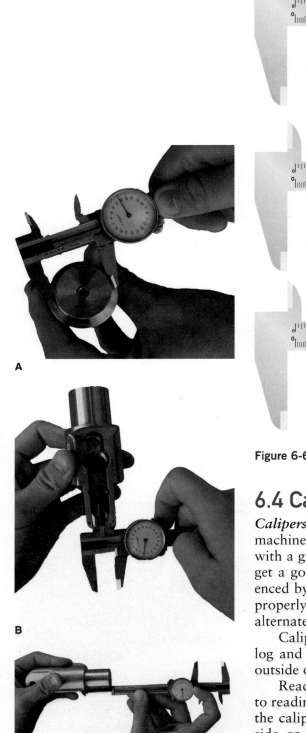

Figure 6-7. Caliper measurements. A—Outside. B—Inside. C—Depth.

6.4 Calipers

Calipers are one of the most used and most versatile measuring tools in a machine shop. Unfortunately, they are often misused to measure features with a greater accuracy than the tool allows. It takes time and practice to get a good feel for using calipers. The actual measurement can be influenced by the operator squeezing the calipers too tightly or not aligning it properly. Use care with your calipers and double-check dimensions with alternate measuring tools when necessary.

Calipers can come in multiple sizes and configurations, including analog and digital varieties, **Figure 6-6**. This versatile tool can measure the outside of a part, the inside of a part, or a depth or step, **Figure 6-7**.

Reading calipers is pretty simple. There are three main components to reading calipers: the *edge*, *scale*, and *dial*, **Figure 6-8**. The reading from the calipers is always taken from the edge, whether reading inside, outside, or depth. The reading indicates how much of the scale is exposed to the left of the edge. The large numbers on the scale represent inches, and the smaller numbers in between the inch designator represent .100″ each. That means the 2 is .200″, the 3 is .300″, and so on. If you look closer at **Figure 6-8**, you can see that the edge is past the 4″ line and lined up on

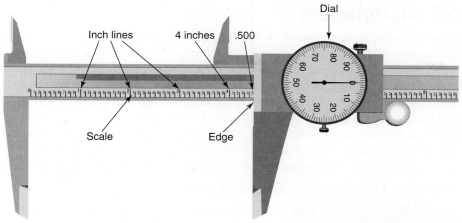

Goodheart-Willcox Publisher

Figure 6-8. The components of dial calipers.

the 5, or .500″ line. To this point, the reading is 4.500″. The last consideration is the the dial. Wherever the needle on the dial lands, you must add that number to the reading. In **Figure 6-8**, that dial is on 0, so the measurement is 4.500″.

Test your knowledge by working through two more examples of caliper readings, **Figure 6-9**.

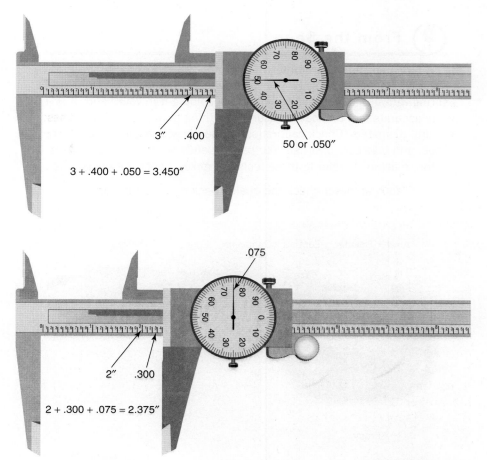

Goodheart-Willcox Publisher

Figure 6-9. Two measurements taken from dial calipers.

6.5 Micrometers

A *micrometer*, also known as a *micrometer screw gage*, is a measuring device that uses a rotating scale on a high-precision screw thread to precisely measure objects. The micrometer has been in use for over 150 years, and forms a cornerstone of modern manufacturing. Because micrometers are precise and portable, they can make extremely accurate measurements right at a machine with repeatable results. Like other tools, micrometers come in multiple configurations, including the outside micrometer, inside micrometer, and depth micrometer. All micrometers work in essentially the same way, off a rotating scale of a precision thread, and share common parts, **Figure 6-10.**

- **Frame.** The main single-piece body of the micrometer.
- **Anvil.** A fixed stop that contacts one side of the piece being measured.
- **Spindle.** Rotating part that screws up and down to secure the second side of a piece being measured.
- **Locknut.** Prevents the spindle from rotating.
- **Barrel.** Fixed in place and provides measurement scale.
- **Thimble.** Rotates around the barrel and has a secondary measuring scale numbered 1 to 24 in US Standard micrometer.
- **Ratchet.** Indicates ("clicks") when tension is applied to the spindle to create even and repeatable force when measuring.

 From the Shop

The Indispensable Micrometer

The micrometer needs to be your best friend in the shop. Practice measuring with it, and get really comfortable holding it in your hand. It is a very accurate tool that can measure everything from material thickness to drill diameters. Check calibration frequently with a gage block or pin gage, and take care of it like your part depends on it—because it does. Often, machinists refer to these tools as "mikes," short for micrometer.

"You can never check the quality back into a scrap part."

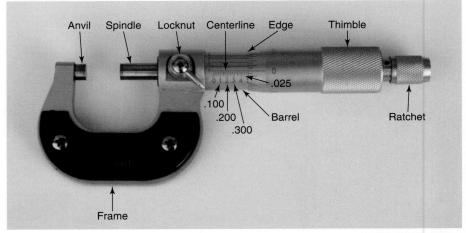

Goodheart-Willcox Publisher

Figure 6-10. A 0–1″ micrometer and its components. The edge is exactly on the number 5, and the thimble is lined up with 0 on the centerline. The reading on this micrometer is .500″.

US customary micrometers are sized in 1-inch increments and measure to thousandths of an inch. The micrometer in **Figure 6-10** is a 0–1″ outside diameter micrometer, so it only measures objects up to 1″. When measuring objects from 1–2″, you would use a 1–2″ micrometer and scale up as needed. In **Figure 6-10**, you will notice that micrometers have an edge, similar to calipers. This edge is used to take a measurement with the micrometer. Also notice the large numbers on the barrel. These numbers represent .100″ each. The 1 represents .100, 2 represents .200, 3 represents .300, and so on. Between each numbered line is a smaller hash mark that represents .025″ each. If the edge is past the first hash mark, it is .025. If it is past the second, it is .050, and past the third it is .075. When the edge reaches the next line, it is at the next larger number because it has rotated .100″.

The last number in a micrometer measurement comes from the thimble. The thimble is numbered from 1–24, representing .001–.024″, and rotates around the barrel in increments of one-thousandth of an inch. Once the thimble passes 24, it returns to 0 and reaches another small hash mark.

Consider another example with a different size micrometer, **Figure 6-11**. Note that this is a 2–3″ micrometer. As in the previous example, start by locating the edge. It is not past a hash mark or any number on the barrel. Next, observe that the centerline is lined up on number 4 of the thimble. The calculation would then be as follows:

$$2.000 + 0 + .004 = 2.004''$$

 Tech Tip

Depth micrometers can be calibrated using a gage block and a granite surface plate. By placing a gage block on the granite plate, the depth micrometer can measure from the top of the block down to the surface plate. The measurement on the micrometer should be the same as the gage block size.

6.5.1 Depth Micrometer

A *depth micrometer* is a common tool used when measuring from one face or surface of a part to another face. Depth micrometers are mechanically similar to outside diameter micrometers, with two main differences. In a depth micrometer, the spindle extends down from the anvil, or fixed base, so the scale is inverted. Depth micrometers are also outfitted with a series of interchangeable rods to allow for measurements in 1″ increments, **Figure 6-12**.

Depth micrometers are often used when measuring a step block or when measuring inside a hole with a flat bottom, **Figure 6-13**.

6.5.2 Inside Micrometer

Inside micrometers are used when measuring internal features such as slots, grooves, or large bore diameters, **Figure 6-14**. The two extending faces should be placed inside the feature being measured and extended out to achieve the measurement. Alignment is critical with these tools. They must reach directly across the feature being measured at 180° from the opposing face to achieve accurate results.

Goodheart-Willcox Publisher

Figure 6-11. A standard 2–3″ micrometer.

The L.S. Starrett Company

Figure 6-12. A 0–1″ depth micrometer with an analog readout system.

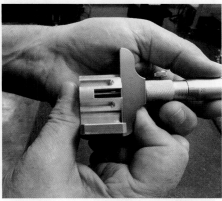
Goodheart-Willcox Publisher

Figure 6-13. In this example, we are measuring from a machined face down the inside of a flat-bottomed hole.

Two extending faces

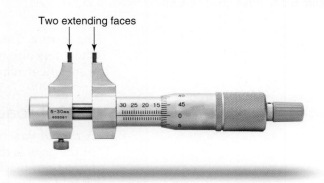
Freedom Life/Shutterstock.com

Figure 6-14. An internal micrometer with extending faces.

2-Point ID Measurement

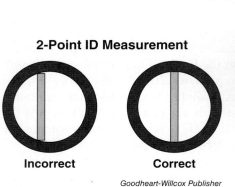

Incorrect   Correct

Goodheart-Willcox Publisher

Figure 6-15. A typical 2-point bore measurement must pass through the centerline of the hole.

Goodheart-Willcox Publisher

Figure 6-16. A 3-point bore measurement ensures the measurement point is in the center of the hole.

6.5.3 Bore Micrometer

Inside micrometers are valuable for certain measurements, but there is an inherent issue with measuring an inside diameter with two points of contact. Those two points must cross through the centerline to achieve an accurate measurement, **Figure 6-15**.

When measuring a large bore diameter or close tolerance bore, a *bore micrometer* is better suited. Bore micrometers have the capacity to make three-point measurements, overcoming the issue in **Figure 6-15**. The third locating point centers the micrometer in the bore, **Figure 6-16**.

Bore micrometers are made in small increments, only measuring a range of around .200″ each. They can be very expensive, but they are a very accurate and repeatable measuring device, **Figure 6-17**.

Kimtaro/Shutterstock.com

Figure 6-17. Three-point bore micrometers in various sizes.

6.6 Comparison Gages

A *comparison gage* is a hard, fixed gage used to compare to a machined feature. These gages can be *go/no-go* style, where the gage either fits the feature or it does not, or an even simpler version is just compared to an existing feature. There are many types of standard comparison gages. A machine shop may also make a custom comparison gage, often called a *template*, for a specific machined feature.

6.6.1 Thread Gages

A *thread gage* is a go/no-go-style gage used to measure threads. One end of the gage is the "go" end and the other is the "no-go" end. After threads have been machined, the "go" end of an appropriately sized thread gage should screw into or onto the machined threads. If the thread gage does not fit, the threads are not cut deep enough and need to be remachined.

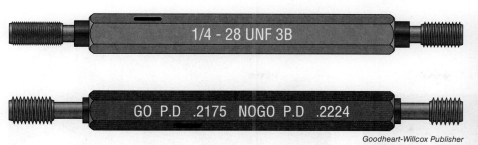

Goodheart-Willcox Publisher

Figure 6-18. Internal thread go/no-go gages. Each tool is marked with thread pitch, class, and go/no-go.

Goodheart-Willcox Publisher

Figure 6-19. External thread go/no-go gages. Each tool is marked with thread pitch, class, and go/no-go.

Once the "go" end fits onto the piece, the "no-go" end should not. If the no-go end still fits, the threads are cut too deep.

Figure 6-18 illustrates an inside diameter (ID) thread gage. Often, the ends are colored green for "go" and red for "no-go" for easy identification. The gage should also be calibrated and stamped with the correct thread size, pitch, and class.

Figure 6-19 shows an outside diameter (OD) thread gage. This gage is used when making external thread features. In this case, two separate gages are required, one each for "go" and "no-go." OD gages should also be clearly marked with pitch, size, class, and either "go" or "no-go."

6.6.2 Feeler Gages

Feeler gages come in sets of shims, or thin pieces of metal, divided by size, to determine a gap or space in a part or between two parts, **Figure 6-20**. One common form of feeler gage is in the spark plug "gap gage." Feeler gages are often used during setup or inspection. Again, these are not precision tools, but they assist in difficult inspection areas.

Goodheart-Willcox Publisher

Figure 6-20. Thickness or feeler gages. Each is marked with the correct thickness.

6.6.3 Fillet Gages

A *fillet gage* is a simple comparison gage that is precut with radii or partial arcs to compare against a machined surface, **Figure 6-21**. The proper gage is selected depending on desired radius, and the gage is simply held against the machined surface to compare the gage to the feature.

6.6.4 Screw Pitch Gages

A *screw pitch gage* is used to verify pitch, measured in threads per inch (TPI), on a threaded piece. Like fillet gages, screw pitch gages come in sets with a variety of sizes all in one tool, **Figure 6-22A**. TPI, or pitch, is given in single-unit increments using whole numbers, so these small variances require engaging multiple teeth of the gage into the threads to verify pitch, **Figure 6-22B**. Ensure that all the teeth of a gage fit evenly into the threads you are attempting to measure.

TLaoPhotography/Shutterstock.com

Figure 6-21. A typical set of fillet gages that ranges from 1/8″ radius up to a 1″ radius. This tool does not give precision measurement of an arc, but it can be used to quickly verify an estimate of arc or radius size.

A

B

Goodheart-Willcox Publisher

Figure 6-22. A—Typical set of comparison thread gages, measuring threads per inch. B—Thread gage comparing threads per inch.

A

B

Goodheart-Willcox Publisher

Figure 6-23. A—Pin gage box set. B— Two pin gages installed in a holder for go/no-go measurement.

6.6.5 Pin Gages

Pin gages are precision round pins that come in sets with .001″ increments, **Figure 6-23A**. These pins come in multiple grades that define the level of accuracy, similar to gage block sets. Pin gages have a variety of uses, including checking the size of a round hole, checking the width of a slot, or even checking a groove on a round part. Pins are often mounted in a holder set with red and green handles for no-go and go, respectively, **Figure 6-23B**.

6.7 Height Gages

Height gages, also called *vertical comparison gages*, are used to compare a known height to an unknown height. The known size can be gage blocks (discussed later), a sample part, or premachined inspection gage. Height gages can also be used to check some geometric dimensioning and tolerancing (GD&T) features like flatness, parallelism, perpendicularity, and concentricity.

6.7.1 Vernier Height Gages

A *vernier scale* is a scale made to slide along divisions of a larger graduated scale and take measurements between the larger divisions. Vernier measurement tools have no dial, readout, or electronics. The reading is taken from a series of scales and matching lines to determine dimensions. Vernier tools are normally classified as semiprecision because it is extremely difficult to achieve an accurate reading of more than a few thousandths of an inch.

Vernier height gages are often used to compare one height to another, **Figure 6-24**. They can also be fitted with a carbide attachment to scribe or mark a surface at a given dimension. Height gages should always be placed on a granite plate to ensure flatness and perpendicularity to the workpiece.

Golffy/Shutterstock.com

Figure 6-24. A height gage with attached drop indicator.

6.7.2 Height Gages with Indicators

A height gage is most commonly used with an indicator attached, **Figure 6-25**, to "set a zero" on top of a known size and measure the difference to a machined feature. With the addition of an indicator, the measurement becomes very precise. The indicator can measure across surfaces for such features as parallelism, flatness, and perpendicularity. The indicator also can be used to analyze differences in dimensions.

6.7.3 Digital Height Gages

Some of the most accurate height gages are digital. The improved accuracy comes from the electronic touch probe sensor that allows for consistent pressure on the indicator. The result is a reading that removes human error and inconsistencies. Digital height gages can be zeroed off a granite plate and then moved in a straight line to measure linear dimensions. Alternatively, a zero can be set from one part feature and measure to a second feature. Most digital height gages also have the capability to set multiple zeroes from different faces or features. This function allows a machinist to measure more than one feature at a time.

High-tech digital devices, including digital height gages, are more susceptible to errors from dust or dirt than their analog counterparts. Temperature and humidity are always a factor in high-precision measurements. Often, precision tools must remain in quality control areas with controlled humidity and temperature.

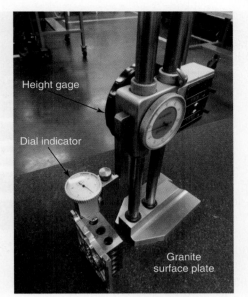

Goodheart-Willcox Publisher

Figure 6-25. A height gage with attached dial indicator.

6.8 Gage Blocks

Gage blocks are precision-ground rectangular blocks used as comparison gages. That is, the blocks do not actually take measurements, but they allow you to compare a feature or part to the known size of a gage block. The gage block was developed by Swedish inventor Carl Johansson in 1901 as a precision method a machinist could use to combine blocks to make up any measurement needed. Today, many machinists still refer to these blocks as *Johansson blocks* or *Jo blocks*.

Gage blocks are available in multiple *grades*. Grades establish and identify the precision and quality of the block set. Gage blocks are precisely machined and lapped to maintain high levels of parallelism and flatness. There are multiple grading systems, depending on the desired accuracy of a measurement, **Figure 6-26**.

Gage Block Grading Systems			
Grade	**Purpose**	**Accuracy**	**Description**
AAA	Reference	± .05 microns (a micron is .000039 of an inch)	These blocks are used to establish the standards.
AA	Calibration	+.1 microns to –.05 microns	Used to calibrate inspection blocks and very high precision gaging.
A	Inspection	+.15 microns to –.05 microns	A toolroom standard for setting up other gages.
B	Workshop	+.25 microns to –.15 microns	Used as shop standards and the most common set seen on a machine shop floor.

Goodheart-Willcox Publisher

Figure 6-26. Chart of inspection grading for gage blocks.

Figure 6-27. An 81-piece gage block set.

Goodheart-Willcox Publisher

Gage blocks come in a box set of 81 pieces as seen in **Figure 6-27**. These blocks can be stacked together to make any size from .100″ to almost 15″ in .0001″ increments. The process of stacking gage blocks together is known as *wringing*. Wringing relies on the molecular attraction that occurs when two very flat surfaces are rubbed together to create a vacuum between the two faces. Wringing causes the blocks to stick together and provides no measurable difference in size between individual blocks and multiple blocks.

It is best practice to use the fewest possible blocks when stacking. For example, to make a 1.725″ stack of blocks, use the 1″, .600″, and .125″ blocks. It is possible to assemble multiple combinations and still achieve the 1.725″ height, but fewer blocks yield better results. Once gage blocks are assembled, they can be used to set up a height gage with an indicator, measure a slot width, or even calibrate other tools, such as micrometers or calipers.

6.9 Sine Bars

Sine bars are used in conjunction with gage blocks to measure angles. A sine bar consists of a precision-ground, hardened body with precision-ground cylinders fixed at each end, **Figure 6-28**. Recall the term *sine* from Chapter 4. Sine describes the mathematical relationship between the opposite side and hypotenuse of a right triangle:

Figure 6-28. A typical 5″ sine bar.

Goodheart-Willcox Publisher

$$\text{Sine} = \frac{\text{Opposite}}{\text{Hypotenuse}}$$

A sine bar of known length can be used to measure or verify an angle. The length of a sine bar is defined as the center-to-center distance between the two cylinders, **Figure 6-28**. This distance becomes the hypotenuse of a right triangle. In the following example, a 5″ sine bar is used, but sine bars come in multiple sizes in 1″ increments, including 5″, 7″, and 12″ sine bars.

If you need to measure a 10° angle, use the trigonometric formula for sine to calculate the opposite side with a 5″ hypotenuse. Recall the calculator functions you learned in Chapter 4 and apply those skills here.

$$\text{Sine} = \frac{\text{Opposite}}{\text{Hypotenuse}}$$
$$\text{Opposite} = \text{Hypotenuse} \times \text{Sine}$$
$$\text{Opposite} = 5″ \times \sin(10°)$$
$$\text{Opposite} = .8682″$$

The result of these calculations, .8682″, gives the length of the opposite side, which is created with gage blocks. Set up gage blocks to equal .8682″, or a .1002″, .118″, and a .650″ block. Wring those blocks together and set them underneath one of the cylinders, **Figure 6-29**. The top surface of the sine bar now sits on the gage blocks at exactly 10°.

When a part with a machined 10° angle is placed upon this sine bar setup, the machined angle should now lie parallel

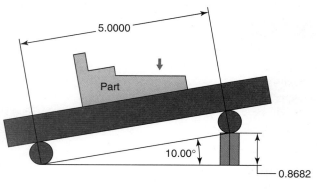

Goodheart-Willcox Publisher

Figure 6-29. Display of a 5″ sine bar set to a 10° angle.

to the surface plate on which the bar and blocks are sitting. By using an indicator on a granite plate, you can measure the amount of deviation in the part surface from a perfect 10° angle. The amount of acceptable deviation is determined by the print tolerance of the angle. This same method can be used for machining that angle or set up on a machine to establish an angled cut, **Figure 6-29**.

6.10 Telescoping Gages

Telescoping gages are another type of comparison gage. By definition, telescoping gages do not measure any feature independently. However, with proper use and application, telescoping gages can be beneficial in measuring semiprecision bore diameters.

Telescoping gages have a set of spring-loaded ends that can be locked into place by turning the lock on the handle end of the gage, **Figure 6-30**. The gage is placed inside a hole or bore diameter and turned on a slight angle to allow the spring-loaded ends to extend out to a size larger than the bore. The end cap is then lightly tightened to hold the extended position, and the telescoping gage is slowly rotated through the hole, **Figure 6-31**. As the gage rotates, it "squeezes" down to the same size as the hole. The gage now has a size equal to the bore diameter and can be measured with a micrometer to determine bore size.

Telescoping gages take some practice to rotate smoothly and appropriately through the bore to attain the right dimension. Always double-check your readings. Also, remember that this is a semiprecision tool—it cannot provide greater accuracy than .001–.002″.

6.11 Indicators

Indicators are widely used comparison gages. Indicators cannot measure dimensions themselves, but they do indicate from a known to an unknown surface. The indicator consists of a dial, which shows the incremental values of measurement, and a spring-loaded "finger." Using a known size, such as a gage block or a sample part, the finger is set to touch the known size and the dial is set to zero. When the inspection part is placed under the finger, the difference between the known and unknown size can be measured from the dial reading. An indicator can be as accurate as .0001″ and range in size from .008″ all the way up to 1″.

Recall the earlier example of an indicator used with a height gage, **Figure 6-25**. In this illustration, a *dial indicator* is attached to a height gage and can be moved up and down. If you were measuring for a specific size, a set of gage blocks would be used to set the zero on the indicator face, and then the indicator would be moved to the part feature. The difference in the indicator reading from zero provides an easy calculation of the actual part size.

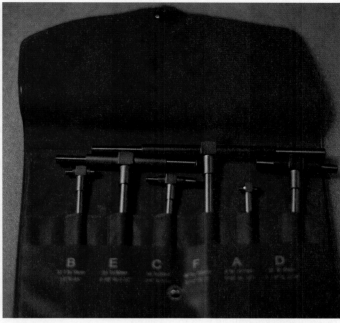

Goodheart-Willcox Publisher

Figure 6-30. A telescoping gage set.

Rotate through bore

Goodheart-Willcox Publisher

Figure 6-31. The proper way to rotate a telescoping gage through a bore.

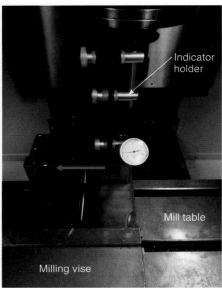

Goodheart-Willcox Publisher

Figure 6-32. Using a dial indicator to traverse across a milling vise to measure straightness.

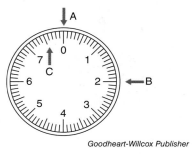

Goodheart-Willcox Publisher

Figure 6-33. A dial indicator face.

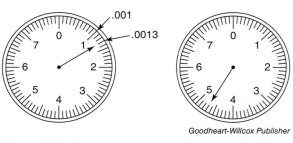

Goodheart-Willcox Publisher

Figure 6-34. Two sample readings from a dial indicator face. In the first example, the needle is pointing past the .001″ line to .0013″. In the second example, the needle is pointing to .0047″.

6.11.1 Dial Indicator

Although analog indicators all have some type of "dial" face, smaller, generally more accurate, indicators are collectively called *dial indicators*. When used correctly, these indicators are extremely accurate and versatile. A dial indicator can be attached to other accessories, such as a magnetic base, height gage, tramming bar, or mill indicator holder.

Consider a common example: dial indicators are used every day in CNC milling machines to align machine vises, **Figure 6-32.** When any vise is installed in a milling machine, that vise needs to be installed directly in line with one axis of machine travel, either the X or Y axis, depending on installation and intended use. To ensure the vise is aligned, follow these steps:

1. Use an indicator holder and lock it onto the spindle or a tool holder.
2. Move the machine into position so that the indicator tip is touching one extreme end of the vise fixed jaw.
3. From here, steadily move or sweep the machine along that axis and measure the difference in alignment.
4. Adjust the vise so the indicator has little to no difference from end to end.

This process requires some practice and patience to master. The alignment of the vise is critical to the manufacturing of a quality part that meets customer specifications.

Reading the face of a dial indicator is similar to reading other precision measuring tools, **Figure 6-33.** The dial is adjustable, so the dial can be rotated to align the needle with the "0" on the indicator, **Figure 6-33A.** Dial indicators have a large set of numbers on the face with small tick marks between those numbers. Recall from the earlier math chapter (Chapter 4) that machinists maintain a standard level of accuracy by reporting measurements to the *thousandths* place, or .001″. On a dial indicator, the numbers are listed in thousandths, where the large number 1 is .001″, 2 is .002″, and so on, **Figure 6-33B.** The small lines, or tick marks, between each large number each measure 1/10 of one thousandth, or .0001″, **Figure 6-33C.** Practice reading a dial indicator by working through the two examples in **Figure 6-34.**

6.11.2 Drop Indicator

A *drop indicator* is an indicator with a vertically dropping stem controlled by a series of springs and gears. Drop indicators generally have extended travel, as much as 2″, but lack the accuracy of a dial indicator. Remember that accuracy is defined by the smallest increment of measurement and reading repeatablity.

A drop indicator can be attached to various holders, but it is most commonly attached to a magnetic base used in CNC lathes or held on its own stand. When used in a lathe, a drop indicator measures the concentricity of a cylindrical part in the chuck jaws. By attaching the indicator to the machine and spinning the chuck, a machinist can measure the difference from the machine spindle to the center of rotation of a part. The part can be lightly tapped with a soft mallet to place the center of part rotation onto the center of spindle rotation and achieve concentricity.

When attached to its own granite base stand, **Figure 6-35**, a drop indicator can be mobile enough to use near a machine and give accurate readings of dimensions that are often difficult to measure, such as while a part is mounted into a machine.

Reading a drop indicator is similar to reading a dial indicator. The difference is the overall travel distance and accuracy level, **Figure 6-36**. The "10" on the face refers to .010″ and the small lines between the larger numbers each represent .001″. The numbers on the inner circle count the revolutions of the dial. Each revolution is .100″, so if the needle rotates completely around, the small needle will count the .100″ revolutions of travel.

 **Thinking Green**

Resurfacing Granite Plates
Machine shops often use solid granite plates to perform inspections because of their wear resistance and the ability to manufacture them extremely flat. Granite plates that have become worn can be resurfaced through a hand "lapping" process. This process restores the granite plate back to original specifications and saves the costs and use of natural resources in replacing it.

6.11.3 Coaxial Indicators

Coaxial is defined as having a common axis or two or more coincident axes. A *coaxial indicator* is a special indicator used in machining that spins around the spindle centerline and is used to find bore or boss features. Coaxial indicators provide an extremely fast way to find the center of a round feature, but they are not particularly accurate. These indicators should not be used if the accuracy level needs to exceed ± .002″. Often, coaxial indicators are used to find an approximate center. Then, a dial indicator is used for final adjustment.

The coaxial indicator, or coax indicator, comes with multiple tips and attachments, **Figure 6-37**. The various tips allow for multiple bore or boss

Goodheart-Willcox Publisher

Figure 6-35. A common configuration of a drop indicator with mounting arm and small granite plate.

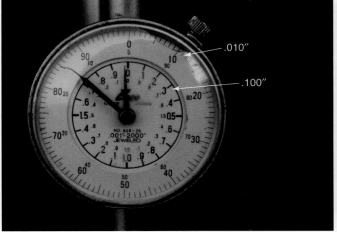

Goodheart-Willcox Publisher

Figure 6-36. Getting readings from a drop indicator face.

Goodheart-Willcox Publisher

Figure 6-37. A coaxial indicator with accessories.

Safety Note

Do not turn on the spindle until you are absolutely positive it will not rotate faster than 100 rpm. Any faster can cause a safety hazard for both the operator and indicator.

Goodheart-Willcox Publisher

Figure 6-38. A coaxial indicator rotates around a spindle center to locate the bore centerline.

sizes and give the capability to indicate small circles as well as much larger diameter circles. Because the indicator rotates around a centerline to function, it should only be used to indicate round features.

When the coaxial indicator is fully assembled, the stationary bar prevents the face from spinning while indicating in the desired workpiece. The coaxial indicator is then placed in a tool holder that is mounted in the machine. The tool holder can be a drill chuck, solid body holder, or a collet-style holder, as shown in **Figure 6-38**. Once in the machine, the spindle can be started at no more than 100 rpm to allow the indicator to slowly rotate around the spindle centerline.

Gently hold the stationary bar to keep the face in place. The needle on the indicator should fluctuate left and right as the spindle rotates. The key to measuring with a coax indicator is to stop the needle from rotating. Do this by moving the axes of the machine until the spindle is aligned with the center of the circle. Once the spindle is aligned, the needle should stop moving. However, remember that coaxial indicators are not high precision. If a setup requires greater than a .002″ tolerance, a more accurate measuring tool must be used.

6.12 Advanced Gaging and Inspection Tools

As machining has evolved over time, so has inspection equipment. The advancement of machining has always been paced by the ability to inspect.

6.12.1 Coordinate Measuring Machine (CMM)

A *coordinate measuring machine (CMM)* works much like a CNC milling machine, but it inspects and does not cut. A CMM uses the Cartesian coordinate system to drive around an electronic probe to find part features, **Figure 6-39**. The "magic" of the CMM is in its software and accuracy.

Goodheart-Willcox Publisher

Figure 6-39. A small, work-area coordinate measuring machine (CMM).

Electronic touch probing is very accurate. In addition, the software can store feature data and create GD&T measurements, such as planes and datums. Often, the repeatability of this tool is better than .00005″ (50 millionths).

6.12.2 Optical Comparator

An *optical comparator* is a measuring device that uses the shadow of a part to magnify features and measure them on screen, **Figure 6-40**. Optical comparators have been around for a long time and provide the ability to measure intricate or oddly shaped parts. A part magnified 20 times or larger becomes relatively easy to see and measure. The screen of the comparator calls out sizes and radii. By scanning across one side of a part to the opposite side, a semiprecision measurement can be made.

With modern technology, optical comparators have become increasingly more accurate and easy to use. Currently, most optical comparators offer digital readouts (DROs) that add to ease of use and accuracy, as well as function keys that do instant comparison math and measurement.

6.12.3 Air Gaging

Air gaging describes inspection tools that are set to a standard size, followed by the use of light air pressure to determine size. The air probe is set using a calibrated ring gage of a known size. As the ring is placed over the air probe, the readout dial is manually set to zero. When the inspection part is placed over the air probe, it measures the difference in airflow and converts it to a difference in size. Air gaging is most common in inside diameter inspection where tolerances are often .0001″ or smaller, such as in bearings and seal diameters.

One benefit of air gaging is that it is a noncontact inspection technique. This means the gage never touches the part, so human error is not a factor. However, air gages are reliant on using a standard, or known part sample, and therefore must be constantly monitored and adjusted for reliability.

6.12.4 Electronic Gaging

Electronic gaging is rapidly becoming the normal inspection method in most machine shops. It is fast, accurate, and quickly becoming more cost-effective for everyday shop owners. The accuracy of electronic gaging comes from light contact with electronic triggers. These sensitive triggers provide great repeatability. Electronic gages also have external convenience features, such as built-in calculations. Electronic gaging tools can include micrometers, indicators, calipers, or even height gages.

Goodheart-Willcox Publisher

Figure 6-40. An optical comparator using a part shadow to measure features.

Chapter Review

Summary

- Measuring tools can be classified as either precision or semiprecision tools. Precision is generally classified as a tool that can measure closer than .005″ or 1°, and semiprecision has incremental measurements of more than .005″ or greater than 1°.
- Calipers are the most versatile measuring tool. They can measure depths, outside diameters, and inside diameters.
- There are seven components to a micrometer: the frame, anvil, spindle, thimble, locknut, thimble, and barrel.
- A height gage is used on a granite surface plate to check geometric features, like parallelism, perpendicularity, or flatness. It can also be used in conjunction with gage blocks or inspection standards to measure deviations.
- Gage blocks are precision-ground rectangular blocks, the known size of which can be compared to a feature or part. They come in sets of 81 blocks and can be stacked together to make multiple sizes.
- A sine bar can be used in conjunction with gage blocks to measure or verify angles.
- An indicator is a comparison gage that indicates the difference in measurement between a known and unknown surface.
- A coaxial indicator attaches to the machine tool spindle and rotates around its axis to measure a bore or boss feature.
- A coordinate measuring machine (CMM) uses an electronic probe to accurately measure features as close as .00005″.

Review Questions

Answer the following questions using the information provided in this chapter.

Know and Understand

1. Generally, _____ measurement is defined as any measurement smaller than .005″ (five-thousandths of an inch) or less than 1° of accuracy.
   - A. coarse
   - B. fine
   - C. precision
   - D. semiprecision

2. _____ are precision-made rulers, usually laser engraved, and manufactured especially for the machining environment.
   - A. Calipers
   - B. Micrometers
   - C. Protractors
   - D. Steel rules

3. *True or False?* Calipers are one of the most used and most versatile measuring tools in a machine shop.

4. There are three main components to reading calipers: the _____, _____, and _____.
   - A. scale, edge, dial
   - B. scale, ratchet, dial
   - C. spindle, edge, dial
   - D. spindle, scale, edge

Match each description with the correct micrometer component.

5. Fixed in place and provides measurement scale.

6. Rotating part that screws up and down to secure the second side of a piece being measured.

7. The main single-piece body of the micrometer.

8. Rotates around the barrel and has a secondary measuring scale numbered 1 to 24 in US Standard micrometer.

9. A fixed stop that contacts one side of the piece being measured.

A. Thimble

B. Barrel

C. Anvil

D. Locknut

E. Frame

F. Spindle

10. A(n) _____ micrometer is used when measuring from one flat surface to another flat surface.

    A. outside diameter   C. inside micrometer
    B. depth   D. bore

11. _____ micrometers can be used when measuring internal features like slots, grooves, or large bore diameters.

    A. Outside diameter   C. Inside
    B. Depth   D. Bore

12. Height gages are a vertical _____ gage, which means they are used to compare a known size to an unknown size.

    A. standard   C. indicator
    B. micrometer   D. comparison

13. *True or False?* Some of the most accurate height gages are digital. The improved accuracy is a result of the electronic touch probe sensor that allows for consistent pressure on the indicator.

14. _____ are precision-ground rectangular blocks that are used as comparison gages.

    A. Gage blocks   C. Johansson blocks
    B. Jo blocks   D. All of the above.

15. A _____ is a precision-ground and hardened body with two precision-ground cylinders fixed at each end.

    A. sine bar
    B. cosine bar
    C. coaxial indicator
    D. hypotenuse bar

16. Telescoping gages have a set of _____ that can be locked into place by turning the lock on the handle end of the gage.

    A. adjustable ends
    B. spring-loaded ends
    C. round ends
    D. parallel ends

17. *True or False?* The term *indicator* refers to the definition that this tool only indicates from a known surface to an unknown surface.

18. The term *coaxial* means to have a common axis or _____ axes.

    A. multiple
    B. singular
    C. designated
    D. coincident

19. A thread gage is a(n) _____ style gage.

    A. outside diameter
    B. bore
    C. go/no-go
    D. +/−

20. _____ are a set of shims or thin pieces of metal, divided by size, to determine a gap or space in a part, or between two parts.

    A. Feeler gages
    B. Thread gages
    C. Air gages
    D. Go/no-go gages

21. *True or False?* A thread micrometer is used to verify the threads per inch, or pitch, on a threaded piece.

22. CMM is an acronym for _____.

    A. centralized measuring machine
    B. coordinate measuring machine
    C. coordinate making machine
    D. centrifugal maintenance machine

23. Air gaging is a term used for inspection tools that are set to a standard size and then _____ is used to determine size.

    A. fitting pressure
    B. weight
    C. oil pressure
    D. light air pressure

24. What type of tool can also be found in an electronic-style tool?

    A. Indicator
    B. Calipers
    C. Micrometer
    D. All of the above.

Apply and Analyze

1. As a machinist who manufactures parts, why is it important to know and understand measuring tools?

2. What are some factors to consider when selecting the correct inspection device to measure a part or feature?

3. The text states that calipers are widely used and versatile. What features of calipers make them so versatile and widely used?

4. Micrometers are extremely accurate measuring tools. Name some of the different types of inspection tools that are classified in the family of micrometers.

5. What are some of the ways to use a set of gage blocks, and what additional inspection tools are used in conjunction with the gage blocks?

6. Telescoping gages are used to measure hole diameters or slot widths. What are some possible issues with using telescoping gages?

7. Indicators can be used directly on a machine or attached to other tools, like a magnetic base or a height gage. Name some of the different types of indicators discussed.

8. There is a group of gages called comparison gages. Name some of those gages and explain why they are classified as comparison.

9. What is a CMM and what makes it accurate?

10. An optical comparator, air gaging, and electronic gaging are considered noncontact inspection techniques. What are some benefits of noncontact gaging?

Critical Thinking

1. Measurement tools and measurement techniques have evolved over many years and have been developed as needed by manufacturing companies. Discuss in detail how the pace of measuring tool development has influenced the development of machinery and engineering designs.

2. Often on machined parts, it is necessary to check the part while it is still in the machine. Is it possible to use all the inspection tools while the part is still in the machine? What are some features that can or cannot be checked in a machine?

3. The future of machining is faster production, closer tolerance parts, and more exotic metals. What new inspection hurdles might this create, and how can new technology influence inspections in the future?

North-gear/Shutterstock.com

This engineer is using a height gage.

7

CNC Machinery

Chapter Outline

7.1 Introduction to CNC Machinery

7.2 CNC Machining Centers

    7.2.1 Components of CNC Machining Centers

    7.2.2 Vertical Machining Center

    7.2.3 Horizontal Machining Center

    7.2.4 Multiaxis Machines

7.3 CNC Turning Centers

    7.3.1 Components of the CNC Turning Center

    7.3.2 2-Axis Turning Center

    7.3.3 CNC Turning Operations

    7.3.4 Multi-Turret Turning Center

    7.3.5 Multi-Spindle Turning Center

7.4 Mill-Turn Machines

7.5 Swiss Machines

7.6 CNC Grinders

7.7 Water Jet

7.8 Plasma Cutters

7.9 CNC Lasers

7.10 CNC Routers

Learning Objectives

After reading this chapter, you should be able to:

- Define the term CNC.
- Explain the difference between a CNC mill and a CNC machining center.
- Describe the application of an automatic tool changer.
- Name the four main components of a CNC machining center.
- Explain the Z axis orientation in vertical and horizontal CNC machining centers.
- Explain what makes a mill a multiaxis machine.
- Describe the difference between integrated multiaxis and axis addition.
- Define a CNC turning center.
- Explain the application of a tailstock.
- Define the turning center turret and axis orientation.
- Identify the Z axis orientation in a turning center with relation to tool direction.
- Explain the three major machining cycles in turning.
- Describe a mill-turn machine.
- Describe the differences between Swiss turning machines and conventional turning and milling machines.
- Name two types of CNC grinding.
- Explain the applications of the water jet.
- Explain the applications of plasma cutting.
- Explain the applications of laser cutting.
- Explain the applications of the router.

Key Terms

automatic tool changer	fixturing/fixture	spindle
boring	grinding	Swiss screw machine
chuck	headstock	tailstock
CNC laser	horizontal machining center	trunnion
CNC machining center		turning
CNC mill	machine controller	turret
CNC router	mill-turn machine	vertical machining center
CNC turning center	multiaxis machine	
CNC turning center spindle	multi-turret turning center	water jet
facing	plasma cutting	workholding table

7.1 Introduction to CNC Machinery

Computer numerically controlled (CNC) is a broad term describing any machine tool that uses a computer to control and operate the machine. The computer age and modern technology changed how people learn, communicate, and do business, impacting nearly every aspect of modern life. While this was devastating to certain business sectors, it also opened up new horizons for imaginative and forward-thinking innovators.

The machining industry evolved throughout this time of technological upheaval with a lot of growing pains and resistance. Machinists are hardworking, skilled professionals, many of whom could not imagine any computer ever making parts as accurately and quickly as they could. However, as they embraced change and new technology, many of those skilled manual machinists have since become great CNC operators and programmers.

Today, manual machines are quickly being replaced by high-tech, high-speed CNC machines that require a new type of skilled machinist. CNC machinists must have machining skills, computer skills, and a willingness to push traditional machining boundaries in speed and materials. To put this shift in perspective, some modern CNC milling machines can operate at 30,000 revolutions per minute (rpm) and move at speeds above 2,000 inches per minute (ipm). Parts that used to require five or six operations and 10 hours of machining time can now be made with a single setup in a few minutes. Machine builders, tooling engineers, and fixturing designers are working nonstop to stay on the leading edge of technology.

This chapter provides an overview of some of the machines you may operate or program in your career. Remember, today's CNC machines are likely to continue changing rapidly.

7.2 CNC Machining Centers

Recall from earlier chapters that a *mill* by definition is a machine where the material or workpiece is secured to a work surface and the cutting tool spins. The cutting tool is then engaged into the workpiece by moving the table or spindle. *CNC mill* is often used as a generic term for any CNC milling-type machine. However, a CNC mill is actually a milling machine with CNC controls, but without an automatic tool changer. In a CNC mill, tools are loaded and unloaded one at a time in a manual operation. Early in the development of CNC, many companies retrofitted their manual machines with CNC controls, servo motors, and encoders. It was a great crossover solution for shops that had a lot of equipment but wanted the new technology. However, the piece that could not be retrofitted was an automatic tool changer.

The addition of an automatic or programmable tool changer converts a CNC mill into a *CNC machining center*, **Figure 7-1**. The tool changer can be

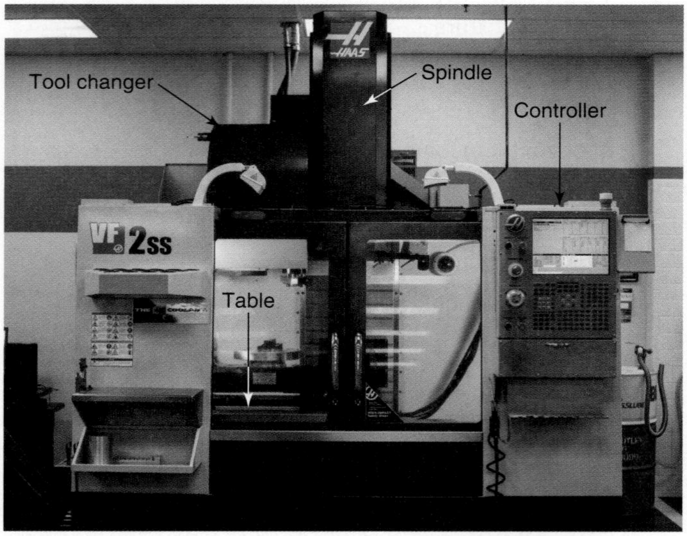

Goodheart-Willcox Publisher
Figure 7-1. A Haas VF2SS Machining Center with an automatic tool changer.

a chain-driven stand-alone unit, an internally mounted umbrella-style unit, or a side-mounted carousel unit as seen in **Figure 7-2**.

7.2.1 Components of CNC Machining Centers

There are multiple types of CNC machining centers, but they all have the same basic components:

- Spindle
- Automatic tool changer
- Controller
- Workholding table

Regardless of configuration, setup, or add-on features, every CNC machining center has these four components.

Spindle

A *spindle* secures a tool holder in place and rotates cutting tools, **Figure 7-3**. The tool holder is secured by a hydraulic or pneumatic pull bar that pulls the tool holder into a tapered spindle. The spindle housing holds multiple pieces, including the bearings the spindle rotates around. On high-speed spindles, these bearings can be steel or ceramic. Many spindles have the capability to allow cutting fluid to flow through the spindle center and directly out of the tool. This allows fluid to be dispersed directly onto the cutting edge. The entire housing, including the spindle motor, pull bar, and bearings, constitute the attachment called the spindle.

Workholding Table

A *workholding table* secures fixtures, vises, and material in place for machining, **Figure 7-4**. Workholding tables can be replaced with a variety

Goodheart-Willcox Publisher

Figure 7-2. A side-mounted tool carousel with 24-tool capacity storage.

Goodheart-Willcox Publisher

Figure 7-3. A vertical machining center spindle assembly.

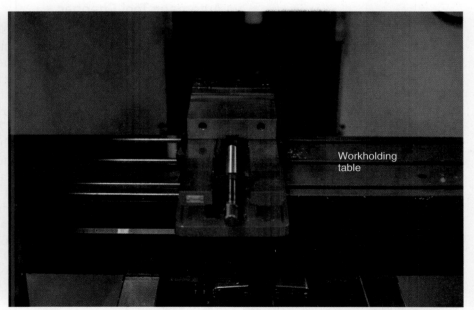
Goodheart-Willcox Publisher

Figure 7-4. A 6″ machining vise securely mounted to a workholding table.

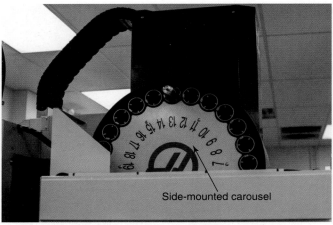

Figure 7-5. A side-mounted tool carousel. This carousel is capable of tool changes in less than one second.

Figure 7-6. An umbrella-mounted tool carousel with a 10-tool storage capacity. This carousel is mounted directly inside the machine.

of fixtures or tooling plates and can be modified to fit any workholding requirements. No matter the configuration of a particular machine, each machine must have a place to secure the material being machined.

 Tech Tip

Take care in preventing loose chips or clamps from damaging the workholding table. A small fragment of steel under a vise or fixture will influence the material being held and can leave permanent damage on the workholding table.

Automatic Tool Changer

An *automatic tool changer* stores tools not currently being used and then brings them to the ready position when a tool change is required. The automatic tool changer can hold as few as six tools or more than 300. Sometimes a tool changer can be a separate freestanding device, but most commonly it is built directly into the machine. The two most common tool changer configurations are the carousel side mount, **Figure 7-5**, and the umbrella mount, **Figure 7-6**. In an umbrella mount, the tool changer is inside the machine with tools located in numbered stations, or pockets. This configuration requires the machine to put tools in the correct station or pocket and then advance to the next station to load the subsequent tool. Carousel mounts have much faster tool-change times, in some cases as fast as 0.4 seconds. In a carousel mount, the computer keeps track of which tool is in what station, and the next tool for use is advanced to the tool-ready position. As soon as a tool change is commanded, the machine can instantly change tools, thus providing a faster change time.

Controller

A *machine controller* is an onboard computer where edits, program uploads, and all tooling and work offsets are made, **Figure 7-7**. This is the central processing area of a CNC machine. Programming commands and all machine functions and movements are controlled from this location.

Figure 7-7. Haas CNC machine controller.

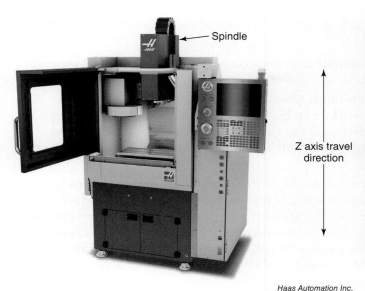

Figure 7-8. A compact Mini Mill from Haas Automation.

Haas Automation Inc.

Goodheart-Willcox Publisher

Figure 7-9. Illustration of machine axis of travel relative to work table and spindle, in a vertical machining center.

The controller has changed and evolved as fast as any other computer component over the last 40 years. Today, these controllers can have massive storage capacity, direct Internet and Ethernet capabilities, and external monitoring capabilities. The base components of the machine tool have not changed that much, but the addition of servo motors and CNC control is essential to the current manufacturing revival.

7.2.2 Vertical Machining Center

In a *vertical machining center*, *vertical* refers to the orientation of the spindle, or Z axis. In a vertical machine, the spindle travels vertically, or up and down toward the floor and ceiling, **Figure 7-8.** The Z axis in all milling machines is through the centerline of the spindle, and other axes are designated by spindle orientation. In a 3-axis vertical mill, for example, the Z axis is up and down through the spindle, the X axis is to the left and right, and the Y axis is forward and backward, as in **Figure 7-9.**

Vertical machining centers range in size from 12″ × 16″ in X and Y directions up to machines large enough to drive a car onto. Some of the largest machines have a special design to allow maximum clearance and are given special classifications, which are discussed later in this chapter.

7.2.3 Horizontal Machining Center

A *horizontal machining center* is easily recognizable, as in **Figure 7-10,** by the spindle orientation. Notice that the spindle comes from the wall of the machine, not the top as in vertical machining centers. It is typical in horizontal machining centers for the table or workpiece to move into the spindle, rather than having the

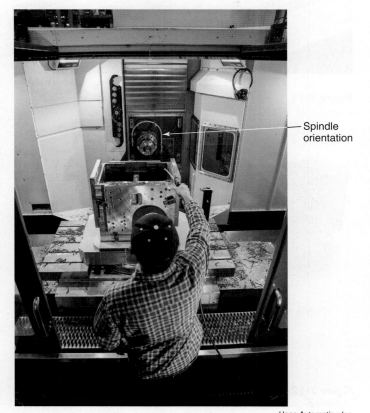

Haas Automation Inc.

Figure 7-10. A horizontal machining center with a multiface rotary fixture.

spindle itself extend or move. There are three major advantages to this configuration:

- The machine spindle does not extend or hang away from its support system, so it is more rigid in design.
- Chip removal is much more efficient than in vertical machining centers. In a vertical machine, chips stay inside the work envelope and can interfere with the cutting edge. In a horizontal machine, gravity and right-angle cutting allow chips to fall free away from the cutting area.
- Horizontal machines can be fitted with optional accessories, such as pallet changing stations and vertical workholding fixtures. Although the configuration of a horizontal machine may vary, the Z axis direction is always oriented through the spindle centerline. Because of this configuration, the tool change usually happens through a door opening and can be faster than in a vertical machine.

Haas Automation Inc.

Figure 7-11. A rotary directly built into a machine table is an example of a fully integrated multiaxis machine.

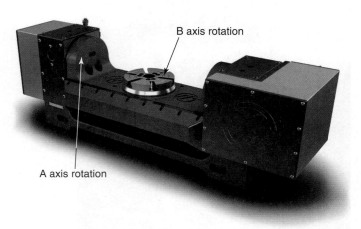

Haas Automation Inc.

Figure 7-12. A TR160 Haas trunnion assembly. This can be mounted on a machine work table to add two additional axes of rotation.

7.2.4 Multiaxis Machines

A *multiaxis machine* is a machine with more axes of travel than a standard 3-axis machining center. Multiaxis machines are rapidly becoming more common as lower cost and higher availability have made them more accessible to machine shops. Multiaxis machines can be integrated into a single configuration or retrofitted by adding equipment to a 3-axis machining center. Integrated multiaxis machines are the best solution for multiaxis machining, but they typically have smaller work envelopes that can make simpler 3-axis work more difficult to set up and operate, **Figure 7-11.**

Converting a standard 3-axis machine to a multiaxis setup requires adding axes by means of a trunnion or rotary, typically referred to as A and B axes. Additional equipment is also needed for this conversion, including an additional servo drive system and a parameter change to designate to the machine controller that more axes are available for rotation.

A *trunnion* is a multiaxis attachment with multiple configurations that can be mounted directly to a machine table, **Figure 7-12** and **Figure 7-13.** The term trunnion originates from the cradle in which a cannon sat, which also allowed directional aiming. Trunnions can rotate in two directions, converting a 3-axis machine to a 5-axis machine.

Figure 7-12 shows a Haas Automation TR160 trunnion. This trunnion is supported on both ends, which allows the center body to rotate (A axis) as well as the table (B axis). The extended distance between supports allows for fairly large workpieces to be secured.

A   *Haas Automation Inc.*   **B**   *Goodheart-Willcox Publisher*

Figure 7-13. A—TRT100 compact designed trunnion. Allows for more work space. B—TRT100 mounted inside a vertical machining center with two additional axes of rotation.

Figure **7-13A** shows a TRT100 unit with a much more compact design. This still gives us 5-axis capability, but it reduces the amount of machining envelope required for mounting. This unit can be mounted on one end of the machine while leaving space for normal 3-axis work on the table, **Figure 7-13B.**

Figure **7-14** shows another variety of 5-axis trunnion using a multiple 5C colleting system. This indexer uses three collet holders that can rotate around a B axis and which can also be tilted front to back in the A axis. This gives the programmer the option of making three identical parts at the same time, or three different parts in one set up.

There are many 4-axis options available as well that may provide all the axes required, depending on the type of manufacturing. One of these is the rotary, which only turns or rotates in one direction. These can be mounted vertically or horizontally to allow for different angles of machining. **Figure 7-15** shows one variety of 4th axis platter-style rotary.

Haas Automation Inc.

Figure 7-14. A Haas T5C3 dual-axis tilting rotary unit.

Haas Automation Inc.

Figure 7-15. A 4-axis platter style rotary axis.

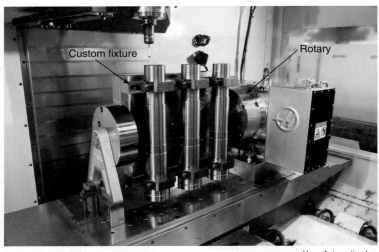

Haas Automation Inc.

Figure 7-16. Custom fixturing can be mounted on a rotary for multiple-part production.

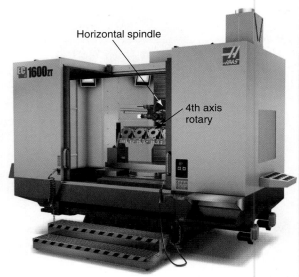

Haas Automation Inc.

Figure 7-17. A horizontal CNC machining center with a 4th axis rotary installed.

In addition, *fixturing* (to include machine vises or custom fixtures) can be mounted to any of these rotary tables and mounts to hold a variety of part shapes and sizes, **Figure 7-16.**

Vertical machining centers, horizontal machining centers, and multiaxis machining centers provide multiple ways to manufacture parts, depending on the desired specifications. For example, **Figure 7-17** shows a horizontal machining center with a 4th axis platter integrated into a 5th axis rotary table, giving it 30° of tilt. This is just one possible combination for multiaxis machining.

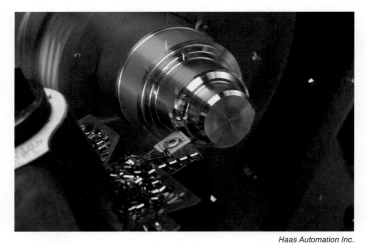

Haas Automation Inc.

Figure 7-18. A CNC turning center machining a steel part.

7.3 CNC Turning Centers

A *CNC turning center* is a computer-controlled lathe fitted with an automatic tool changer. These machines can be referred to as *CNC lathes*. A lathe is a machine tool where the material or cutting piece is rotated and the cutting tool is fixed, or nonrotational. Because the material is rotating, these machines make round or cylindrical parts, **Figure 7-18.** Remember from earlier chapters that the lathe was the earliest machine tool, and, in many industry sectors, it is still the primary machining process.

Most CNC turning centers are oriented horizontally with the material going from left to right of the operator, but, as in other machines, this configuration can change. Just keep in mind the definition of the lathe as a spinning workpiece with a fixed tool.

Thinking Green

Reducing Chips

Managing waste from chips is critical for eco-conscious machine shops. Optimum cutting produces long, curling chips. Leaving these chips intact makes collection more difficult and wasteful. Cut the chips at intervals to make gathering and recycling easier. Chips can then be repurposed into new objects, like wire brushes and scouring pads, or pressed into bricks for pick-up by industry recyclers.

Before collection or bricking, treat all chips to remove excess cutting fluid. To remove cutting fluid, drain the chips or use centrifugal force to separate the fluid from the chips. Careful removal of cutting fluid from chips may allow the fluid to be reused as well.

7.3.1 Components of the CNC Turning Center

There are multiple types and configurations of CNC turning centers. All CNC turning centers have four major components:

- Spindle
- Automatic tool changer
- Tailstock
- Controller

Spindle

A *CNC turning center spindle* is made up of two subcomponents: the chuck and headstock, **Figure 7-19.** *Chuck* generally refers to any workholding

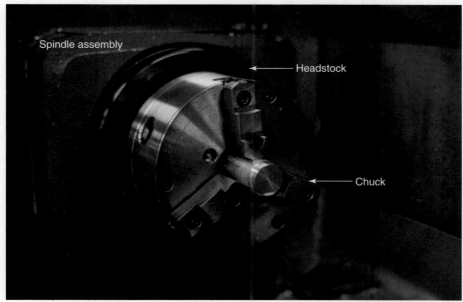

Goodheart-Willcox Publisher

Figure 7-19. A manual chuck mounted in a CNC turning center.

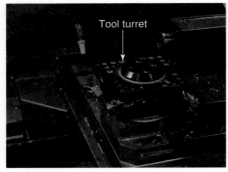

Tool turret

A

Slant turret

B

Goodheart-Willcox Publisher

Figure 7-20. A—4-tool position turret on a Haas TL-1. B—The most common style of tool turret is the slant-style turret.

device, including 3-jaw chucks, 4-jaw chucks, collet closers, faceplates, and any fixtures or devices that secure a workpiece for machining. The *headstock* includes all components that secure a chuck or open and close a chuck. The housing, bearings, and manual, pneumatic, or hydraulic chuck closing components are all part of a headstock.

Automatic Tool Changer

The automatic tool changer on a CNC turning center is called a *turret*. Turrets are where tools are fixed and stored in place when not in use and secured for cutting, **Figure 7-20**. A turret can hold as few as four tools in small, toolroom-style lathes or up to 20 in larger machines. Some exceptional specialty machines have off-machine add-ons that can hold 100 tools if desired.

Tailstock

A *tailstock* is a moveable or programmable device that provides longitudinal support to a workpiece, **Figure 7-21**. In bar stock machining, tailstocks may be required to support the bar at the point farthest from the chuck to eliminate unwanted surface finish issues or avoid tapered cuts caused by the bar being pushed away during manufacturing. The general rule is 4:1. That is, if the length is four times the diameter, use a tailstock. True turning centers must have tailstocks. Machines without tailstocks are called "chuckers."

Controller

As in CNC machining centers, a machine controller in a CNC turning center is an onboard computer where edits, program uploads, and

Goodheart-Willcox Publisher

Figure 7-21. A CNC machining center with a tailstock. In this illustration the tailstock is manually positioned and locked in place, but tailstocks can be fully programmable.

all tooling and work offsets are made, **Figure 7-22**. The machine controller interprets programming commands, and all machine functions and movements are controlled from the controller. There is virtually no difference in appearance between machining center controllers and turning center controllers, but there are differences in machine functions and processing code.

7.3.2 2-Axis Turning Center

A standard CNC turning center has only two axes of movement, the X axis and Z axis. The direction of the Z axis is defined based on the orientation of the cutting tool. Going toward the spindle is defined as the negative Z or Z– direction, and going away from the spindle is defined as the positive Z or Z+ direction. The origin or Z0. position is usually set at the face of the finished part. This will simplify the calculations and the CNC program, but it can be set at any location corresponding with the appropriate program. The origin on the X axis or X0. is a reference to the centerline of the spindle, which is the center of rotation for any part on a lathe. Moving away from that centerline is the X+ direction, and going toward that centerline is the X– direction. Consider the following examples where the tool is oriented on different sides of the spindle.

In **Figure 7-23**, the turret is closest to the operator. The dashed line is the machine spindle centerline, or center of rotation. Everything from that centerline to the operator is X+, and everything going toward the chuck is Z–.

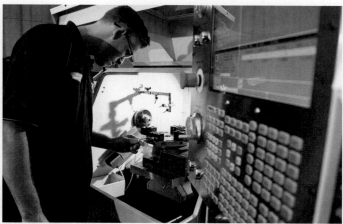

Haas Automation Inc.

Figure 7-22. A CNC operator is setting up tools. The controller is in the foreground.

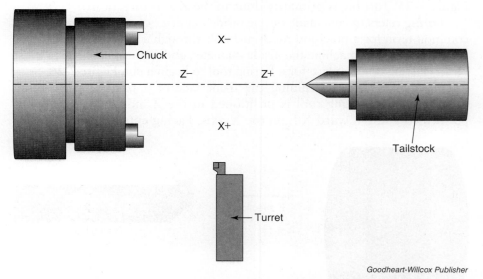

Goodheart-Willcox Publisher

Figure 7-23. Illustration showing axis direction relative to chuck and tool position. The turret is closest to the operator.

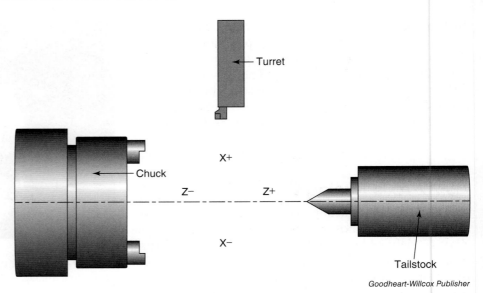

Figure 7-24. Illustration showing axis direction relative to chuck and tool position. The turret is farthest from the operator.

Figure 7-24 shows a more common turret configuration, with the tool farthest away from the operator. The X+ direction is defined based on the orientation of the tool, so it is now on the opposite side of centerline.

7.3.3 CNC Turning Operations

There are three basic operations performed in a turning center: turning, boring, and facing. You may also encounter other, less common operations, such as threading and grooving.

Turning refers to a machining operation performed on the outside of a part, or the outside diameter. Any straight-line cut, tapered diameter, or radius process performed on the outside diameter is called turning, **Figure 7-25.** Turning is primarily done in the Z axis direction.

Boring refers to cuts made on the inside diameter of a part. *Bore* is also a common term for a precision-machined hole through a part. A boring bar or tool is required to machine the inside diameter, and in most cases there must be an existing hole in place for a boring tool to begin a cut, **Figure 7-26.**

Facing describes machining the front face of a part. During facing operations, the cutting tool is positioned in the Z axis, while the cut is performed going toward X0. on the X axis. Facing cuts are performed to

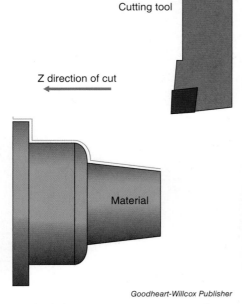

Figure 7-25. A typical turning operation showing a taper, radius, and straight-line cut. The tool is traveling in a Z– direction.

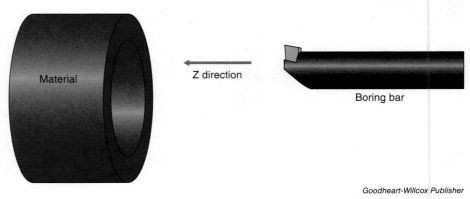

Figure 7-26. A typical boring operation showing an inside diameter operation. The tool is traveling in a Z– direction.

remove any excess material on the front face or to establish a consistent part length through multiple parts, **Figure 7-27**.

7.3.4 Multi-Turret Turning Center

A CNC turning center with more than one turret is called a ***multi-turret turning center***. Multi-turret turning centers are used in operations where it is advantageous to machine from both sides of a part or perform multiple operations simultaneously. For extended length parts, multi-turret turning centers can perform pinch turning, a process where tools cutting on opposite sides of a part can prevent part vibration or chatter. The trick to pinch turning is to have one tool slightly ahead of the other tool and cutting at a slightly deeper depth, or X axis position. Multi-turret machines can also complete more than one operation at a time. For example, the outside of a part can be machined while a drill is making a hole in the center. This combines operations and reduces the total machine time, or cycle time.

Machinists using additional turrets must exercise extreme caution when programming to prevent possible part or machine damage. Each turret is programmed individually with specific timing for operations. As one turret completes an operation, the second can begin a new operation. This swapping of operations increases efficiency and prevents the turrets from a possible collision with each other. Each turret is also bound by the same coordinate restraints, **Figure 7-28**.

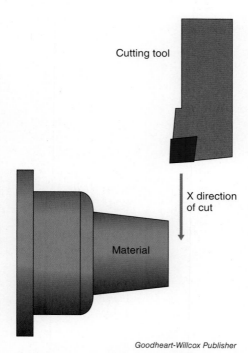

Goodheart-Willcox Publisher

Figure 7-27. A typical facing operation. The tool is traveling in an X– direction.

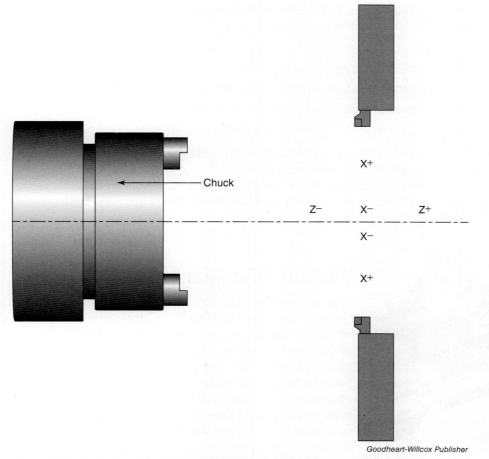

Goodheart-Willcox Publisher

Figure 7-28. A dual-turret machine and the axis travel direction relative to the cutting tool direction. Cutting toward the centerline will be an X– directional cut.

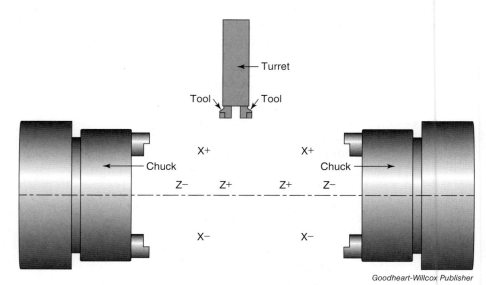

Figure 7-29. A single-turret, dual-spindle machine and the axis travel direction relative to the cutting tool direction.

7.3.5 Multi-Spindle Turning Center

In addition to multiple turrets, a turning center can be equipped with multiple spindles, **Figure 7-29**. Multiple spindles allow for machining to be accomplished on both ends of the part. In this configuration, the turret must be adapted to hold tools facing both left and right. The basic procedure for multi-spindle machining is as follows:

- One spindle advances.
- Spindle speeds are timed together.
- The first machining operation is completed.
- The part is transferred from one spindle to the other.
- The second side is finished in a single operation.

Multi-spindle turning machines can also be programmed like traditional, single-spindle turning centers.

As the demand for more versatility increases every year, so does the variety of CNC turning center configurations. There are even dual-spindle, dual-turret machines.

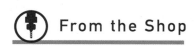

From the Shop

Building Our Future
There are a lot of different CNC machines and many variations of those types of machines. The great irony is that as manufacturing comes up against manufacturing problems, we build our own machines to solve those problems. As more machines are developed, more machinists and CNC operators will have to learn this new equipment to compete in the market.

7.4 Mill-Turn Machines

One of the revolutionary advances in machining over the last few years has been the *mill-turn machine*. A mill-turn machine is a true hybrid of a machining center and a turning center, fully capable of both turning and milling, **Figure 7-30**. For many years machining centers have been able to make round parts in center setups, and turning centers have had the ability to accomplish some drilling and minor mill-type operations. However, only recently has it been possible to do full turning and full milling in a single machine.

Although mill-turn machines may resemble lathes, they have the horsepower and capabilities to complete

Figure 7-30. A mill-turn machine with full turning and full milling capability.

full milling operations. The spindle is programmable to allow orientation at specific angles, or it can be slowly rotated for full multiaxis machining. Previously, a part requiring both lathe and mill work took multiple setups on multiple machines. Now, these difficult parts can be machined in a single setup on a single machine. Mill-turn machines often have software specific to their controllers to allow custom programming, or they support more advanced computer-aided machining (CAM) software.

7.5 Swiss Machines

CNC *Swiss screw machines* were first developed for producing tiny, high-precision internal watch and clock components, **Figure 7-31**. A Swiss screw machine is an automatic lathe that has a sliding headstock and a guide bushing. An entire day's production from a Swiss machine might be as many as 3,000 pieces that fit in the palm of your hand.

Figure 7-31. Internal watch components made on a CNC Swiss screw machine.

Swiss screw machines differ from other milling and turning machines in several ways:

- Tools are fixed in place on a slide and do not move to cut material. Instead, the bar is pushed into the tool. This allows the cut to be very rigid and does not allow for the material to flex or push away.

- The only way to make parts from a Swiss machine is to load full-length round bars into the machine. The bars are supported by bushings and collets, and the machine can be outfitted with a multi-bar loading system to allow for very high production runs, often up to 500,000 pieces.

- Swiss machines routinely have multiple spindles and milling capabilities.

The parts produced by a Swiss screw machine are so small that it is impractical to do secondary operations on a separate machine. The largest Swiss machines only run a 3/8″ diameter bar size. If a machine shop has to make a high-volume, small-diameter part, a Swiss machine can be an appropriate choice.

7.6 CNC Grinders

Grinding is an abrasive rather than a cutting process, where material is removed in small increments of less than .005″. In its most basic form, a grinding wheel is made from a natural or synthetic abrasive mineral that is bonded together in a matrix. In more advanced applications, a wheel can be made from steel and then vitrified bonded with an industrial diamond edge. The grinding process works best with harder materials, generally above 40 Rockwell (RC) hardness, and can produce very fine surface finishes with tolerances as accurate as ± .0001″. The application of CNC controls has made grinding a valuable process in high-production precision machining.

A — *oYOo/Shutterstock.com* B — *Goodheart-Willcox Publisher*

Figure 7-32. A—An OD grinding operation. B—An illustration of a formed grinding wheel capable of OD and face grinding simultaneously.

The two basic types of grinding are outside diameter (OD) grinding and inside diameter (ID) grinding. An OD grinder, **Figure 7-32**, works by spinning a grinding wheel against the outside of a part. The part must be turning slowly in the opposite direction of the grinding wheel, which is spinning much faster. In a CNC grinder, the wheel is profiled with a diamond-tipped tool to cut a form into the wheel. The formed wheel can now be brought against the material to grind both faces and outside diameters simultaneously. Parameters are set to re-form the wheel after a given number of plunges. Because of the heat and abrasion created in grinding, the part should be flooded with coolant.

 Safety Note

Grinding wheels are subject to cracks that can cause catastrophic wheel failure. Always inspect a grinding wheel for cracks or damage before installation, and discard any wheel that appears damaged. All grinding wheels come with a listed maximum rpm. Never exceed maximum rpm, as this will cause the wheel to fail. Every time you double the speed of a grinding wheel, you multiply the centrifugal force on that wheel by four times.

ID grinders work on the inside of a bored hole, **Figure 7-33**. Grinding an inside diameter requires a slightly different setup. The workpiece is held in a lathe-style chuck, and the grinding wheel is mounted on a shaft that extends parallel to the machine centerline. As the grinding wheel engages the material, it begins a reciprocation cycle that traverses the wheel along the bore diameter. This cycle allows for a consistent finish and diameter size.

Figure 7-33. An ID and face-grinding operation on an ID grinder.

7.7 Water Jet

A water jet is another abrasive machine that is not often the primary machine used in CNC manufacturing. A *water jet* uses a mixture of abrasives and high-pressure water, up to 100,000 PSI, to cut out intricate shapes in various materials, including aluminum, steel, marble, wood, glass, and even plastics, **Figure 7-34.** Water jets have advanced to the point where they can cut up to 6″-thick steel plates.

In a water jet, water is pressurized and directed through a small diameter nozzle programmed to cut a desired shape. Water jets are typically equipped with proprietary programming software that can read CAD files directly for 2-dimensional shapes. Water-jet programming is similar to programming for a CNC machining center, where the water stream acts as a cutting tool. Early versions of water jets were used primarily for sign-making and decorative artwork. Now, water jets are used to cut out large

Figure 7-34. A high-pressure water-jet operation.

amounts of material in unconventional shapes, which makes them important machines in conventional machine shops and manufacturing.

7.8 Plasma Cutters

Plasma cutting involves forcing an electrically charged, superheated gas through a nozzle. Because this is an electrical process, only ferromagnetic materials—those that are capable of carrying an electrical charge—can be cut, including aluminum, steel, stainless steel, brass, and copper. An electrical arc is created by touching the nozzle to the grounded material. Then, the arc is enveloped by a compressed gas, such as air, oxygen, or an inert gas. The gas surrounds the arc and focuses it into a directed beam. The compressed gas flow blows the molten material and debris away from the work area, allowing for a clean cut, **Figure 7-35**.

Andrey Eremin/Shutterstock.com

Figure 7-35. A CNC plasma cutting operation on sheet metal.

Handheld torches can cut materials up to 1.5″ thick. CNC plasma cutters can gouge through 6″-thick steel. Plasma cutters are regularly used to complement or replace saws, snips, cutoff wheels, and oxygen acetylene cutting torches. Plasma cutters are faster and cleaner for cutting steel than a gas torch. CNC plasma machines are built on the same type of platform as water jets and lasers and are often used for sheet metal and steel plate cutouts. Often, many parts of different shapes and sizes can be arranged or *nested* in a CAD system and cut out in a single operation by tracing the outlines.

7.9 CNC Lasers

CNC laser cutting technology may be the fastest growing field in CNC machinery, especially in sheet metal applications, **Figure 7-36**. New laser technology can engrave or cut almost any shape with tolerances as close as .001″.

Prasit Rodphan/Shutterstock.com

Figure 7-36. A high-efficiency laser operation.

Unlike plasma cutters and routers, laser cutters were built specifically for CNC operations. Plasma and router machines originally were handheld before they were added to a head and table to create CNC variations. Lasers have never been a handheld cutting option. Laser beams are focused streams of monochromatic coherent light, which are directed toward the workpiece material by mirrors or fiber optics.

7.10 CNC Routers

Routers have been used as handheld power tools for many years for wood or plastic cutting, edge formations, and engraving. A router is a relatively simple machine consisting of a motor spinning a tool in a collet holder, **Figure 7-37**.

The first **CNC routers** were 2-axis machines that drove a router bit around a profile or line. The depth of cut was preset before machine operation. Today, CNC routers are full 3-dimensional milling machines, and they are the standard in any production woodworking facility. They have also been implemented in cutting plastic, plexiglass, and composites, such as polyurethane.

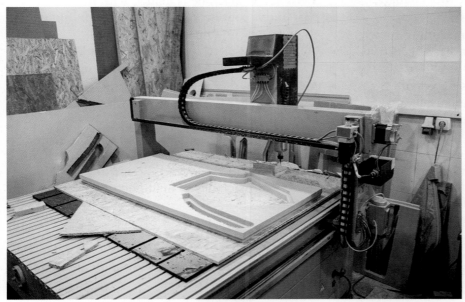

olgers/Shutterstock.com

Figure 7-37. A CNC router used to machine wood for cabinetmaking.

Chapter Review

Summary

- CNC, or computer numerically controlled, is a broad term referring to any machine that uses a computer and a control to operate a machine.

- A CNC machining center is a CNC controlled mill with an automatic tool changer. A CNC mill has tools that can only be changed manually.

- An automatic tool changer stores tools not being used currently and then brings them to the ready position when a tool change is required.

- A CNC machining center has four major components: the spindle, automatic tool changer, controller, and a workholding table.

- The Z axis orientation of a CNC machining center is through the centerline of the spindle. This is true in a vertical or horizontal machining center.

- The standard machining center has 3 axes of movement, the X, Y, and Z. Machining centers with additional axes are referred to as multiaxis machines. These axes of travel are known as A and/or B axis.

- A machining center that was designed and built with additional axes from the machine builder is known as an integrated multiaxis machine. Machining centers can be retrofitted with post-production equipment installed to add these axes.

- A rotary is a type of retrofitted device that only moves in one direction, making the machining center a 4-axis machine.

- A trunnion is a type of retrofitted device that moves in two directions, making the machining center a 5-axis machine.

- A CNC turning center is a computer-controlled lathe with an automatic tool changer.

- The use of a tailstock in a turning center provides stability to long workpieces and increases machinability.

- A turret is the automatic tool changer on a CNC turning center. The orientation of the turret dictates X and Z axis directions.

- The Z axis direction in a turning center is directly toward or away from the spindle. The X axis direction travels toward or away from the spindle centerline.

- The three major cycles in a CNC turning center are turning, boring, and facing.

- A CNC mill-turn machine is a fully integrated machining and turning center. It has full capabilities of both milling and turning.

- A CNC Swiss screw machine is designed for ultra-high production. The tools are fixed in position and the material is advanced past the cutter. The Swiss screw machine is loaded with full-length bars and is routinely equipped with multiple spindles and milling capabilities.

- Two types of CNC grinding are outside diameter (OD) and inside diameter (ID) grinding.

- A CNC water jet uses high-pressure water and an abrasive to cut intricate shapes through various materials.

- Plasma, laser, and router technology is used and implemented with CNC technology. Different industries with specific requirements use CNC-controlled machines to manufacture parts for their industries.

Review Questions

Answer the following questions using the information provided in this chapter.

Know and Understand

1. The term CNC, or _____, is a broad term describing any machine tool that uses a computer control to operate the machine.

   A. computer normally controlled
   B. computer numerically controlled
   C. completely numerically controlled
   D. completely not controlled

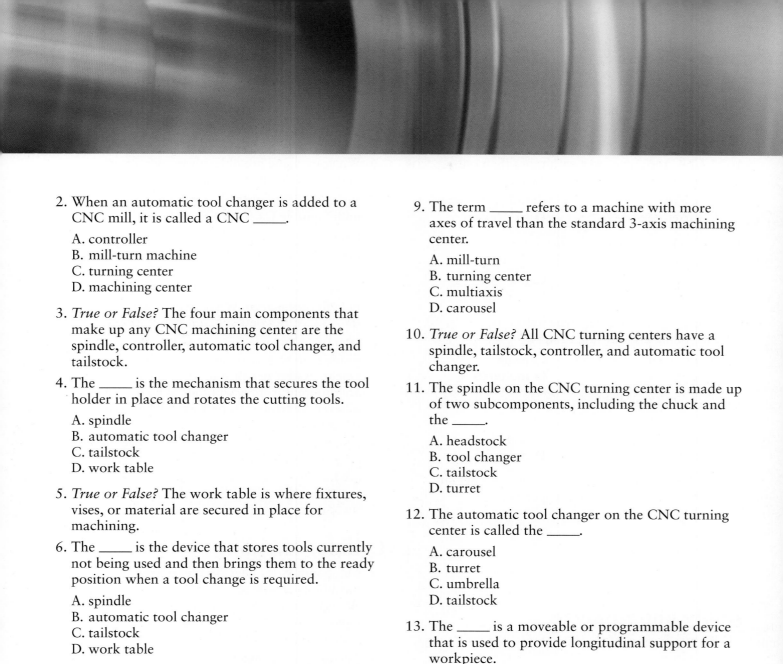

2. When an automatic tool changer is added to a CNC mill, it is called a CNC _____.

A. controller
B. mill-turn machine
C. turning center
D. machining center

3. *True or False?* The four main components that make up any CNC machining center are the spindle, controller, automatic tool changer, and tailstock.

4. The _____ is the mechanism that secures the tool holder in place and rotates the cutting tools.

A. spindle
B. automatic tool changer
C. tailstock
D. work table

5. *True or False?* The work table is where fixtures, vises, or material are secured in place for machining.

6. The _____ is the device that stores tools currently not being used and then brings them to the ready position when a tool change is required.

A. spindle
B. automatic tool changer
C. tailstock
D. work table

7. The machine _____ is the onboard computer where edits, program uploads, and all tooling and work offsets are made.

A. center
B. work coordinate
C. tailstock
D. controller

8. *True or False?* In a horizontal machining center, the spindle direction is vertical.

9. The term _____ refers to a machine with more axes of travel than the standard 3-axis machining center.

A. mill-turn
B. turning center
C. multiaxis
D. carousel

10. *True or False?* All CNC turning centers have a spindle, tailstock, controller, and automatic tool changer.

11. The spindle on the CNC turning center is made up of two subcomponents, including the chuck and the _____.

A. headstock
B. tool changer
C. tailstock
D. turret

12. The automatic tool changer on the CNC turning center is called the _____.

A. carousel
B. turret
C. umbrella
D. tailstock

13. The _____ is a moveable or programmable device that is used to provide longitudinal support for a workpiece.

A. turret
B. trunnion
C. rotary
D. tailstock

14. In a turning center, going toward the chuck is the Z– direction and going away from the chuck is the _____ direction.

A. X–
B. X+
C. Z+
D. A+

15. *True or False?* The three basic machining operations performed in a turning center are milling, boring, and facing.

16. A CNC turning center with more than one _____ is called a multi-turret turning center.

   A. spindle   C. tailstock
   B. turret   D. tool holder

17. Multiple _____ allow for machining to be accomplished on both ends of the part.

   A. spindles   C. tailstocks
   B. turrets   D. tool holders

18. The _____ machine is a true hybrid of the machining and turning centers.

   A. mill-turn   C. prius
   B. laser   D. multiaxis

19. The act of grinding is not a cutting process, but a(n) _____ process.

   A. abrasive   C. high-pressure
   B. erosion   D. hand

20. Unlike other cutting processes like plasma and router, the _____ was built specifically for a CNC operation.

   A. Swiss screw machine
   B. trunnion
   C. laser
   D. grinder

Apply and Analyze

1. What are some of the advantages of CNC machines?

2. A machining center must have an automatic tool changer. What are the three different types of tool changers? List a possible advantage of each one.

3. A rotary and a trunnion are two supplementary pieces of equipment that can be installed in a machining center to provide additional axes of movement. What are the possible advantages of adding these to standard 3-axis machines?

4. A fully integrated multiaxis machine is built as one unit from the machine builder. What are some advantages to integrated multiaxis machines?

5. The machine controller is the control center of the CNC machine. What are some modern advantages built into the CNC controller?

6. What are some possible advantages of using a multi-turret CNC turning center?

7. What are some possible advantages of using a multi-spindle CNC turning center?

8. How does a CNC mill-turn maximize machining operations and minimize setups and machining time?

9. Swiss screw machines are designed for small parts in high production. What characteristics make these machines best suited for this application?

10. CNC grinding can be accomplished on the inside diameter or on an outside diameter. The machines that perform these two types of grinding are different from each other. Why does it take different machines to perform OD grinding and ID grinding?

11. CNC laser and CNC plasma machines are generally used to cut large sheets of steel that are less than 1/2″ thick. What are some of the industries that might benefit from implementing these types of machines?

Critical Thinking

1. Many industries have been impacted by technological advances in CNC machine building. What industry do you think might have seen the most significant improvements in manufacturing capabilities by adding CNC machinery?

2. The early CNC machines were very expensive and only available from foreign markets, leading to long wait times for delivery. What major changes in part manufacturing have occurred since CNC machines have become affordably priced and readily available in local markets?

3. CNC machines are still advancing in technology and integration. What possible advances do you imagine might come next? Consider robotics, 3D printing, computer technology, social media, or any other modern advances.

Close-up of CNC milling machine spindle.

8

Cutting Tools and Inserts

Chapter Outline

8.1 High-Speed Machining

8.2 Inserts

8.2.1 Insert Naming System

8.2.2 Carbide

8.2.3 Coatings

8.2.4 Diamond Cutter

8.2.5 Ceramics

8.2.6 PCBN

8.3 Drills

8.3.1 Twist Drills

8.3.2 Center Drills

8.3.3 Spot Drills

8.3.4 Indexable Drills

8.4 Taps

8.4.1 Straight Flute Taps

8.4.2 Spiral Flute Taps

8.4.3 Bottom Taps

8.5 Countersinks and Counterbores

8.6 End Mills

8.6.1 Edge Cutting Geometry

8.6.2 Roughing End Mills

8.6.3 Inserted End Mills

8.7 Face Mills

8.8 Toolholders

8.8.1 Milling Toolholders

8.8.2 Lathe Toolholders

8.9 Machinability Ratings

Learning Objectives

After reading this chapter, you should be able to:

- Explain the benefit of insert tooling.
- Summarize the ISO numbering system.
- Explain the evolution of carbide tooling.
- Describe the benefits of PVD and CVD coating.
- Explain the possible uses for polycrystalline diamond tooling inserts.
- Explain the possible uses for ceramic tooling inserts.
- Describe the benefits of polycrystalline cubic boron nitride inserts.
- Explain the different components on a twist drill.
- Differentiate between a center drill and a spot drill.
- Describe the benefits of insertable drilling tools.
- Identify a spiral flute, straight flute, and bottom tap.
- Understand the use of counterbores and countersinks.
- Identify the cutting-edge geometry of an end mill.
- Explain the usage of roughing end mills.
- Explain the usage and limits of face mills.
- Outline the difference in CAT and BT toolholders.
- Identify the different styles of lathe toolholders.

Key Terms

bottom tap

center drill

chemical vapor deposition (CVD)

counterbore

countersink

end mill

face mill

insert

physical vapor deposition (PVD)

polycrystalline cubic boron nitride (PCBN)

polycrystalline diamond (PCD)

roughing end mill

spiral flute tap

spot drill

straight flute tap

tap

twist drill

8.1 High-Speed Machining

The machining process usually involves taking a piece of metal and slamming another piece of cobalt or carbide into it to shear away unwanted material. This does not always go as planned, and the intense forces involved mean that cutting tools and materials can literally explode into pieces. In modern machine shops, the materials that CNC machinists are asked to cut are more exotic and challenging than ever before, requiring more advanced machines and more diverse cutting tools. Machining could not continue to evolve without the engineering advances seen in the cutting tool industry.

Only a few years ago, machining exclusively used high-speed steel tools with a piece of carbide brazed to the end. High-speed steel (HSS) is a high-performance, hard metal that can contain cobalt, tungsten, vanadium, or molybdenum to add wear resistance to a cutting tool. HSS drills are suitable for running 300 rpm and .003″ feed per revolution, or 5 inches per minute in a mill. When newer machines can reach 15,000 rpm and 150 or 200 inches per minute, the added force and heat cause tool fracturing and ultimately catastrophic tool failure in high-speed steel.

Once tooling requirements increased, tool designers and engineers went to work. The implementation of solid carbide tools and their continued improvement have made high-speed machining a reality, **Figure 8-1**.

Haas Automation Inc.

Figure 8-1. A lathe turning operation using inserted tooling.

8.2 Inserts

An *insert* is a small, clamped-in cutting tool. When only one edge is cutting, it is most practical and cost efficient to make a screw-in or clamp-in carbide insert. This allows the insert to be changed as needed without having to replace the entire cutting tool. Almost all turning tools are carbide inserts, and many milling cutters use insert technology. Solid carbide end mills are common in milling and will be discussed in a later section. This section focuses on insert naming theory, composition of carbides, insert coatings, and inserts that are not mainly composed of carbide.

 From the Shop

Check the Data

There are a lot of different insert manufacturers to choose from. Some operators prefer one brand over another, and some operators use certain brands for certain applications. The real key to the right insert is to do some testing on your machine and your operation and chart the life of the tool. Choosing the right insert for the right job can be the difference between making a profit or taking a loss. The truth is in the data.

8.2.1 Insert Naming System

The International Organization for Standardization (ISO) provides a standard naming system for a multitude of different products, including inserts, in 162 different countries throughout the world. Standardizing insert names and types makes it easier to talk to multiple companies providing similar tooling as well as differentiate the shape and size of tools. For example, a CNMG432 insert from any tooling supplier should be the same shape, size, and angle as any other supplier's brand. However, that does not mean it is identical in function, as different tools are made to different specifications and the material composition or carbide quality can vary.

Naming conventions for inserts call for a series of four letters and three numbers, as in CNMG432. Milling and turning inserts have a slightly different system.

The first letter identifies the outside shape of a tool, **Figure 8-2**. The shape of the insert used is influenced by the type of cut being made. For example, a heavy straight roughing cut is often performed with a "C" or "W" type insert because of its strength and support behind the cutting edge. Finishing cuts can be done with "D" and "V" style inserts because the clearance behind the cutting edge gives a better surface finish. There are many outside shapes, and it will take some experience and trial-and-error to select the adequate tool for any cutting application. Inserts are most often some type of parallelogram, such as a diamond, but they may also be a number of other polygons. In the given example, "C" designates an 80° diamond.

First Letter Designates Outside Shape in Insert Naming System			
ISO Code Letter	**Description**	**Shape**	**Shape Angle**
A	85° parallelogram		85°
B	82° parallelogram		82°
C	80° diamond		80°
D	55° diamond		55°
E	75° diamond		75°
H	Hexagon		120°
K	55° parallelogram		55°
L	Rectangle		90°
M	86° diamond		86°
N	55° parallelogram		55°
O	Octagon		135°
P	Pentagon		108°
R	Round		360°/Full radius
S	Square		90°
T	Triangle		60°
V	35° diamond		35°
W	Trigon		80°
X	Special parallelogram		85°

Goodheart-Willcox Publisher

Figure 8-2. Chart for first letter designator in ISO numbering system defining shape.

Second Letter Designates Relief Angle in Insert Naming System		
ISO Code Letter	Angle	Diagram
N	0°	
A	3°	
B	5°	
C	7°	
P	11°	
D	15°	
E	20°	
F	25°	
G	30°	

Goodheart-Willcox Publisher

Figure 8-3. Chart for second letter designator in ISO numbering system defining relief angle.

Third Letter Designates Tolerance in Insert Naming System			
ISO Code Letter	Cornerpoint (inches)	Thickness (inches)	Inscribed Circle (inches)
A	.0002	.001	.001
C	.0005	.001	.001
E	.001	.001	.001
F	.0002	.001	.0005
G	.001	.005	.001
H	.0005	.001	.0005
J	.002	.001	.002–.005
K	.0005	.001	.002–.005
L	.001	.001	.002–.005
M	.002–.005	.005	.002–.005
U	.005–.012	.005	.005–.010

Goodheart-Willcox Publisher

Figure 8-4. Chart for third letter designator in ISO numbering system defining overall tolerance of insert dimensions.

The second letter designates the relief angle on the front edge of a cutting tool, **Figure 8-3**. A smaller relief angle makes the cutting edge of the insert more stable or rigid, allowing for heavier cuts. A steeper relief angle creates a sharper edge and can improve surface finish or be used in softer materials. The selection of the appropriate relief angle will be dependent on the type of material and amount of stock to be removed. For example, "N" corresponds to a 0° relief.

The third letter designates the tolerance or accuracy of a cutting tool, **Figure 8-4**.

The fourth letter designates the hole or chipbreaker on a cutting tool, **Figure 8-5**. Inserts can have a hole through the center to secure the insert

Fourth Letter Designates Hole Shape in Insert Naming System				
ISO Code Letter	Shape	Hole	Hole Shape	Chipbreaker Type
None		No		None
A		Yes	Cylindrical	None
B		Yes	70–90° double countersink	None
D		Yes	Cylindrical	None
E		No		None
F		No		Double-sided
G		Yes	Cylindrical	Double-sided
H		Yes	70–90° single countersink	Single-sided

(continued)

Figure 8-5. Chart for fourth letter designator in ISO numbering system defining hole shape.

Fourth Letter Designates Hole Shape in Insert Naming System *(continued)*

ISO Code Letter	Shape	Hole	Hole Shape	Chipbreaker Type
M		Yes	Cylindrical, or double countersink	Single-sided
N		No		None
P		Yes	Cylindrical	Hi-double positive
Q		Yes	40–60° double countersink	None
R		No		Single-sided
S		Yes	Cylindrical	Hi-double positive
T		Yes	40–60° double countersink	Single-sided
U		Yes	40–60° double countersink	Double-sided
W		Yes	40–60° double countersink	None
Z		Yes	Cylindrical	Double-sided Hi-double positive

Goodheart-Willcox Publisher

Figure 8-5. *(Continued)*

to the toolholder or no hole, requiring a top clamp to secure the insert in place. A chipbreaker is a relief behind the cutting edge that helps peel the removed metal away from the cutting edge. For example, "G" represents a double-sided insert with a cylindrical hole.

After the four letters, three numbers complete a conventional insert name. The letters represent insert shape, and the numbers represent insert size. The first number designates the inner-circle size of a tool, **Figure 8-6.** The inner circle is the largest diameter circle that can fit inside the shape of the top of the tool. The larger the number, the larger the insert size. For example, "4" designates a 1/2″ inner circle (IC).

The second number designates the thickness of a tool, **Figure 8-7.** The correct thickness of tool must be chosen based on the toolholder size. The tool tip must be on the centerline of the machined part. For example, "3" designates a 1/8″ thickness.

The third and final number designates the radius on the nose of a tool, **Figure 8-8.** A larger radius will add strength to the cutting edge, while a small radius is subject to fracturing. Often the radii designation on the print will determine the radius of the tool. Roughing tools generally have a larger radius, and finishing inserts have a smaller radius. For example, "2" designates a 1/32″ radius.

First Number Designates Inner-Circle Size in Insert Naming System

ISO Code Number	Inscribed Circle Size	
	Decimal Inches	Fractional Inches
0.5	.0625	1/16
1.2 (5)	.15625	5/32
1.5 (6)	.1875	3/16
1.8 (7)	.21875	7/32
2	.25	1/4
2.5	.3125	5/16
3	.375	3/8
3.5	.4375	7/16
4	.5	1/2
4.5	.5625	9/16
5	.625	5/8
5.5	.6875	11/16
6	.75	3/4
6.5	.8125	13/16
7	.875	7/8
8	1	1
10	1.25	1 1/4

Goodheart-Willcox Publisher

Figure 8-6. Chart for first number designator in ISO numbering system defining inner-circle size of insert.

Second Number Designates Tool Thickness in Insert Naming System		
ISO Code Number	Decimal Value (inches)	Fractional Value (inches)
—	.03125	1/32
T0	.040	—
1	.0625	1/16
T1	.078	5/64
2	.094	3/32
T2	.109	7/64
3	.125	1/8
T3	.156	5/32
4	.187	3/16
5	.219	7/32
6	.25	1/4
7	.313	5/16
9	.375	3/8

Goodheart-Willcox Publisher

Figure 8-7. Chart for second number designator in ISO numbering system defining insert thickness.

Third Number Designates Radius of Tool Nose in Insert Naming System		
ISO Code Number	Decimal Value (inches)	Fractional Value (inches)
0	.004	—
0.5	.008	—
1	.016	1/64
2	.031	1/32
3	.047	3/64
4	.062	1/16
5	.078	5/64
6	.094	3/32
7	.109	7/64
8	.125	1/8

Goodheart-Willcox Publisher

Figure 8-8. Chart for third number designator in ISO numbering system defining tool tip radius.

As you can see, there are many variables and combinations of inserts to meet all kinds of machining needs.

8.2.2 Carbide

Carbide is one of the most important materials in inserts and cutting tools. Carbide tools are actually made of tungsten carbide, which is a mix of carbon and tungsten. Tungsten carbide by itself is very brittle because of its hardness.

In the 1930s, it was discovered that adding some titanium to the mix of carbon and tungsten gives tools longer life at higher speeds. Together, the titanium, carbon, and tungsten are reduced to a granular powder form. Then, that mixture is added to a cobalt solution, known as a *binder*, and pressurized to create tools and inserts used in machining, **Figure 8-9**. Typically, an insert is about 80% cobalt and 20% tungsten carbide.

The process for making carbide starts with mining. Tungsten is mined from the ground and then treated with chemicals to create a tungsten oxide. The tungsten oxide is then mixed with carbon and heated to around 2,200° Fahrenheit. At this temperature the oxygen is removed and we are left with tungsten carbide. This compound is crushed into granules much smaller than a grain of sand, around 1 to 10 microns, or .00001 to .00010 inches. The different grain sizes give different mechanical properties to the carbide. Those granules are then added to the cobalt solution to create the tool or insert. Inserts are made in forms, or molds, from a powder form, and then pressurized to bond together. Round tools, such as end mills, can be made as small as 1/32″ or as large as a 55-gallon drum. New technology has allowed carbide tool and insert manufacturers to control carbide granules in the cobalt solution to create increased strength in specific areas of a tool and improve tool life tremendously.

Haas Automation Inc.

Figure 8-9. A carbide insert performing a threading operation on a CNC turning center.

8.2.3 Coatings

Coatings on carbide tools and inserts were introduced to create harder edges, provide better surface finishes, and prevent material buildup on cutting tool edges. To increase tool life under tough cutting conditions, many types and combinations of coatings have been developed. They can be applied in two ways: *chemical vapor deposition (CVD)* or *physical vapor deposition (PVD)*. Both types are applied in furnaces.

CVD coatings are usually 5–20 microns thick and are made up of multiple layers of different composition. A typical CVD coating has three layers: a layer of titanium carbon nitride, a layer of aluminum oxide, and a layer of titanium nitride. Each layer produces a different characteristic for cutting or preventing edge deformation. Many tool manufacturers use different recipes with different materials to create their specific coating. Tools are put into a furnace, reduced to a vacuum, and then coating materials are added as gases that bond to the insert, **Figure 8-10**.

PVD coatings are applied through electrical methods. Negatively charged inserts have positively charged coating material applied to them. Titanium carbonitride, titanium nitride, and titanium aluminum nitride are widely used as PVD coatings. Tools and inserts with PVD coatings are usually used for machining alloys, such as materials containing nickel, cobalt, or titanium, at high temperatures.

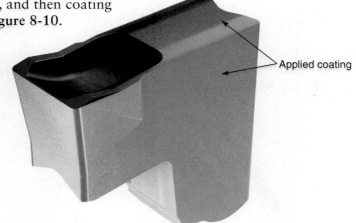

Applied coating

Iscar

Figure 8-10. A grooving insert with applied CVD coating.

8.2.4 Diamond Cutter

Polycrystalline diamond (PCD) is a manufactured diamond material made in a factory using extremely high heats and pressures. PCD is especially effective in applications involving high speeds, high volumes, or certain materials such as aluminum or graphite. However, PCD does have limitations. It generally can't cut ferrous materials—those that contain iron—because iron causes a chemical reaction with PCD.

PCD is available in multiple diamond grades to fit different applications. A fine-grain diamond is used for less-abrasive applications requiring better surface finishes. Medium-grain diamond is used for most general purpose machining applications. Coarse-grain diamond is used in heavy machining and for abrasive materials where surface finish may not be as important. PCDs do not generally handle interrupted cuts, as they tend to fracture. Most PCD applications also require very high spindle speeds.

8.2.5 Ceramics

Excessive heat is a common challenge in machining with carbide tools. Cobalt's melting temperature is around 2,700°F, so machining in a high-temperature environment can cause the binding agent of carbide to begin to break down and deform. This breakdown can lead to catastrophic failure of a tool or insert. High-temperature machining usually occurs in materials with high hardness levels above 50 Rockwell C scale.

Unlike carbide tools, ceramic tools require heat to make cuts. High spindle speeds with balanced feed rates heat the material in front of the ceramic cutting tool and allow free machining of hardened materials.

8.2.6 PCBN

Polycrystalline cubic boron nitride (PCBN) is the ultimate hard-turning insert. PCBN inserts provide superior edge integrity in materials above 45 Rockwell C scale (RC) hardness. PCBN cutters are able to cut up to 68 RC hardness in ferrous materials. The application of PCBN requires very rigid machines with complete workholding. In modern machine shops, it is possible to hard turn with PCBN inserts, where it used to be possible only to grind.

8.3 Drills

Drills are cutting tools used to create holes quickly and easily. Drills can be solid body, inserted, or point interchangeable. They can be made from high-speed steel, cobalt, or carbide. Different types of drills offer many options for making holes.

8.3.1 Twist Drills

The term *twist drill* generally refers to any solid body drill bit. The name comes from the manufacturing process in which straight relief grooves are created, then the drill is literally twisted around its center to create the common twist drill, **Figure 8-11**. Each part of the drill has a specific function that plays a vital role in the function of the drill bit:

- **Body.** The body is the outside diameter of a drill. The body determines the final size of a drilled hole.
- **Flute.** The flute is a relief inside the body diameter that allows chips to be removed during the drilling process. Flutes also allow coolant to flow to the cutting edge.

 **Tech Tip**

The *Rockwell scale* is a hardness scale based on hardness of a material. The Rockwell test determines the hardness by measuring the depth of penetration under a load compared to the penetration made by a preload. There are different scales, denoted by a single letter. Hardened steel is measured on the Rockwell C scale.

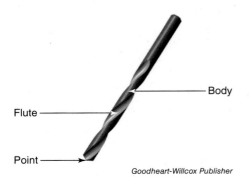
Goodheart-Willcox Publisher

Figure 8-11. A twist drill showing the body, flute, and point.

- **Cutting edge.** The cutting edge is the lip that allows a drill to enter a part and remove material. This is where cutting occurs.
- **Point.** Although technically no cutting happens at the point, it is a critical component for guiding a drill into the material being machined.

8.3.2 Center Drills

A *center drill* is a special drilling tool that is used only to create a starter hole prior to drilling. A center drill creates a small clearance hole that allows a drill tip to enter a cut straight. Center drills come in a variety of sizes. Choose a center drill that is smaller than the desired finished hole diameter. See **Figure 8-12.**

Because of the small tip on a center drill, care must be taken not to break or chip the tip during the drilling process. Center drills are standard in manual machining but should not be used in CNC mills or lathes.

8.3.3 Spot Drills

Like center drills, *spot drills* are also used to start holes. Spot drills have a short flute with a large tip and a 90° point, **Figure 8-13.** These features offer distinct advantages that make spot drills the drill of choice in CNC applications. The larger tip is sturdy enough not to break at higher spindle speeds or feed rates. The 90° point allows machinists to create a small edge break on the finished hole diameter.

8.3.4 Indexable Drills

Indexable drills, or insertable drilling tools, offer the same features and advantages of insert tooling in drilling applications. Indexable drills are becoming more prevalent in CNC machines as technology and tool diversity improves. In the past, indexable drilling was limited by the size of inserts to only large diameters, but new interchangeable drill heads have made it possible to use indexable drills on work as small as 3/8″ diameter.

Figure 8-14 shows a drill body with an inserted end. Notice that the tool geometry is somewhat different from a twist drill in that there is no real tool point. Using offset inserts and a center cutting technique allows drilling with no pilot hole and high feed rates.

An inserted drill with an interchangeable head looks more like conventional drills, **Figure 8-15.** Each drill body can take a small range of interchangeable heads and create extremely accurate hole sizes over extended lengths.

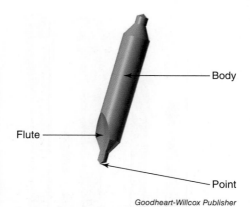

Goodheart-Willcox Publisher

Figure 8-12. A double-ended center drill.

Goodheart-Willcox Publisher

Figure 8-13. A 90° spot drill placing a starter hole in a piece of material.

Tech Tip

When programming these tools, calculate recommended speeds and feeds carefully and do not use any peck drilling cycles.

Iscar

Figure 8-14. An insertable drill body used for deep hole drilling or alloy materials.

Iscar

Figure 8-15. An insertable drill with a twist-in style drill insert.

Figure 8-16. An assortment of taps, including bottom, spiral flute, and straight flute.

Iscar

8.4 Taps

A *tap* is a tool used for creating internal threads in a drilled hole, **Figure 8-16**. Common tap sizes are outlined as Unified National Coarse (UNC) and Unified National Fine (UNF). Taps are also defined by their size and threads per inch (TPI). For example, a coarse threaded 1/4″ tap is called a 1/4-20 tap, meaning the outside diameter is 1/4″ and TPI is 20. A tap cannot create a hole, so it can only be used after another drilling operation, such as spot drilling. For a 1/4-20 tap, the common hole size is .201″ diameter, or a #7 drill. This will give you the correct internal diameter for the threaded hole.

8.4.1 Straight Flute Taps

A *straight flute tap* has no twist in the clearance flutes, **Figure 8-17**. This is optimal for tapping holes that go through a part and do not need chip evacuation. You will notice this tap is also coated to prevent chip buildup on the cutting edge.

8.4.2 Spiral Flute Taps

A *spiral flute tap* has a flute twist similar to a twist drill, **Figure 8-18**. The twist allows chips to be pulled out of the existing hole. This type of tap is optimal for tapping holes that do not go completely through a piece of material.

8.4.3 Bottom Taps

A *bottom tap* is used when tapping a blind hole, or a non-through hole, where the print requires threads to extend down the entire hole depth, **Figure 8-19**. Without a tapered lead-in, tap teeth are under additional stress and subject to chipping or breaking. Often, bottom taps are used after a spiral or straight tap has already been used.

Tech Tip

To find the correct drill size for a tapped hole, refer to the *Machinery's Handbook* or a tap-drill chart.

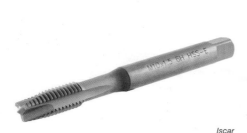

Iscar

Figure 8-17. A straight flute tap that has been coated to prevent chip buildup.

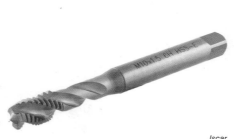

Iscar

Figure 8-18. A spiral flute tap that allows for optimal chip removal from the hole.

Iscar

Figure 8-19. A bottom tap that allows for full-depth tapping.

8.5 Countersinks and Counterbores

Countersinks and *counterbores* are tools specifically designed to create relief angles or clearance diameters for bolt and screw heads, **Figure 8-20**. Countersinks create clearances for screw heads and are typically cut at 82° or 100°, depending on the type of screw being used in the assembly. Counterbores are used for clearances of bolt heads and most socket head cap screws. Having sufficient clearance for bolt installation or matching angles on screw thread assemblies is critical to meeting engineering requirements for assembling components. There are various standards, such as ANSI or ISO, governing screw and bolt head clearances. Usually, clearances are defined on the print, which determines allowable tolerance for clearances.

A   **B**

Iscar

Figure 8-20. An interchangeable countersink. A—Tip. B—Body.

8.6 End Mills

End mills are the most common cutting tool in milling. They are used for roughing, finishing, slotting, contouring, and profiling. They can be solid body or inserted. They come in various forms, including square shoulder, ball type, tapered, bull nose, or dovetailed. End mills are very versatile tools that come in sizes from .015″ all the way up to 6″ or greater, depending on the application.

 Thinking Green

Recycling Cutting Tools

Cutting tools can be recycled in several ways. Many can be redesigned or reshaped for additional use after they are initially worn out. Reusing tools is both environmentally and economically sensible, since reuse saves the materials used in the cutting tools themselves and the cost to purchase the tools. If the tool cannot be reshaped, sometimes the tool holding can be redesigned to fit a partially worn tool. Tools that are too worn to reuse may be recycled into new tools. Properly recycled materials require substantially less energy to craft new tools, as compared to crafting new tools from raw materials.

8.6.1 Edge Cutting Geometry

Typical end mills have several main parts: the body, flutes, and cutting edges, **Figure 8-21**. The body is the outside diameter of the end mill. This is the reference used in naming an end mill. For example, an end mill with a 1/2″ body diameter is referred to as a 1/2″ end mill. Cutting edges are formed by the lip just inside the body diameter on the bottom face. Cutting edge placement gives the end mill the ability to perform side cutting or face cutting with the tool end face. Flutes are the internal relief clearance, and they have a significant impact on surface finish and feed rates. End mills with four or six flutes are typically used for finishing operations.

The number of flutes affects the performance of an end mill, including feed and speed rates. The center of an end mill is referred to as the *web*. End mills with fewer flutes are thicker in the web, and end mills with more flutes are thinner. A thinner web can create vibration in heavy cuts. Recall

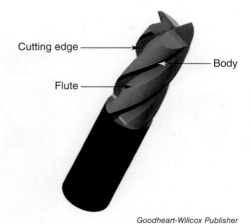

Cutting edge

Body

Flute

Goodheart-Willcox Publisher

Figure 8-21. A typical four-flute, non-center cutting end mill.

Goodheart-Willcox Publisher

Figure 8-22. A roughing end mill with varied edges.

from Section 4.7, "Calculating Speeds and Feeds," that one factor in the calculation is number of flutes:

Speed (rpm) × Chip Load × Number of Flutes = Feed Rate

For example, calculate feed rate in inches per minute (ipm) given a speed of 7,000 revolutions per minute (rpm) and .005″ chip load.

For a two-flute end mill,

7,000 rpm × .005 × 2 = 70 ipm

For a six-flute end mill,

7,000 rpm × .005 × 6 = 210 ipm

Observe that the feed rate is faster for an end mill with more flutes, but machinists must also consider web differences. Depending on the application, it might be more efficient to make smaller cuts with more flutes at higher feed rates.

8.6.2 Roughing End Mills

Roughing end mills have a special cutting-edge geometry that allows for varying chip formation and removing large volumes of material. The edges in a roughing end mill are serrated and vary in frequency, **Figure 8-22.** Roughing end mills are used to remove a lot of material in a short amount of time. Their cutting-edge formation allows for large step-overs and prevents chipping of the cutting edge.

8.6.3 Inserted End Mills

Iscar

Figure 8-23. A four-flute inserted end mill.

Technological advances in inserts and cutting tools have impacted end mills as well as many other tools. With the right application, inserted end mills can be even more cost-effective than solid body end mills. Inserts for end mills can have two, four, six, or even eight cutting edges each. See **Figure 8-23.** This means that once an edge dulls, the insert can be rotated to a new edge. Inserted end mills can have between one and six flutes, much like solid end mills. With the correct insert style and grade, an end mill body can be the most versatile tool for any machining application or material.

8.7 Face Mills

Face mills are often confused with end mills, but face mills are specifically designed to machine the top face of a workpiece or part. Face mills do not side cut or plunge into material. They typically have 4 to 12 flutes and range in size from 2″ to 12″ in diameter. Because of their large diameter and specific engineering, face mills create a fine surface finish with no steps or lips in the face of a part.

Inserts for face mills are designed for improved surface finish and mild cutting. Inserts can be *wiper inserts* (**Figure 8-24A**), with no relief clearance for a smooth finish, or *round inserts* (**Figure 8-24B**), which create very small tool engagement surfaces. In certain applications there can be a need for heavier roughing operations in

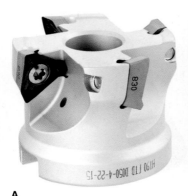
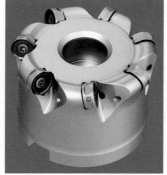

A **B** *Iscar*

Figure 8-24. Two examples of multi-flute facing mills.

a face mill. Specialty face mills for heavier roughing, **Figure 8-25**, have the same serrated edges seen in roughing end mills.

8.8 Toolholders

After selecting the right tool and machine, machinists must also select the right toolholder. It is possible to select the right tool for a job, but putting the right tool in the wrong toolholder can cause failure. Extending a tool out too far from the spindle can cause inferior surface finishes, machining harmonics, or tool edge breakdowns. Toolholders without enough clearance can cause the holder to engage the stock material and damage the spindle and workpiece.

8.8.1 Milling Toolholders

Milling toolholders have more diversity and options than turning toolholders. There are some odd variations in machining center toolholders, but most machines come with two standard taper types. The US standard is called a *CAT taper*, and non-US machines come with a *BT taper*. CAT taper is short for "Caterpillar V Flange Taper" and refers specifically to the starting diameter and angle of the tapered end that fits into the spindle. BT stands for "British Taper" and also refers to the starting diameter and taper angle.

CAT Standards

CAT taper toolholders are available in many different sizes. The smaller the size, the easier it is on the spindle to produce very high speeds. Larger tapers correspond to more rigid toolholders. These are usually found in larger machines with heavy machining and slower spindle speeds. Familiarize yourself with the sizes, specifications, and distinctions between the three most common CAT toolholders. See **Figure 8-26**. Note that the 60° angle relief for the tool changer arm is standard on all sizes.

Iscar

Figure 8-25. Iscar™ Cutting Tools produces a ShredMill™ specifically designed for heavy facing operations.

 Tech Tip

There are more sizes available, but CAT 30, 40, and 50 can cover almost every machine on the market. If there is a question about toolholder spindle size, refer to the machine user's manual.

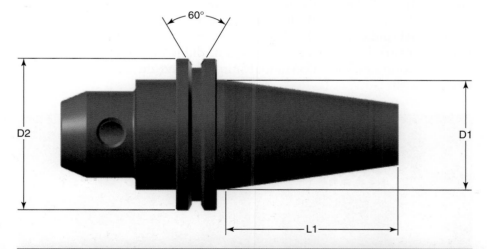

CAT Toolholder			
Toolholder Size	**D1**	**L1**	**D2**	**Taper Angle**
CAT 30 Toolholder	1.250″	1.875″	1.812″	16.260°
CAT 40 Toolholder	1.750″	2.687″	2.500″	16.260°
CAT 50 Toolholder	2.750″	4.000″	3.875″	16.260°

Goodheart-Willcox Publisher

Figure 8-26. Chart of CAT-style toolholders and dimension for each size.

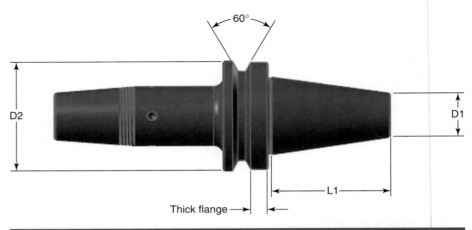

BT Toolholder			
Toolholder Size	D1	L1	D2	Taper Angle
BT 30 Toolholder	1.250″	1.906″	1.811″	16.260°
BT 40 Toolholder	1.750″	2.575″	2.480″	16.260°
BT 50 Toolholder	2.750″	3.937″	3.937″	16.260°

Goodheart-Willcox Publisher

Figure 8-27. Chart of BT-style toolholders and dimension for each size.

BT Standards

BT taper toolholders are also available in a variety of sizes. As with CAT tapers, smaller BT tapers make it easier on the spindle to produce very high speeds, whereas larger tapers require more rigid toolholders in larger machines with slower spindle speeds. Though CAT and BT holders are similarly sized, they are not identical. Tooling will not properly seat if they are interchanged. The easiest way to differentiate the two different holders is the thick flange on the BT holder, **Figure 8-27.**

Both BT and CAT holders have a screw-in end, called a *retention knob*, which can vary from machine brand to machine brand, **Figure 8-28.** The CAT retention knob has US standard threads, and the BT retention knob has metric threads.

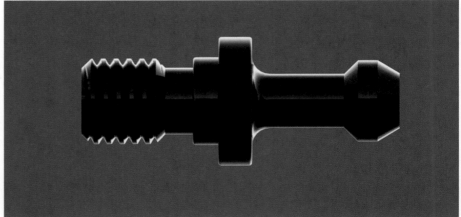

Haas Automation Inc.

Figure 8-28. Tooling retention knob.

Solid Body Holders

The term *solid body holder* refers to a toolholder built for a specific-size tool. For example, a 1/2″ solid body holder only holds a 1/2″ diameter tool. Usually, tools are secured in place by a setscrew. See **Figure 8-29**. Another type of solid body holder is called a *shrink fit holder*. These holders are heated to expand (about .0005″), then the tool is inserted. As the holder cools, it shrinks into place. Shrink fit holders tend to be very accurate and secure.

Collet Holders

Collet holders use a colleting system, usually a double-angled collet, to secure a tool in place, **Figure 8-30**. Collets provide some flexibility with tool diameter. By changing out the collet size, machinists can hold a variety of different-size tools, usually ranging from 1/32″ up to at least 1″ in diameter. Collet holders are available in a variety of sizes and lengths to fit drills, spot drills, and end mills.

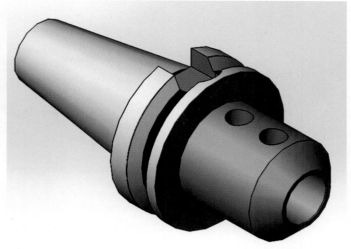

Goodheart-Willcox Publisher

Figure 8-29. Solid body toolholder, built for a specific-size tool.

8.8.2 Lathe Toolholders

Lathe tools are significantly different from milling tools. Most outside diameter (OD) and inside diameter (ID) turning tools are inserted. Turning inserts, by design, must be set into a matching toolholder, **Figure 8-31**. For example, a CNMG-style insert will only fit into a CNMG toolholder.

A

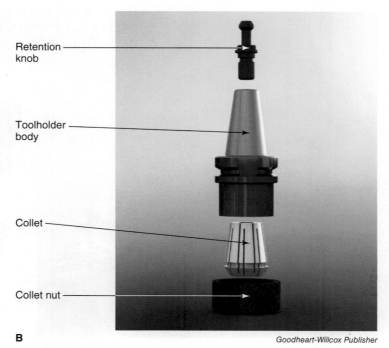

Retention knob

Toolholder body

Collet

Collet nut

B

Goodheart-Willcox Publisher

Figure 8-30. A—Collet holder. B—Exploded view.

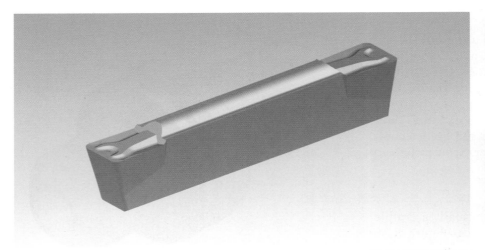

Goodheart-Willcox Publisher

Figure 8-31. A grooving insert used in a lathe groove or cut-off operation.

Haas Automation Inc.

Figure 8-32. A CNC turning center toolholding turret.

The variables in lathe toolholding are determined by machine design. Some tools mount directly into a fixed turret, **Figure 8-32**. These turrets rotate around a center shaft and are locked into place with a hydraulic or pneumatic cylinder. It is critical that the correct shank size is used to maintain the appropriate centerline for turning. An incorrect shank size will force the tool to cut off-center.

Another type of lathe uses a *gang tool* configuration to hold tools, **Figure 8-33**. Gang tool turning centers can perform extremely fast tool changes, but the gang tool setup cannot accommodate a tailstock and can limit the stock size of material being machined. Gang tool turning centers are most often seen in small-part, high-volume machining.

A third type of turning center uses a turret-mounted, quick-change tooling body called *VDI tooling*. VDI stands for the German name for the Society of German Engineers. VDI tooling is similar to turret-mounted tooling, but it gives a much faster change over time between setup operations, **Figure 8-34**.

Haas Automation Inc.

Figure 8-33. A gang tool configuration for toolholding in a CNC lathe.

Goodheart-Willcox Publisher

Figure 8-34. A model of a VDI toolholder.

8.9 Machinability Ratings

Once a machinist has selected proper tooling, the next consideration is how fast to machine with the selected tooling and inserts. Remember the formulas for calculating speeds and feeds in milling:

$$\frac{3.82 \times \text{Surface Feet per Minute (sfm)}}{\text{Tool Diameter}} = \text{Speed (rpm)}$$

$$\text{Speed (rpm)} \times \text{Chip Load} \times \text{\# of Flutes} = \text{Feed Rate (ipm)}$$

For a given tool setup, the unknown quantity is surface feet per minute, which measures how many linear feet a tool can travel per minute. All tool distributors provide different ranges for sfm based on their cutter engineering. Recommendations from different distributors for identical inserts with the same shape, nose radius, and material configuration can often give different results in parts cut per insert. The tables given in **Figure 8-35** provide a starting point for most materials to be machined using carbide inserts. Consult with your local tool distributor for the most accurate estimate of sfm and adjust as needed for optimal part manufacturing.

Surface Feet per Minute (sfm) for Common Materials Machined with Carbide Inserts					
Material Grade	Hardness Brinell	sfm Carbide	Material Grade	Hardness Brinell	sfm Carbide
Stainless			**Super Alloys**		
304	135–185	325	A286	250–320	150
316	185–250	315	L605	180–230	70
403/410	175–225	475	Nickel 200	80–170	400
420	175–225	475	Nickel 400	115–240	270
430	135–185	525	718	200–300	80
440	225–275	325	Waspaloy	200–300	80
15-5PH	150–200	350	4750	135–185	350
15-7PH	150–200	350	Kovar	135–185	525
Light Metals			**Copper Alloys**		
Wrought Alum	N/A	1,200	Free Mach Brass	N/A	650
Sandcast Alum	N/A	1,400	Low Lead Brass	N/A	525
Diecast Alum	N/A	500	Beryllium Copper	N/A	225
All Magnesium	N/A	2,000	Alum Bronze	N/A	225
Aircraft Alloys			**Cold-Rolled Carbon**		
4130	172–225	400	1018	125–175	485
4140	225–275	350	1045	225–275	375
4340	225–300	350	1117	150–200	600
52100	275–325	350	1144	270–350	525
9310	175–225	440	1215	150–200	700
Tool Steels					
A2, A3, A6, S7	200–275	250			
O1, O2, O6, O7	175–225	250			
W1, W2, W5	150–200	325			

Goodheart-Willcox Publisher

Figure 8-35. A reference chart of surface feet per minute (sfm) for some materials with standard carbide tools.

Chapter Review

Summary

- Insert tooling provides a cost-effective solution to replacing worn or damaged tooling.

- The ISO numbering system provides a standard insert and tool designation that allows for consistent nomenclature for similar tooling, and it also provides specific information on inserts' size and shape.

- The science of carbide tooling has evolved into a structurally designed formation of various binder metals and coatings that are application specific.

- Chemical vapor deposition and physical vapor deposition are coating application processes that allow for micro-coating of inserts to improve lubricity and tool life.

- Polycrystalline diamond is a manufactured diamond insert used to cut nonferrous materials at high speed.

- Ceramic inserts are used in high-temperature machining of hard materials such as steel or alloys.

- Polycrystalline cubic boron nitride (PCBN) is a hard-turning insert. These PCBN inserts provide superior edge integrity in materials above 45 Rockwell C scale hardness.

- The body, flute, cutting edge, and point are the main components of a twist drill.

- Center drills and spot drills are used to create a small starter hole before drilling. These ensure drill hole location and straightness. Center drills are used in manual machining, while spot drills are the drill of choice in CNC applications because their larger tips are sturdy enough not to break at higher spindle speeds or feed rates.

- Insertable drills allow a cost-effective, high-feed drilling application in CNC machines.

- A straight flute tap has no twist in the clearance flutes, a spiral flute tap has a flute twist similar to a twist drill, and a bottom tap does not have a tapered lead-in.

- Countersinks and counterbores are tools specifically designed to create relief angles or clearance diameters for bolt and screw heads. Countersinks create clearances for screw heads, and counterbores are used for clearances of bolt heads and most socket head cap screws.

- Cutting-edge placement on an end mill gives it the ability to perform side cutting or face cutting with the tool end face. Flutes are the internal relief clearance, and they have a significant impact on surface finish and feed rates. End mills with four or six flutes are typically used for finishing operations.

- Roughing end mills have a special cutting-edge geometry that allows for varying chip formation and removal of large volumes of material.

- Face mills are designed to machine the top face of a workpiece or part. Face mills do not side cut or plunge into material.

- Standard CNC machining centers use Caterpillar V Flange (CAT) and British Taper (BT) toolholders. Sizes can vary between 30, 40, and 50, depending on machine size and maximum spindle rpm.

- There are turret, gang tool, and VDI tooling turning toolholders.

Review Questions

Answer the following questions using the information provided in this chapter.

Know and Understand

1. Until just a few years ago, machining exclusively used _____ lathe tools with a piece of carbide brazed to the end.

   A. manual
   B. high-speed steel
   C. inserted
   D. facing

2. The term _____ refers to a small, clamped-in cutting tool.

   A. insert
   B. roughing
   C. finishing
   D. ISO

3. ISO is the _____

   A. International System of Organization
   B. Industrial Standard Organization
   C. International Society of Organizers
   D. International Standards Organization

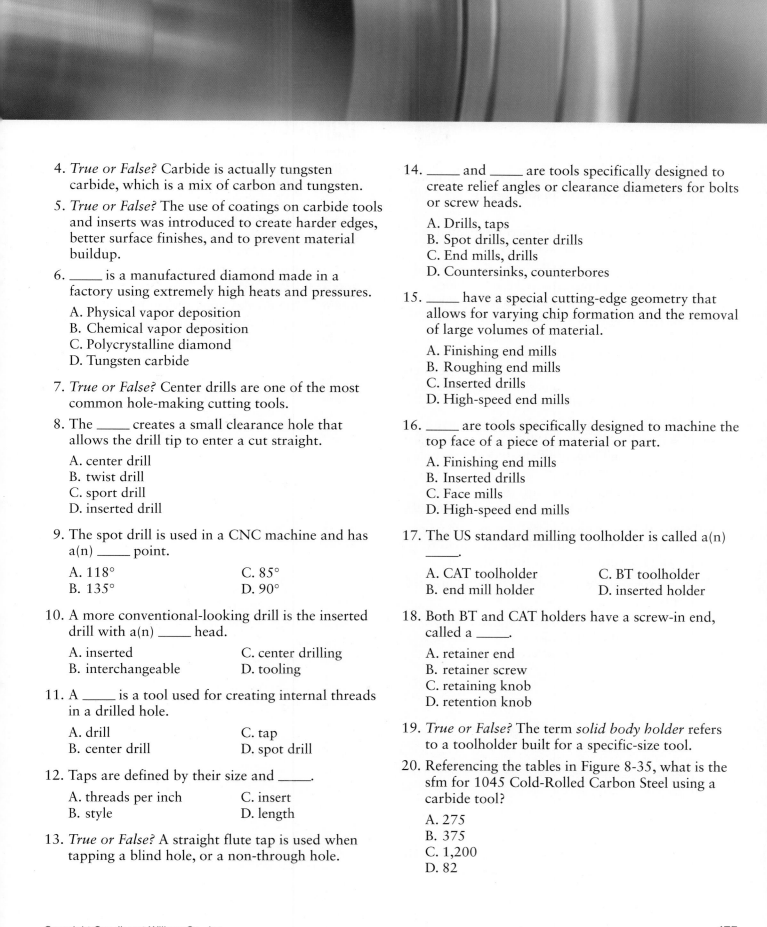

4. *True or False?* Carbide is actually tungsten carbide, which is a mix of carbon and tungsten.

5. *True or False?* The use of coatings on carbide tools and inserts was introduced to create harder edges, better surface finishes, and to prevent material buildup.

6. _____ is a manufactured diamond made in a factory using extremely high heats and pressures.

   A. Physical vapor deposition
   B. Chemical vapor deposition
   C. Polycrystalline diamond
   D. Tungsten carbide

7. *True or False?* Center drills are one of the most common hole-making cutting tools.

8. The _____ creates a small clearance hole that allows the drill tip to enter a cut straight.

   A. center drill
   B. twist drill
   C. sport drill
   D. inserted drill

9. The spot drill is used in a CNC machine and has a(n) _____ point.

   A. 118°   C. 85°
   B. 135°   D. 90°

10. A more conventional-looking drill is the inserted drill with a(n) _____ head.

    A. inserted   C. center drilling
    B. interchangeable   D. tooling

11. A _____ is a tool used for creating internal threads in a drilled hole.

    A. drill   C. tap
    B. center drill   D. spot drill

12. Taps are defined by their size and _____.

    A. threads per inch   C. insert
    B. style   D. length

13. *True or False?* A straight flute tap is used when tapping a blind hole, or a non-through hole.

14. _____ and _____ are tools specifically designed to create relief angles or clearance diameters for bolts or screw heads.

    A. Drills, taps
    B. Spot drills, center drills
    C. End mills, drills
    D. Countersinks, counterbores

15. _____ have a special cutting-edge geometry that allows for varying chip formation and the removal of large volumes of material.

    A. Finishing end mills
    B. Roughing end mills
    C. Inserted drills
    D. High-speed end mills

16. _____ are tools specifically designed to machine the top face of a piece of material or part.

    A. Finishing end mills
    B. Inserted drills
    C. Face mills
    D. High-speed end mills

17. The US standard milling toolholder is called a(n) _____.

    A. CAT toolholder   C. BT toolholder
    B. end mill holder   D. inserted holder

18. Both BT and CAT holders have a screw-in end, called a _____.

    A. retainer end
    B. retainer screw
    C. retaining knob
    D. retention knob

19. *True or False?* The term *solid body holder* refers to a toolholder built for a specific-size tool.

20. Referencing the tables in Figure 8-35, what is the sfm for 1045 Cold-Rolled Carbon Steel using a carbide tool?

    A. 275
    B. 375
    C. 1,200
    D. 82

Apply and Analyze

1. What are the limitations of using high-speed steel tools in CNC machining?

2. What are some of the benefits of the ISO tooling numbering system?

3. Using the chart in Figure 8-2, describe the outside shapes of inserts with the first letters of "C," "W," and "D."

4. What is the tip radius on an insert designated as CNMG431? How about WNMG333?

5. Carbide is made from two different elements and then mixed with a binder to make a "carbide insert." What are the two elements that make carbide, and what is the most common binder element?

6. Chemical vapor deposition (CVD) and physical vapor deposition (PVD) are two processes of insert coating. What is the difference in the application of the coating type in these two processes?

7. Center drills and spot drills are used to create a small starter hole before a drilling operation. What is the purpose and benefit of creating a start hole before a drilling operation?

8. When tapping a blind hole with threads the full depth of the hole, which type of tap is best suited?

9. Calculate the feed rate in inches per minute where the spindle speed is 5,250 rpm, the chip load is .006", and the cutter is a four-flute end mill.

10. What are some of the advantages to using inserted end mills?

11. CAT and BT toolholders both come in multiple sizes. What are the applications and advantages and disadvantages to the different sizes?

Critical Thinking

1. Over the last few years, CNC machines have increased spindle speeds and horsepower, and demands for faster machining and more exotic materials have increased. What are the tooling challenges that had to be solved to allow for this type of machining?

2. There are a variety of different insert makers and tooling companies. Although these inserts have the same ISO designator, they do not all machine with the same results. What are some practical ways you can identify the "right tool for the right job"?

3. Different types of operations require different tooling. Who do you think should be responsible for selecting the appropriate tooling for a machining operation before the job begins?

Close-up of an industrial cutter tool.

9

CNC Mill Programming

Chapter Outline

9.1 Introduction to CNC Mill Programming

9.2 Program Planning

9.2.1 Print Review

9.2.2 Part Workholding

9.2.3 Tool Selection

9.3 Program Format

9.3.1 Opening Statement

9.3.2 Program Body

9.3.3 Closing Statement

9.4 Address Codes

9.4.1 G Codes

9.4.2 M Codes

9.5 Contouring

9.5.1 Tool Start Position

9.5.2 Cutter Compensation

9.6 Radius and Angle Milling

9.6.1 G02 and G03 Radial Movements

9.6.2 I, J, and K Method

9.6.3 Programming Angles

9.7 Pocketing

9.7.1 Cutter Entry

9.7.2 Roughing Method

9.8 Canned Cycles

9.8.1 G83 Cycle (Full Retract Peck Drilling)

9.8.2 G73 Cycle (Chip Break Peck Drilling)

9.8.3 G81 Cycle (Spot Drilling)

9.8.4 G82 Cycle (Counterbore Cycle)

9.8.5 G84 Cycle (Tapping Cycle)

9.8.6 G85 Cycle (Fine Boring Cycle)

9.8.7 G86 Cycle (Spindle Stop Boring Cycle)

9.9 Tool Setting

9.9.1 Length Offsets

9.9.2 Diameter Offsets

9.10 Work Coordinate Offsets

9.10.1 Finding WCS on Material

9.10.2 Setting WCS on Control

Learning Objectives

After reading this chapter, you should be able to:

- Describe the process of program planning.
- Explain how workholding affects programming.
- Define the three parts of a CNC mill program.
- Explain the use of G codes.
- Describe when to use an M code.
- Specify when a pocketing toolpath would be used.
- Explain the contour toolpath.
- Describe the benefits of canned cycles.
- Explain the use of cutter compensation.
- Use a G02 and G03 arc command.
- Program angle moves.
- Describe at least two different tool entry techniques.
- Enter tool offsets on the appropriate offset page.

Key Terms

address code

arc

body

canned cycle

closing statement

contouring

cutter compensation

default block

design intent

diameter offset

hard tool

I, J, and K method

incremental position

machine function (M code)

machine home

modal command

opening statement

peck drilling

pocket milling

preparatory command (G code)

threads per inch (TPI)

tool offset

work coordinate system (WCS)

workholding

zero return

9.1 Introduction to CNC Mill Programming

In this chapter you will learn how a print, read correctly (Chapter 5), can tell you exactly how to machine a part. You will also discover how complex operations and calculations are made simple through a variety of *canned cycles* in which the computer does all the work. You will learn how workholding impacts part production. Think of the many aspects of machining a part that must be planned and communicated to the machine: What tools are available? How do you control where each tool goes, how fast it gets there, and how fast you can machine a piece of material? These are things that are all controlled by the CNC program, and ultimately by the programmer.

This chapter will cover codes specifically for FANUC- and Haas-controlled machines. These two controller types, and all their different models, cover most machine controllers in the market. They use a type of programming referred to as G and M code. While other types of controls can also use G and M code, the codes used for FANUC and Haas are by far the most common, and in most cases interchangeable.

Writing a CNC program is as simple as telling a story one tool at a time. Your most important tool is the creativity and planning you bring to the project before making even one keystroke toward writing a program.

9.2 Program Planning

The most critical, and often overlooked, step in creating a successful CNC part program is the planning stage. Weak programmers will grab a print and start producing code, adding tools that make sense to them, which often results in programs that have not considered the complex issues of workholding or fixturing. A great programmer will spend time considering the machines and tooling available, the most efficient workholding method, and the customers' requirements.

It is always a good idea to assemble a team when preparing to program a part and move it to production. A customer representative—possibly a salesperson or manager—someone from the engineering staff, a setup specialist, and an operator could all have helpful feedback in the planning process to make a quality finished part. Every program produced and every part manufactured should provide end users a piece that fits their needs in a cost-effective manner, while giving your company a profitable solution. This quality customer service and smart planning takes a team of highly skilled and organized professionals.

9.2.1 Print Review

Interpreting **design intent** is an important first step in reviewing prints. By examining the tolerances of features and the position from which tolerances are established, a print reader can gain an understanding of part function, what features are critical in manufacturing, and how this part can be machined. The simple print in **Figure 9-1** provides an example.

The block size is dimensioned as 6″ by 3″, with a total tolerance of .020″. A reader can determine from the print that this block could be manufactured from 3″-wide stock and that the 6″ length cannot be left as saw cut, but the

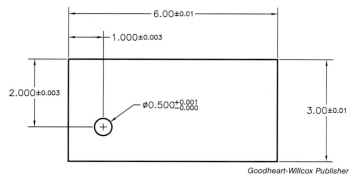

Goodheart-Willcox Publisher

Figure 9-1. A sample drawing of a 6″ × 3″ block with a 1/2″ hole.

block is also not intended for a close fit into another assembly. To meet these requirements, saw the 3″-wide material about 6.1″ long, then machine both ends to 6.00″. In the hole position, the tolerances are a little closer. Those tolerances reference the top-left corner of the part. It is a good idea to establish the work coordinate system (WCS) origin—the point where both X and Y values are zero—at the same corner on the top-left edge. In the Cartesian coordinate system, this would be in quadrant 4, or the X+, Y– quadrant.

An important feature to this part's engineering is the hole diameter, which has a very close tolerance of .500″/.501″. This tolerance callout actually tells you how to machine this feature. There are a variety of ways to make a hole in a part, including drilling, milling, reaming, and boring. A drilled hole must be spot drilled first and then drilled, resulting in a hole no closer than ±.003″ of a specified size that is not perfectly round or accurately positioned. A milled hole is accurate in position and roundness, but due to machining technique, these holes can have some taper from top to bottom, or a cylindricity issue. A third option is a reamed hole, which must be spot drilled, drilled undersize, and then reamed. A reamer is considered a *hard tool*, or a tool that only cuts on the sides and must enter through a preexisting hole. For the hole in **Figure 9-1**, an adequate machining technique would be to drill this hole .485″ in diameter and then ream with a .5005″ reamer. The last option is to bore this hole, accomplished by drilling a hole and then using a boring head to machine a round, close-diameter hole. Boring achieves the best results—holes accurate in cylindricity, position, and diameter—but it is slower than the previous options. Boring this hole is not the most cost-effective operation, but it is sometimes required to meet print specifications.

In communicating where dimensions are established and which features are most important, the **Figure 9-1** print tells us how to machine this part. In sum, use 3″ stock, cut it to 6.1″, place the part in a vise where both ends can be machined, establish the top-left corner as the origin of the WCS, use an end mill to cut to 6.00″, and then spot drill, drill, and ream a .5005″ hole. By carefully examining the print, a good machinist gains a much better understanding of the designer's intent for a part, which can then inform the best machining process.

9.2.2 Part Workholding

After reading and understanding the full print requirements, a machinist must decide how to hold the part. Holding the part securely during the machining process is critical, but machinists also need to have access to machine as many features as possible in a single holding to reduce setup and overall cycle time. For some projects, *workholding* will be as simple as placing a piece in a vise, others will require the production of a specialty fixture, and some projects may require specialized devices like magnetic or vacuum chucks (see **Figure 9-2** and **Figure 9-3**). The objective of workholding is to maximize machining and minimize setups or operations.

inTer_b/Shutterstock.com

Figure 9-2. A magnetic chuck used to secure ferrous metals in place.

Goodheart-Willcox Publisher

Figure 9-3. A vacuum chuck that uses negative air pressure to secure the workpiece.

Workholding is a critical decision that has to be made in the preprogramming phase. If you do not know how you are going to hold the part, there is no way to determine how to machine it.

9.2.3 Tool Selection

The next important decision before writing a program is determining what tools to use. It is impossible to write a program without knowing what tools are available. To reduce tool changes and cycle time, try to minimize the number of tools used. Always evaluate the possibility of using the same tools for multiple features. For example, if cutting a .625″-wide slot and a .750″-wide slot on one part, a 1/2″ end mill can accomplish both without using a second tool. It is also best practice to spot drill every hole before drilling. If any specialized tooling is required, it is best to identify this in the planning phase so that it can be accounted for while programming.

9.3 Program Format

After analyzing the print, workholding, and tooling, it is time to generate code. Writing a program is simply telling a story. It has an *opening statement*, *body*, and *closing statement*. The code for each tool used within the program will have these same components. Writing programs consistently with this format will simplify the process and spotlight any missing information.

9.3.1 Opening Statement

Each program should have a consistent opening statement based on the machine for which it is written. The opening statement will contain the program number and a default block that cancels any previous cycles that may still be active.

These are the first few lines of a sample program code:

```
O3283
G20 G17 G40 G80
G91 G28 Z0.
```

In this example, **O3283** is the program number. It starts with the letter "O." This line will be the only instance of using the letter O in a CNC program. The program can carry any number, but it is best practice to use a number that correlates with the part number in some discernible way. Every program must start with the program number.

G20 G17 G40 G80 is often called the *default block*. It cancels commands that may have activated in a previous program. This line can vary widely based on the programmer, but this example is a safe starting point. G codes will be covered in more depth in a later section, but it is worth analyzing the details of this block:

- G20 places the machine in inch mode (as opposed to metric, which uses code G21).
- G17 places all arcs in X,Y planes.
- G40 cancels cutter compensation.
- G80 cancels active canned cycles.

G91 G28 Z0. is a safe *zero return* move in the Z axis. This is a safety move to prevent any initial movement of the spindle or tool that is close to the fixture, part, or clamping. It will return the spindle to its Z-axis home position.

- G91 places the machine in incremental mode.
- G28 returns the axis to zero.
- Z0. defines the axis position to which the machine will return. In this instance, the G28 will return the Z axis to its Z0. position.

9.3.2 Program Body

The body is the heart of the program, defining all of the tools and movements to create a part according to the print. The program for each tool will also have an opening statement, body, and closing statement within the larger program. Consider the following example:

T1 M6	
G0 G90 G54X_____ Y_____ M3 S5000	**OPENING STATEMENT**
G43 H1 **Z2.** M8	

T1 M6 calls tool #1 and tool change.

G0 G90 G54 X_____ Y_____ M3 S5000 executes a rapid move in absolute mode to XY location and turns on the spindle clockwise to 5,000 rpm.

G43 H1 Z2. M8 turns on tool height offset for tool #1, positions the tool at Z2., and turns on coolant.

G01 Z−.5 F30.	
X_____	**BODY: ALL MOVEMENTS**
Y_____	

G01 Z−.5 F30., in a feed-rate mode, moves the tool to Z−.5 at 30 inches per minute.

X_____ moves the tool to an X position (still in a feed-rate mode).

Y_____ moves the tool to a Y position (still in a feed-rate mode).

G80	
G91 G28 **Z0.**	**CLOSING STATEMENT**
G28 Y0.	

G80 turns off canned cycles.

G91 G28 Z0., in incremental mode, return to Z0.

G28 Y0., in incremental mode, return to Y0.

Any additional tools should follow this repetitive format for each tool's code. Consistently programming in this manner will allow you to easily scan your work and see any deficiencies, preventing missing information and ensuring a successful program.

9.3.3 Closing Statement

The closing statement contains the last lines of the program and prepares the machine to run the next part in a safe manner. It should also define the optimum position to remove the part from the fixture and load new material. For example:

```
M9
G91 G28 Z0. M5
G28 Y0.                                    CLOSING STATEMENT
M30
```

M9 turns off coolant.

G91 G28 Z0. M5 incrementally moves the tool to Z0. home and turns the spindle off.

G28 Y0. moves the tool to Y0. location.

M30 rewinds the program.

The story is now complete. This is the format this text will use going forward to create all programs.

9.4 Address Codes

Address codes are single-letter characters at the beginning of each word that define what the computer should do with the numerical data that follows. They are used to designate commands or machine functions. **Figure 9-4** lists the available address codes and their functions.

9.4.1 G Codes

The program address G is commonly referred to as the ***preparatory command***. The G code puts the CNC control in a specific mode or state of operation. For example, G00 places the machine in a rapid mode, and G01 places the machine in a feed-rate mode. G codes can also be used to command **canned cycles**, or abbreviated multi-movement cycles, such as G83, which commands a drilling cycle but does not produce any machine movement.

The table in **Figure 9-5** lists the most commonly used G codes, which can be used to produce most programs. There are more G codes available, including machine-specific G codes for special functions. For specific codes for different controls, refer to the user's manual for that particular machine.

The *Function* column of **Figure 9-5** can help you decide when to use each G code. Categorizing G codes by function and using them to create some sample programs will help clarify the programming process.

- G00, G01, G02, and G03 are all movement commands: rapid, straight lines, and arcs.
- G28 sends the machine home.
- G40, G41, and G42 are commands for cutter compensation.
- G43 is tool height offset, or how far your tool tip is from home.
- G54–G59 are work coordinate offsets, or where your part is in the machine.
- G73–G86 are canned cycle commands.
- G90 and G91 establish incremental or absolute programming mode.
- G98 and G99 are used in canned cycles to establish safe return planes.

Address Codes

Address Code	Function
A	Angular dimension around X axis.
B	Angular dimension around Y axis.
C	Angular dimension around Z axis.
D	Cutter compensation register number.
E	Angular dimension for special axis.
F	Feed rate.
H	Height offset register number.
I	Alternate X axis used for canned cycles or circular motions.
J	Alternate Y axis used for canned cycles or circular motions.
K	Alternate Z axis used for canned cycles or circular motions.
L	Loop count repeat for canned cycles.
N	Block number.
O	Program number.
P	Dwell time.
Q	Canned cycle repeat dimension.
R	Arc radius or Z axis retract distance.
S	Spindle speed.
T	Tool selection number.
U	Alternate X axis used to specify motion.
V	Alternate Y axis used to specify motion.
W	Alternate Z axis used to specify motion.
X	X-axis motion.
Y	Y-axis motion.
Z	Z-axis motion.

Goodheart-Willcox Publisher

Figure 9-4. A standard address code list.

Commonly Used G Codes

G Code	Function
G00	Move in a straight line at rapid speed.
G01	Move in a straight line at a feed rate.
G02	Clockwise arc at feed rate.
G03	Counterclockwise arc at feed rate.
G04	Dwell; stop for a specified time.
G17	Arc in X-Y plane.
G18	Arc in X-Z plane.
G19	Arc in Y-Z plane.
G20	Program coordinates are inches.
G21	Program coordinates are millimeters.
G28	Return to home position.
G40	Tool cutter compensation off.
G41	Tool cutter compensation left.

(continued)

Figure 9-5. A commonly used G code table. Some machine-specific G codes are not listed.

Commonly Used G Codes *(continued)*	
G Code	**Function**
G42	Tool cutter compensation right.
G43	Apply tool length compensation.
G54	Work coordinate system (1st workpiece).
G55	Work coordinate system (2nd workpiece).
G56	Work coordinate system (3rd workpiece).
G57	Work coordinate system (4th workpiece).
G58	Work coordinate system (5th workpiece).
G59	Work coordinate system (6th workpiece).
G73	High-speed drilling cycle.
G76	Fine boring cycle.
G80	Cancel canned cycle.
G81	Spot-drilling cycle.
G82	Drilling cycle with dwell (counterboring).
G83	Full retract peck drilling cycle.
G84	Tapping cycle.
G85	Boring canned cycle, no dwell, feed out.
G86	Boring canned cycle, spindle stop, rapid out.
G90	Absolute programming of XYZ.
G91	Incremental programming of XYZ.
G98	Return to initial Z plane after canned cycle.
G99	Return to initial R plane after canned cycle.

Goodheart-Willcox Publisher

Figure 9-5. *(Continued)*

9.4.2 M Codes

The program address M is commonly referred to as a *machine function*. The M code can best be described as a switch. For example, M8 turns on the coolant, and M9 turns off the coolant. These commands literally turn on and off the electrical switch for the coolant pump. M codes can be machine or machine-type specific, although many M codes are the same for lathes and mills. Many machines also come with "open M codes," which can be tapped into for optional equipment. Unlike G codes, only one M code can be used on each line of the program.

Figure 9-6 lists the most commonly used M codes and their functions, which can be used to produce most programs. There are more M codes available, including machine-specific M codes for special functions. For specific codes for different controls, refer to your user's manual.

As with G codes, the list of M codes can be sorted by functional category for use with future programs.

- M00 is a mandatory program stop used for forcing a stop mid-cycle.
- M01 is an optional stop turned on and off by a switch.
- M03, M04, and M05 control spindle on and off; M03 will be used in most cases to turn the spindle on and M05 to turn the spindle off.
- M06 initiates a tool change.
- M10 and M11 are only used on multi-axis machines to clamp/unclamp the rotary or table.

Commonly Used M Codes	
M Code	**Function**
M00	Program stop.
M01	Optional program stop.
M03	Spindle on clockwise.
M04	Spindle on counterclockwise.
M05	Spindle off.
M06	Tool change.
M08	Coolant on.
M09	Coolant off.
M10	Chuck or rotary table clamp on.
M11	Chuck or rotary table clamp off.
M30	Program end; return to start.
M98	Subprogram call.
M99	End of subprogram or program restart.

Goodheart-Willcox Publisher

Figure 9-6. A commonly used M code list. Some M codes are not listed.

- M30 stops the program and rewinds to the beginning.
- M98 sends the machine to a subprogram in conjunction with a P address (see **Figure 9-4**).
- M99 rewinds and reruns the program; often called a loop.

9.5 Contouring

Contouring is a commonly used toolpath, typically performed with an end mill, that follows a joined path or single piece of geometry. Simply think of this as cutting down a line around a circle or a rectangular shape. The shape can be much more complex, but the steps to creating the program will be the same. Look at the print in **Figure 9-7** as we walk through a sample program for contouring.

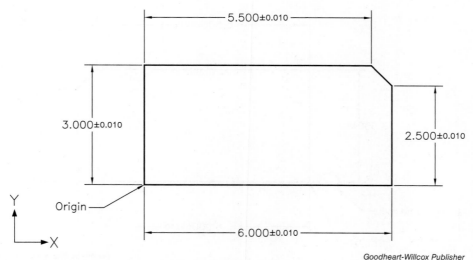

Goodheart-Willcox Publisher

Figure 9-7. A sample print for contour programming.

In the print, the origin is shown as the bottom-left corner, indicating that this print is in quadrant 1, or the X+, Y+ quadrant. With that origin as a starting point, examine each line of the following program to contour this part:

O3283	Program number
G20 G17 G40 G80	Default line
G91 G28 Z0.	Send machine to Z0. home
T1 M6 (1/2" End mill)	Tool change to tool #1
G0 G90 G54 X−.5 Y0. M3 S5000	Rapid, G54 WCS, go to XY position, spindle on to 5000 rpm
G43 H1 Z2. M8	Height offset 1, move to Z2. coolant on
G01 Z−.5 F30.	Feed to Z−.5 at 30 inches per minute
X0.	Move to X and Y positions
Y3.	
X5.5	
X6.0 Y2.5	X and Y on same line create angle
Y0.	
X0.	
X−.5	
G0 Z1.	Rapid to Z1.
M9	Coolant off
G91 G28 Z0. M5	Incremental mode, home in Z, spindle off
G28 Y0.	Home in Y
M30	Program rewind

9.5.1 Tool Start Position

In the preceding program, notice that the starting point of this tool is G90 G54 X−.5 Y0. In the diagram in **Figure 9-8**, that position is shown relative to the WCS zero, or origin. In a contour toolpath, it is important to start the end mill off the part, take the tool down to the initial Z depth, and then move into the part whenever possible. Driving the tool straight down the edge of a contour will create a mark or defect in the finished contour.

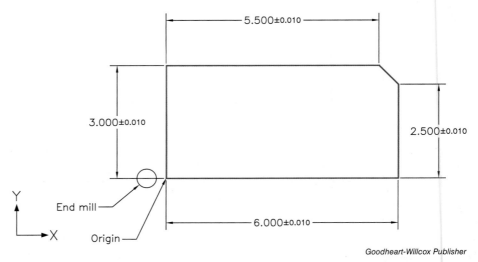

Figure 9-8. The same sample print showing a possible start position for the end mill.

The actual starting position can be anywhere off the part, and it might be dictated by the fixture or holding device. Notice that the endpoint of this program is also off the part to prevent any damage to the finished part.

9.5.2 Cutter Compensation

A contour toolpath program directs the center of the spindle, or tool, through the Cartesian coordinate points of the part. The diagram in **Figure 9-9** represents the path programmed using a 1/2″ end mill from the example code in the previous section. The blue line shows the specified finished dimensions of the part, and the orange shading indicates the tool as it follows that line.

This is not what was intended. The finished part, cut to the size of the white section within the orange toolpath, has been cut smaller than the print requirement. The source of this error is the lack of compensation for the diameter of the tool, or *cutter compensation*.

Refer back to the G code list, **Figure 9-5**. Note the G codes that refer to cutter compensation:

- G40 turns tool cutter compensation off.
- G41 moves tool cutter compensation left.
- G42 moves tool cutter compensation right.

 From the Shop

G41: Left-Hand Compensation

Remember from earlier sections, we always want to *climb cut* with an end mill to stop the tool from grabbing the material, making a rough surface finish, or pulling the part out of the fixture. That means our tool is most often cutting on the left-hand side of the material, or in a G41 cutter compensation. There are a few rare exceptions, but the majority of milling operations use the G41 left-hand compensation.

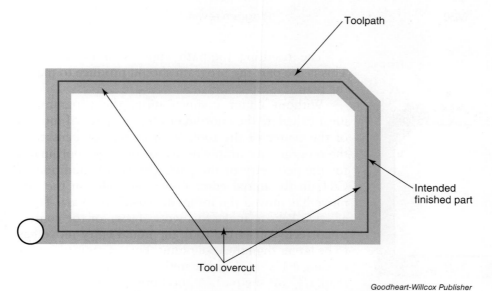

Goodheart-Willcox Publisher

Figure 9-9. An illustration of a contour toolpath. The resulting toolpath violates the finished part boundary.

The toolpath in our example needs to move to the left-hand side of the plan geometry, to compensate for climb cutting with end mills in a contour. There are two ways to make this adjustment. The first way requires recalculating all movements and positions. Since this program uses a 1/2″ end mill, this potential solution requires moving the radius of the tool, or .250″, to the left of all the geometry. This will be time consuming and lead to possible programming errors. The second solution is to let the computer do the calculations by using G code commands. Turn on cutter compensation left (G41) in the first move into the part, and then cancel cutter compensation (G40) on the exit move off the part. Review the adjusted program and new results:

O3283	Program number
G20 G17 G40 G80	Default line
G91 G28 Z0.	Send machine to Z0. home
T1 M6 (1/2″ End mill)	Tool change to tool #1
G0 G90 G54 X−.5 Y0. M3 S5000	Rapid, G54 WCS, go to XY position, spindle on to 5000 rpm
G43 H1 Z2. M8	Height offset 1, move to Z2., coolant on
G01 Z−.5 F30.	Feed to Z−.5 at 30 inches per minute
G41 D1 X0.	Turn on cutter comp left tool 1, move to X position
Y3.	
X5.5	
X6.0 Y2.5	X and Y on same line create angle
Y0.	
X0.	
G40 X−.5	Turn off cutter comp, exit move off part
G0 Z1.	Rapid to Z1.
M9	Coolant off
G91 G28 Z0. M5	Incremental mode, home in Z, spindle off
G28 Y0.	Home in Y
M30	Program rewind

The resulting toolpath appears in **Figure 9-10.** This is the part the program should produce to match the plan.

Without cutter compensation, the programmer must calculate the coordinates for the path of motion of the center of the tool. With cutter compensation, the programmer simply needs to enter the coordinates for the geometry of the part. The simple addition of G41 on the initial entry move and G40 on the exit move has moved the toolpath outside the part. There are two keys to make this work. First, for the control to correctly offset the tool, the initial entry move has to be larger than the tool radius. For example, this part is using a 1/2″ tool, so the initial move must start more than 1/4″ off the part. A good practice is to make the first positioning move the same distance as the tool diameter. This program started at X−.5 and moved

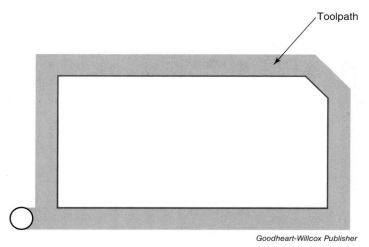
Toolpath

Goodheart-Willcox Publisher

Figure 9-10. This contour toolpath, which includes cutter compensation, will produce the part as specified on the print.

<< PROBING		TOOL OFFSET		TOOL INFO >>	
TOOL 1	COOLANT	H(LENGTH)		D(DIA)	
OFFSET	POSITION	GEOMETRY	WEAR	GEOMETRY	WEAR
1 SPINDLE	10	4.5680	0.	0.	← 0.
2	0	0.	0.	0.	0.
3	0	0.	0.	0.	0.
4	0	0.	0.	0.	0.
5	0	0.	0.	0.	0.
6	0	0.	0.	0.	0.
7	0	0.	0.	0.	0.
8	0	0.	0.	0.	0.
9	0	0.	0.	0.	0.

Goodheart-Willcox Publisher

Figure 9-11. The tool offset screen where the tool diameter is entered.

into X0., while turning on the G41. Second, designate the offset used (D1 in the given example) and enter the tool diameter in the *tool offset* page, **Figure 9-11**. This entry is your *diameter offset*. Some machines may be set to radius instead of diameter. See the user manual to change this setting, or use the radius of the tool as the offset.

Diameter Offsetting

The diameter offset in your control can be a very useful function. The previous example shows how easy it becomes to program a path if there are no calculations needed for the tool diameter. In a machine shop environment, it is sometimes necessary to use a tool with a different diameter than originally specified in the program. If a program is written for a 1/2″ end mill, but the only end mills available are 5/8″, you would have to start a new program to accommodate the difference in size. With the control calculating G41 or G42 positions, it is a simple matter to go back to the tool offset page and change the diameter geometry offset to .625″.

Another benefit to using diameter offsetting is the ability to make small, incremental changes. In our example print and program, this part is defined as a 6.000″ × 3.000″ rectangular part. But what if this piece comes out to 6.010″ × 3.010″? The machinist can make a small change to the .500″ tool diameter offset to make the part either bigger or smaller. Define the tool as .490″ diameter (or smaller) to move the path closer into the part and cut the part smaller. Remember, the machine is cutting lines on both sides of the rectangle, so moving in the tool will affect both sides.

This function can also serve as a setup technique in creating the first part to help prevent cutting that part undersize. Change the tool to .550″ diameter offset so that the machine calculates the path farther away from the programmed dimensions, thus leaving the part larger. After the program is complete, measure the part and offset the tool to reach the desired final size. After the diameter offset is altered and the correct size is achieved, all subsequent parts can be machined at the new diameter offset value.

9.6 Radius and Angle Milling

While contours are the most common toolpath, even in simple contoured parts there are often radii and angles to be programmed. The advent of better control technology has simplified this once complex programming task. This section will discuss the G02 and G03 codes used to create arcs as well as the I, J, and K method (used in older controls or to create circles).

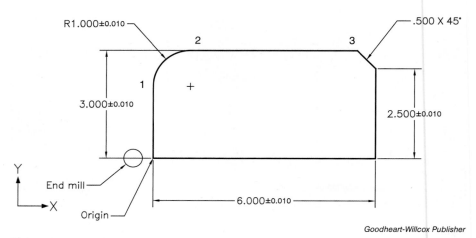

Figure 9-12. A part toolpath that will require an arc command of G02.

9.6.1 G02 and G03 Radial Movements

The G02 and G03 codes are the simplest methods to program arcs. The term *arc* refers to a circular shape less than 360°. At 360°, the arc becomes a circle by definition, not an arc. G0, G1, G02, and G03 are all *modal commands*, meaning once they are turned on, they stay on until they are turned off. Just like a light switch, if you turn on a G0, you will continue in rapid movements until an alternate movement is commanded, or until the modal command is turned off. The difference between G02 and G03 is simply which direction the cutter is headed. G02 will take the tool in a clockwise direction, while G03 takes the tool in a counterclockwise direction. This is true in both lathe and mill programming. Visualize the desired tool direction as it enters and exits the arc to select the appropriate code. See **Figure 9-12** for a modified version of the part from the previous example, **Figure 9-8**, with an arc added.

To program this arc, look at the following code from the origin or WCS, to Point 1, the arc command, and then out to Point 3:

G1 X0. Y0.	Origin
Y2.	Point 1
G02 X1. Y1. R1.	Arc to Point 2
G01 X5.5	Move to Point 3. G1 turns off arc for a straight line.

This method can be used for all arcs less than 180°. For an arc that is more than 180° but less than 360°, things become more complicated. As shown in **Figure 9-13**, there are two mathematical possibilities for this arc.

The 90° arc has the same start and endpoint as the 270° arc, so the control cannot differentiate between the two arcs. By default, the control will do a less than 180° arc. So how can the programmer create a path for a 270° arc? Simply make the R value in the program a negative (–) number.

| G02 X1. Y1. R–1. | Arc to Point 2 with a 270° arc |

As already stated, the G02 and G03 codes can only be used with an arc less than 360°. So, is there a simple way to program a full circle using this technique? Yes—break the circle in half and create two 180° arcs on two separate lines of code, as shown in **Figure 9-14**.

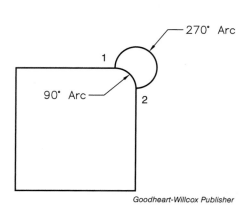

Figure 9-13. There are two possible arcs with the same radius and endpoints.

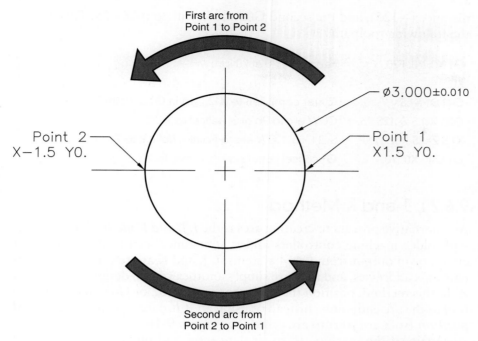

First arc from
Point 1 to Point 2

Ø3.000±0.010

Point 2
X−1.5 Y0.

Point 1
X1.5 Y0.

Second arc from
Point 2 to Point 1

Goodheart-Willcox Publisher

Figure 9-14. Breaking a full circle into 180° arcs allows the use of G02 and G03 commands.

If the tool starts in the center, the code will look like this:

G1 X0. Y0. F20.	Feed to XY 0 at 20 ipm
Z−.25	Feed down to Z−.25
G41 D1 X1.5	Cutter comp left to X1.5 still in G1 at Point 1
G03 X-1.5 R3.	Counterclockwise (CCW) arc to Point 2
X1.5 R3.	Still in G03, CCW arc to Point 1
G1 G40 X0. Y0.	G1, cancel cutter comp, move to XY 0

It is even acceptable to ramp down while cutting the arc by adding Z− moves onto the G03 lines, **Figure 9-15**. In this case, the cutter ends up at Z−.25, so the move to Point 1 will cover half of the Z-depth

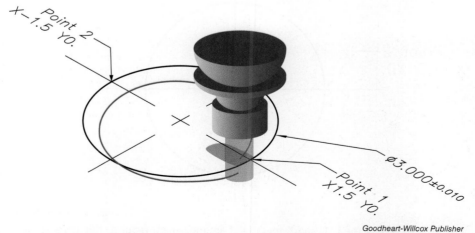

Point 2
X−1.5 Y0.

Ø3.000±0.010

Point 1
X1.5 Y0.

Goodheart-Willcox Publisher

Figure 9-15. While cutting an arc, the tool can also be moved in the Z axis.

distance (−.125), and the second G03 move will go to Z−.25. This creates the following toolpath:

G1 X0. Y0. F20.	Feed to XY 0 at 20 ipm
Z0.	
G41 D1 X1.5	Cutter comp left to X1.5, still in G1 at Point 1
G03 X-1.5 Z-.125 R3.	CCW arc to Point 2, helical to Z-.125
X1.5 Z-.25 R3.	Still G03, CCW arc to Point 1, helical to Z-.25
G1 G40 X0. Y0.	G1 cancel cutter comp, move to XY 0

9.6.2 I, J, and K Method

An alternative process to creating arcs is the *I, J, and K method*. This works with older machine controllers and is sometimes seen in the creation of circles or in output from CAM systems. I, J, and K are referred to as alternate axis addresses, and they are simply another way to designate X, Y, and Z. In this method, position the tool, program a G02 or G03, and move the tool to the arc endpoints. Instead of using an R for arc, use the *incremental position* from arc start to arc center. See **Figure 9-16**.

Again, if the machine starts at the center and moves to Point 1, and then G03 is used to move to Point 2, the code will look like this:

G1 X0. Y0. F20.	Feed to XY 0 at 20 ipm
Z−.25	Feed down to Z−.25
G41 D1 X1.5	Cutter comp left to X1.5, still in G1 at Point 1
G03 X−1.5 I−1.5	CCW arc to Point 2, I−1.5 is distance from start to arc center, incrementally
X1.5 I1.5	CCW arc to Point 1, I1.5 is distance from start to arc center, incrementally
G1 G40 X0. Y0.	G1 cancel cutter comp, move to XY 0.

This method can be useful in creating circles, but it is not necessary on most controls. The "I" was used in this case because the arc center was in

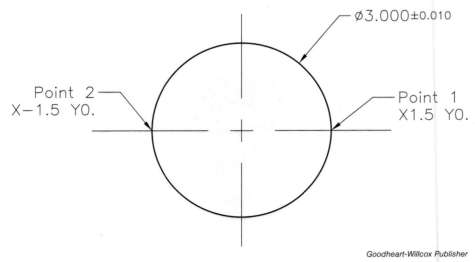

Figure 9-16. Using the I, J, and K method of arc movement requires the incremental position from arc end to arc center.

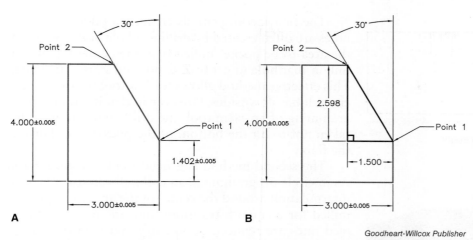

Figure 9-17. A—A print might require some additional calculations to plot toolpath. B—Part shows the resultant right triangle and point coordinates.

the X direction from the start. The "J" would be used if the center was in the Y-axis direction from the start.

9.6.3 Programming Angles

Typically, cutting angles requires moving along two axes simultaneously. This is typically X and Y, but can be performed in X and Z or Y and Z directions. An example of this method was presented earlier in the sample program that created a 1/2″ by 90° angle for the part in **Figure 9-7**. This is all completed with a G1 movement, and no special code is required.

X5.5	
X6.0 Y2.5	X and Y on same line create angle
Y0.	

Many prints will require knowledge of trigonometry to calculate the start and endpoints needed to create a program. Consider the following example:

There appears to be no coordinate for Point 2 in the X axis, **Figure 9-17A**. This omission requires the right triangle calculations, as explained in Chapter 4. There is a triangle highlighted in **Figure 9-17B** with enough measurements for the programmer to calculate Point 2.

9.7 Pocketing

Pocket milling is a fairly common machining operation on CNC machines. A *pocket* is a term for machining inside a closed area or boundary. The inside of a pocket can include islands or internal components that are not meant to be machined, and the walls or bottom of the pocket can be flat or tapered. A typical example of a pocketed part appears in **Figure 9-18**.

9.7.1 Cutter Entry

In the contouring example, the cutter entered and exited the part from outside its boundaries. Obviously, the same strategy will not work for a pocket, so cutter entry is a concern that must be considered in the program. There are three strategies that can be deployed for entering the cutter safely.

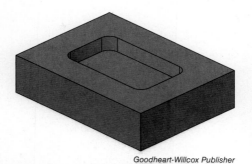

Figure 9-18. A typical part requiring a pocket-cutting cycle.

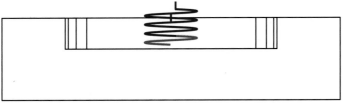

Goodheart-Willcox Publisher

Figure 9-19. A side view of a helical-entry strategy.

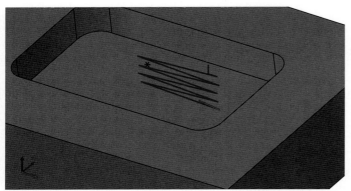

Goodheart-Willcox Publisher

Figure 9-20. A ramped-entry strategy.

The first method entails creating a pilot hole. Pilot holes are often created by drilling a hole in the center of the area to be pocket milled first, then entering an end mill in that hole to get to Z depth and start machining. This effective method allows for less tool pressure upon entry. It has downsides, however, in that it requires programming of an extra tool (the drill), and care must be taken not to hit the bottom of the pocket and damage the floor.

The second method is a helical entry, **Figure 9-19.** As in circle programming, an end mill can use a helical interpolation around the center of the area to be pocket milled for a smooth transition into the pocket. Often, feed rates are reduced upon entry and then turned up once depth is achieved. End mill size and rigidity of the holder may be factors in deciding how fast the tool can enter the pocket area.

The last strategy is a ramp angle entry, **Figure 9-20.** The tool is positioned above the part and then moved in an X,Z line or Y,Z line. Normally, the angle of entry is 3° to allow for the trailing edge of the end mill to clear the cut as it progresses. This is a very effective entry move, but it requires more lines of code. The feed rate can be increased significantly on the ramp and then reduced for the pocket roughing depending on the ramp depth and angle.

9.7.2 Roughing Method

Once entry has been considered, the tool must now be programmed to remove the material from the pocket. This simply involves making straight-line moves in either the X or Y axis, then moving the tool over an amount in the opposite axis, and then traveling back. A good starting number for the step-over amount is the radius of the end mill. It is best not to mill all the way out to the edge of the pocket initially, but instead leave a small amount of material for a finish pass. Once the pocket is roughed out to depth, use a contour path to finish to print specifications. Consider the pocket radii when selecting a tool. The tool radius must be smaller than the part radii or it will not be able to cut the correct corners. Analyze this sample program, which utilizes a ramp-angle entry:

```
O1426 (POCKET)
(MATERIAL - 6061 ALUMINUM)
(T1 | 1/2 FLAT END MILL | H1)
G0 G17 G20 G40 G80
T1 M6 (Rough pocket)
G0 G90 G54 X.05 Y−.4 S3820 M3
G43 H1 Z.25 M8
Z.2
G1 Z.1 F100.
X.1 Z.0967
X−.9 Z.0443
```

```
X.1 Z-.010
X-.9 Z-.060
X.1 Z-.115
X-.9 Z-.165
X.1 Z-.220
X-.9 Z-.270
X.1 Z-.320
X-.9 Z-.375
X-1. Y-.5 F61.12
X1.
Y-.25
X-1.
Y0.
X1.
Y.25
X-1.
Y.4999
X1.
G0 Z.25
G0 X0. Y.25
G1 Z-.375 F60. (Finish contour)
G41 D1 Y.75
X-1.25
Y-.75
X1.25
Y.75
X0.
G40 Y.25
G0 Z.25
M9
G91 G28 Z0. M5
M30
```

9.8 Canned Cycles

Canned cycles reduce repetitive amounts of machine code and make editing repetitive steps easy. For instance, hole making is a common process in milling, and writing identical code for multiple holes in a single part can be tedious and prone to program errors. Canned cycles allow a single line of code to define multiple parameters, and then all subsequent holes can be positioned in the proper X and Y location to repeat the process.

9.8.1 G83 Cycle (Full Retract Peck Drilling)

G83, one of the most common drilling canned cycles, is a *peck drilling* cycle, meaning it should be used whenever the depth of the hole is four or more times greater than the diameter of the drill. An example would be a .250″ diameter hole that is more than 1″ deep. The peck drilling cycle drives the tool down to a partial depth and then fully retracts above the part to remove chips and allow coolant to enter. Then it reenters the hole and goes to a deeper depth. Without the G83, the code looks like this:

```
G90 G54 X1. Y1. M3 S2500
T1 M6
G43 H1 Z.5 M8
G1 Z.1 F10.
Z-.25
G0 Z.1
G1 Z-.5
G0 Z.1
G1 Z-.75
G0 Z.1
G1 Z-1.
G0 Z.1
```

This tool drills −.25″ deep a total of four times and finishes at Z−1 deep. This is not terribly difficult, but imagine if you wanted to drill only .100″ deep with each peck or if you had ten holes to drill. This code gets very long and difficult to edit. Compare this method to one using a G83 cycle:

```
G90 G54 X1. Y1. M3 S2500   (X1. Y1. is location of first hole)
T1 M6
G43 H1 Z1. M8
G0 Z.25
G99 G83 Z-1. R.1 Q.25 F10.   (First hole drilled)
X2. Y2.   (Second hole)
X3. Y3.   (Third hole)
G80
```

By adding the X and Y locations after the G83 line, the same drilling pattern is applied at those locations. If the drilling conditions require changing the peck depths to .1″ instead of .25″, just change the Q to .1″, and all the holes are edited. A G80 cancels this canned cycle. The components to this cycle are as follows:

- G98: Returns to last Z position (.25″ in this case) between holes.
- G99: Returns to reference return position (.1″ in this case) between holes.
- G83: Turns on peck drilling.
- Z: Final depth of hole. Note: Different Z positions can be programmed on subsequent holes.
- R: The rapid position in Z that spindle goes before drilling.
- Q: The peck amount, always a positive number.
- F: The feed rate while drilling.

After learning the basics of when to use canned cycles and why, a good CNC machinist can make use of additional canned cycles. Common cycles are presented in the following sections. A complete list of all canned cycles for a particular machine can be found in the machine's programming manual.

9.8.2 G73 Cycle (Chip Break Peck Drilling)

The G73 canned cycle is similar to G83, but it does not retract the drill completely out of each hole, just a short amount to break off the chip. Here is a G73 program for the same hole we coded before:

```
G90 G54 X1. Y1. M3 S2500   (X1. Y1. is location of first hole)
T1 M6
G43 H1 Z1. M8
G0 Z.25
G98 G73 Z–1. R.1 Q.25 F10.
X2. Y2.   (Second hole)
X3. Y3.   (Third hole)
G80
```

Since the G98 is programmed, the drill will return to Z.25 between holes and not the Z.1.

- G98: Returns to last Z position (.25″ in this case) between holes.
- G73: Turns on chip break peck drilling.
- Z: Final depth of hole. Note: Different Z positions can be programmed on subsequent holes.
- R: The rapid position in Z that spindle goes before drilling.
- Q: The peck amount, always a positive number.
- F: The feed rate while drilling.

9.8.3 G81 Cycle (Spot Drilling)

The G81 is a standard drilling cycle that does not peck; instead, it travels to a programmed Z depth. This cycle is often used for spot drilling or for holes with a depth less than four times greater than the diameter. This cycle should also be used for solid carbide drills, as they will shatter under pecking conditions.

```
G90 G54 X1. Y1. M3 S2500   (X1. Y1. is location of first hole)
T1 M6
G43 H1 Z1. M8
G0 Z.25
G98 G81 Z–1. R.1 F10.
X2. Y2.   (Second hole)
X3. Y3.   (Third hole)
G80
```

Notice the following features:

- G98: Returns to last Z position (.25″ in this case) between holes.
- G81: Turns on spot drilling.
- Z: Final depth of hole. Note: Different Z positions can be programmed on subsequent holes.
- R: The rapid position in Z that spindle goes before drilling.
- F: The feed rate while drilling.

9.8.4 G82 Cycle (Counterbore Cycle)

The G82 is a cycle to create counterbores in a predrilled hole, most often made with an end mill or specific counterboring tool. The purpose is to create a flat-bottomed hole or smooth finish by adding a dwell while the tool is at the finished Z depth.

```
G90 G54 X1. Y1. M3 S2500   (X1. Y1. is location of first hole)
T1 M6
G43 H1 Z1. M8
G0 Z.25
G98 G82 Z–1. R.1 P500 F10.
X2. Y2.   (Second hole)
X3. Y3.   (Third hole)
G80
```

Notice the following features:

- G98: Returns to last Z position (.25″ in this case) between holes.
- G82: Turns on counterboring cycle.
- Z: Final depth of hole. Note: Different Z positions can be programmed on subsequent holes.
- R: The rapid position in Z that spindle goes before drilling.
- P: The dwell time in milliseconds. 500 milliseconds equals .5 seconds.
- F: The feed rate while drilling.

 **Thinking Green**

Recycling Tungsten Carbide Drills
Machine shops collect and store worn-out tungsten carbide drills for resale. Industry recyclers pay high rates for scrap tungsten carbide and recycle it into raw material ready for new uses. Tungsten carbide is one of the most sought-after ceramic scraps because of its strength and suitability for reuse. Reselling these old tungsten carbide drills benefits shops economically while also saving resources.

9.8.5 G84 Cycle (Tapping Cycle)

G84 allows a machinist to tap holes with a single line of programming. This is an amazing feat when you consider the intricate steps in tapping. A tap must enter a hole at a feed rate with an exact relationship to its spindle speed, go to depth, stop and reverse the spindle, and then feed out with the same relationship to spindle speed. This is a hand programming impossibility. Tapping used to be done with a floating tapping head to allow some flex while the machine slowed down and reversed. Now, the controller and servo motors can do the math automatically, once the machinist calculates feed and speed with relationship to the *threads per inch (TPI)* of the tap.

It is easiest to start with the spindle rpm and then calculate the feed rate. Assume a 1/4-20 tap (20 represents threads per inch) and 700 rpm for spindle speed. This is the formula for calculating feed rate:

$$\frac{rpm}{TPI} = \text{Feed rate}$$

$$\frac{700\ rpm}{20\ TPI} = 35\ \text{feed rate}$$

Tech Tip

When using the G84 cycle, spindle and feed rate overrides will be disabled by the control to maintain the spindle-feed relationship.

```
G0 G90 G54 X1. Y1. M03 S700   (X1. Y1. is location of first hole)
G43 H1 Z1. M8
G0 Z.25
G99 G84 Z−.5 R.5 F35.
X2. Y2.   (Second hole)
X3. Y3.   (Third hole)
G80
```

Explanation of cycle:

- G99: Returns to reference return position (.5″ in this case) between holes.
- G84: Turns on tapping.
- Z: Final depth of hole.
- R: Reference plane. Notice a larger .5″ height, which allows tap to get to speed before entering hole.
- F: The feed rate while tapping. Critical feature. Notice 700 rpm above.

 From the Shop

How Fast Can You Tap?

In the G84 example, we used 700 rpm as the spindle speed for the tap. Why? Well, it was an example of what speed we might use if tapping steel. You will find that there are many variables at play in tapping speeds; for example, type of material, type of tap, coating on tap, depth of tap, or even the class of thread. There is not one answer that fits all taps. It will take some experience and some expert input to help you get the best spindle speed. Your best resource is the tooling manufacturer's recommended speed chart for your specific tool, or a tooling engineer from that manufacturer. Start with a slower, conservative speed and then increase speed, with some trial and error. The optimum speed will be fast and provide long tooling life. *But remember: if you change the speed, you have to recalculate the feed.*

9.8.6 G85 Cycle (Fine Boring Cycle)

The G85 cycle is used when boring or reaming holes. The tool will be positioned over an existing hole and feed down to the final Z depth.

Unlike the other canned cycles in which the tool will then rapid out, the tool will exit the hole at the feed rate in a G85 cycle. Note that some small amount of material will be removed on the exit move because of the reduced tool pressure. This cycle will give the best surface finish.

```
G0 G90 G54 X1. Y1. M03 S700   (X1. Y1. is location of first hole)
G43 H1 Z1. M8
G0 Z.25
G99 G85 Z−1. R.1 F5.
X2. Y2.   (Second hole)
X3. Y3.   (Third hole)
```

Explanation of cycle:

- G99: Returns to reference return position (.1″ in this case) between holes.
- G85: Turns on fine boring cycle.
- Z: Final depth of hole.
- R: Reference return plane.
- F: The feed rate while boring, usually slower for boring tools.

9.8.7 G86 Cycle (Spindle Stop Boring Cycle)

The G86 cycle is also used when boring or reaming holes. The tool will be positioned over an existing hole and feed down to the final Z depth. Unlike the other canned cycles, the spindle will stop at final Z depth and the tool will rapid out. Note that this can create a small scratch in surface finish as the tool drags out of the hole. This cycle will usually be performed for rough or semi-finish operations, as it is faster than G85.

```
G0 G90 G54 X1. Y1. M03 S700   (X1. Y1. is location of first hole)
G43 H1 Z1. M8
G0 Z.25
G99 G86 Z−1. R.1 F5.
X2. Y2.   (Second hole)
X3. Y3.   (Third hole)
```

Explanation of cycle:

- G99: Returns to reference return position (.1″ in this case) between holes.
- G86: Turns on spindle stop boring cycle.
- Z: Final depth of hole.
- R: Reference return plane.
- F: The feed rate while boring, usually slower for boring tools.

9.9 Tool Setting

To this point, this chapter has addressed how to plan a program, analyze a print in depth, determine the best workholding option, and create a program. It is time to load the program into the machine and assemble the necessary tools and holders. When building tools, make sure they extend far enough to reach all of the programmed depths. All required tools will be placed in the tool changer according to their programmed tool number. For example, when a program defines tool #1 as a 1/2″ end mill, place that tool in the designated tool #1 position.

9.9.1 Length Offsets

Now it is time to tell the controller the distance from the bottom of the tool at the machine home position to the programmed part Z0. The controller does not know how long the tool extends from the holder without the programmer defining this dimension. This is accomplished using G43, which establishes tool length compensation for each tool.

Within the program, tool #1 will normally be designated as G43 H1, or Height 1. Although not required, it makes sense to match the height offset number to the tool number in this manner. Occasionally you might assign different height offsets to the same tool so that they can be offset independently.

Next, you will need to determine the tool height for your G43 code. If programming from the top of the part, this is the distance shown in **Figure 9-21**. To measure this distance, move the tool down to touch the part Z0 position. While the tool is at this position, push the *Tool Offset Measure* button on the control panel, **Figure 9-22**. This will record the

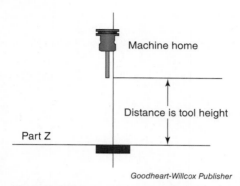

Goodheart-Willcox Publisher

Figure 9-21. The distance from tool end to programmed Z0. is the tool height offset value.

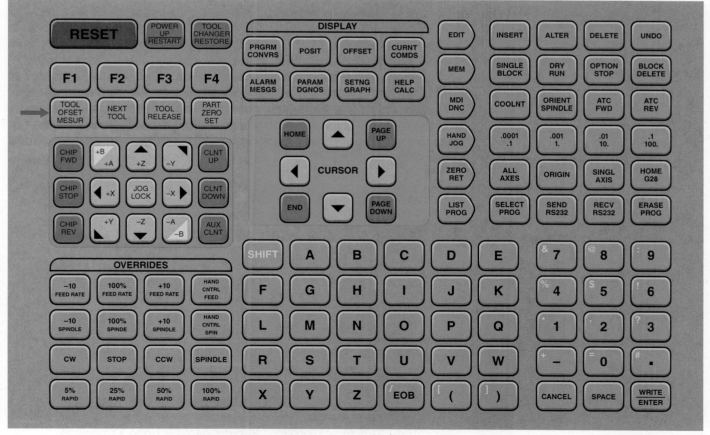

Goodheart-Willcox Publisher

Figure 9-22. The *Tool Offset Measure* button on a Haas controller.

<< PROBING		TOOL OFFSET			TOOL INFO >>
TOOL 1	COOLANT	H(LENGTH)		D(DIA)	
OFFSET	POSITION	GEOMETRY	WEAR	GEOMETRY	WEAR
1 SPINDLE	10	4.5680	0.	0.	0.
2	0	0.	0.	0.	0.
3	0	0.	0.	0.	0.
4	0	0.	0.	0.	0.
5	0	0.	0.	0.	0.
6	0	0.	0.	0.	0.
7	0	0.	0.	0.	0.
8	0	0.	0.	0.	0.
9	0	0.	0.	0.	0.

Goodheart-Willcox Publisher

Figure 9-23. The tool offset page where the tool height is stored or entered.

distance in the *Tool Offset Geometry* page for the corresponding active tool, **Figure 9-23**.

Repeat this process for all tools used in the program. You will notice a *Wear* column next to the *Geometry* column. The *Wear* column is where small adjustments can be made to Z heights after the part is machined and measured.

 Safety Note

Improperly called length offsets (H numbers) can lead to severe machine crashes. If the machine does not know the length of the tool from the home position, it can cause the tool to engage the material in a full rapid movement.

9.9.2 Diameter Offsets

If the G41 or G42 commands have been used in the program, the diameter of those tools must also be defined. Recall how to use those commands. In a previous example, the program cut a 6″ × 3″ rectangle by using G41 (left cutter compensation). The tool we used was a .500″ end mill.

T1 M6 (1/2″ End mill)	Tool change to tool #1
G0 G90 G54 X−.5 **Y0.** M3 **S5000**	Rapid, G54 WCS, go to XY position, spindle on to 5000 rpm
G43 **H1 Z2.** M8	Height offset 1, move to Z2., coolant on
G01 Z−.5 **F30.**	Feed to Z−.5 at 30 inches per minute
G41 **D1 X0.**	Turn on cutter comp left for tool 1, move to X position
Y3.	
X5.5	
X6.0 Y2.5	X and Y on same line create angle
Y0.	
X0.	
G40 X−.5	Turn off cutter comp, exit move off part
G0 G90 G54 X−.5 **Y0.** M3 **S5000**	Rapid, G54 WCS, go to XY position, spindle on to 5000 rpm

<< PROBING	TOOL OFFSET			TOOL INFO >>	
TOOL 1	COOLANT	H(LENGTH)		D(DIA)	
OFFSET	POSITION	GEOMETRY	WEAR	GEOMETRY	WEAR
1 SPINDLE	10	4.5680	0.	0. ←	0.
2	0	0.	0.	0.	0.
3	0	0.	0.	0.	0.
4	0	0.	0.	0.	0.
5	0	0.	0.	0.	0.
6	0	0.	0.	0.	0.
7	0	0.	0.	0.	0.
8	0	0.	0.	0.	0.
9	0	0.	0.	0.	0.

Goodheart-Willcox Publisher

Figure 9-24. When using cutter compensation, the tool diameter is entered into the *Diameter Offset* column.

The G41 was defined with a D1, so the diameter of that tool must be entered in the *Geometry* column of the *Diameter Offset* page under tool #1, **Figure 9-24.** Again, D1 was used with tool #1 to match the diameter number to the tool number, but this is not required.

The *Wear* column is where small adjustments can be made to the offset after the part is machined and measured. Once the part is set up and major adjustments are made, a small adjustment to the wear offset can bring the part dimensions to exacting specifications. These are meant for final, minor adjustments as the part is being manufactured.

9.10 Work Coordinate Offsets

In the same way that the machine does not know how "long" the tools are, it also does not know where the part is on the table or fixture until it is defined by the **work coordinate system (WCS).** A program includes G54–G59 designations, and a specific point on the part is considered to be at X0. and Y0. The programmer must tell the machine where that zero point is, physically, in the work envelope of the machine, **Figure 9-25.**

When a machine is initially started up and zero returned, it travels to its **machine home** position and stores an X, Y, and Z zero. Now, the programmer must physically move the machine to find the same place on the material as the program zero. This zero point can be a corner, the center, or some existing geometry like a hole or boss.

9.10.1 Finding WCS on Material

There are various ways to locate this point, such as an edge finder, an indicator, or a hard post, and each axis position can be found individually

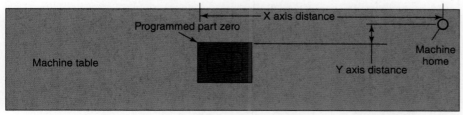

Goodheart-Willcox Publisher

Figure 9-25. The distance from machine home to programmed zero is the work coordinate value.

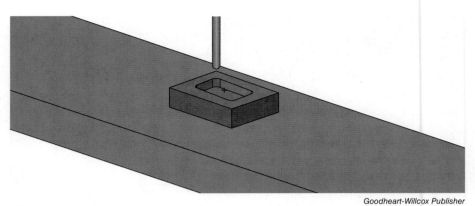

Goodheart-Willcox Publisher

Figure 9-26. The X and Y axes intersection can be determined by finding the edges.

Tech Tip

In the program planning stages, ensure that your workholding is repeatable. If the material can easily be placed back in this same position on the machine with each subsequent part, this WCS only has to be set once.

or in both X and Y simultaneously. The critical note is that the machine controller will record the exact location of the machine when the offset is entered, **Figure 9-26.**

9.10.2 Setting WCS on Control

Once the machine's spindle centerline is sitting directly over the X- and Y-axis programmed zero locations, that location needs to be recorded into the *Work Offset* page.

On this Haas Machine control, **Figure 9-27**, the *Work Zero Offset* page is directly below the *Height Offset* page. Different controllers place

<< TOOL OFFSET		TOOL INFO			PROBING >>
TOOL 1		ACTUAL	TOOL	TOOL	TOOL
TOOL	FLUTES	DIAMETER	TYPE	POCKET	CATEGORY
1 SPINDLE	2	0.5000	4-END MILL	0	x
2	0	0.0700	1-DRILL	1	x*
3	0	0.0860	2-TAP	2	x
4	0	0.	0-NONE	3	x
5	0	0.	0-NONE	4	x
6	0	0.	0-NONE	5	x
7	0	0.	0-NONE	6	x
8	0	0.	0-NONE	7	x
9	0	0.	0-NONE	8	x
ENTER A VALUE					

<< WORK PROBE		WORK ZERO OFFSET		WORK PROBE >>	
G CODE	X AXIS	Y AXIS	Z AXIS		
G52	0.	0.	0.		
G54	0.	0.	0.		
G55	0.	0.	0.		
G56	0.	0.	0.		
G57	0.	0.	0.		
G58	0.	0.	0.		
G59	0.	0.	0.		
G154 P1	0.	0.	0.		
G154 P2	0.	0.	0.		
G154 G3	0.	0.	0.		

Goodheart-Willcox Publisher

Figure 9-27. A Haas controller with the tool offset page and work coordinate offset page.

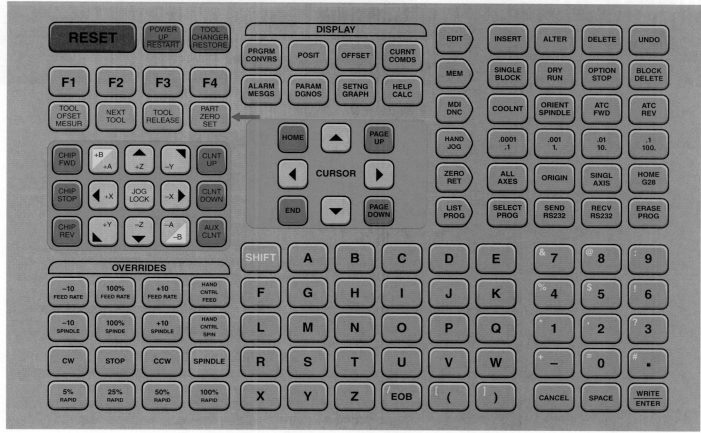

Figure 9-28. The *Part Zero Set* button will record the machine position in a work coordinate setting.

this page in different locations, but the process is similar regardless. Use the cursor to highlight the X-axis box in the correct G54–G59 row and press the *Part Zero Set* button, **Figure 9-28.** This records the machine position in the appropriate offset. Repeat this process for the Y axis.

Notice there is also an offset for the Z axis. Placing a number in that column will change the Z0 plane on which all the tools were established. There are instances where tools should be set to alternate planes. In those cases, a Z offset will be used. For the purposes of this text, this will be left as 0. You may also notice that there are more WCS settings than just G54–G59, and those can be used, if necessary, for multiple parts needing multiple work coordinates, **Figure 9-29.** As a programming note, this system allows for the manufacturing of multiple pieces with the same program just by altering the WCS.

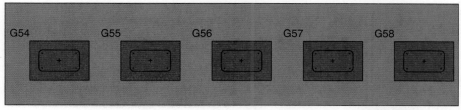

Figure 9-29. The same program can be run on multiple parts at different locations by using various WCS numbers.

Chapter Review

Summary

- The process of program planning should include a thorough print review, consideration for workholding, and tools needed to complete the process. It is best practice to take a team approach to part manufacturing.
- Special review of workholding should be considered to maximize machining and minimize setups and operations.
- A CNC mill program consists of an opening statement, body, and closing statement.
- G codes are called preparatory commands and place the machine in a specific mode or state of operation.
- M codes are called miscellaneous functions and serve as "switches" to turn on and off specific machine functions.
- A pocketing toolpath is used when a large volume of material needs to be removed on the internal boundaries of a part.
- The contour toolpath is used when following a single line of geometry or one specific path.
- Canned cycles allow a single line of code to define multiple parameters. They reduce programming time and programmer errors.
- The use of cutter compensation offsets the tool around the finished geometry by an amount stated in the controller. This allows for programming to part dimensions and making minor adjustments to offset cut dimensions.
- The G02 and G03 commands create an arc movement with the tool. The G02 creates a clockwise arc, and the G03 creates a counterclockwise arc.
- An angle movement is created by using two axis locations on the same line of the program. This can be used with any combination of axes that are available on the machine.
- When entering into material with a cutter, an angled or helical move should be used. Direct plunging into material can cause a tool failure or part damage.

Review Questions

Answer the following questions using the information provided in this chapter.

Know and Understand

1. Try to minimize the number of tools you use to reduce _____ and _____.
   A. tool changes, setup work
   B. tool changes, cycle time
   C. errors, offset changes
   D. canned cycles, work offsets

2. What are the three sections of a CNC program?
   A. Beginning, middle, end
   B. Starting, body, end
   C. Opening statement, tool statement, closing argument
   D. Opening statement, body, closing statement

3. *True or False?* Each program number will be preceded by the letter "O."

4. The program address G is commonly referred to as the _____.
   A. preparatory command
   B. address code
   C. opening statement
   D. rapid movement

5. The G code command for a straight-line feed move is the _____.
   A. G03
   B. G00
   C. G01
   D. G72

6. *True or False?* One of the G code designators for work coordinate systems is G73.

7. The program address M is commonly referred to as a _____.
   A. machine function
   B. miscellaneous function
   C. multiple function
   D. multiple fraction

8. Which of the following M codes can be used to create a program stop?
   - A. M03, M04
   - B. M98, M99
   - C. M03, M05
   - D. M00, M01

9. _____ stops the program and rewinds to the beginning.
   - A. M03
   - B. M30
   - C. M00
   - D. M01

10. _____ is a type of toolpath that is typically performed with an end mill and follows a joined path or single piece of geometry.
   - A. Contour
   - B. Pocket
   - C. Spotfacing
   - D. Pecking

11. Which of the following codes are the simplest method to program arcs?
   - A. G81, G82
   - B. G40, G41
   - C. G02, G03
   - D. G84, G85

12. A term for machining inside a closed area or boundary is _____.
   - A. contour
   - B. pocket
   - C. spotfacing
   - D. pecking

13. The G83 is a _____ cycle.
   - A. spot-facing
   - B. chip-breaking
   - C. spot-drilling
   - D. peck-drilling

14. The G84 is a _____ cycle.
   - A. tapping
   - B. drilling
   - C. boring
   - D. pecking

15. The _____, along with an H number designator, sets the tool height offset.
   - A. G40
   - B. G42
   - C. G43
   - D. G80

16. Once a machine's spindle centerline is sitting directly over the X-axis and Y-axis programmed zero locations, that location needs to be recorded into the _____ page.
   - A. tool height
   - B. work offset
   - C. diameter offset
   - D. tool length

Apply and Analyze

Use the following parameters to write some short programs.

1. Using program #1212, a spot drill tool #1, and drilling .25″ deep, write a program to spot drill two holes at X1., Y−1., and X2., Y−2. Spindle speed is 2,500 rpm and feed rate is 10 ipm.

2. Using program #2156, a 1/2″ drill tool #2, and drilling 2″ deep, write a program to chip break peck drill two holes at X1., Y−1., and X2., Y−2. Spindle speed is 3,500 rpm and feed rate is 18 ipm.

3. Using program #3237, a 1/4-20 tap, tool #3, and tapping .70″ deep, write a program to tap two holes at X1., Y−1., and X2., Y−2. Spindle speed is 1,000 rpm and feed rate is 50 ipm.

4. Write the four lines of programming code needed to move the tool from X0., Y0. to X1.75, Y0., make a .25″ clockwise radius, then travel to X2., Y−1.

5. With the #1 tool starting at X−.5 Y0., use cutter compensation left to move to X0. Y0., an angled move to X3. Y1., and then move to X3.5 Y1.5 while turning off cutter compensation.

Critical Thinking

1. Machine programming is often thought to be straightforward, but it is actually full of variables. What are some ways a programmer can ensure that he or she is cutting the part to print specifications in a safe and economical process?

2. In the program planning stage, it is suggested that a team be built to discuss part production. In different manufacturing plants those team members will vary. Why do you think it might be important to get members' feedback from all different levels of production, such as operators, engineers, the sales team, and customers?

3. The programmer has a vital role in final part production. What are some of the character traits, or soft skills, a successful programmer possesses?

10

CNC Lathe Programming

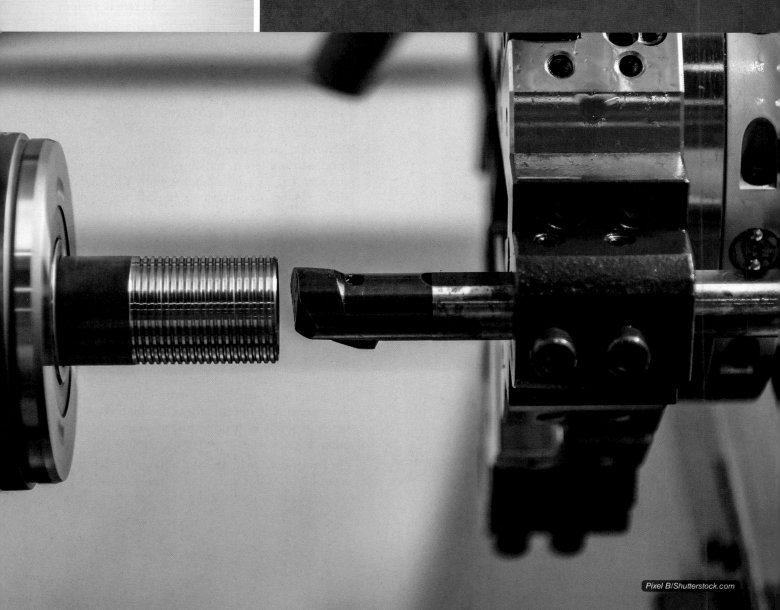

Chapter Outline

10.1 Introduction to CNC Lathe Programming

10.2 Program Planning

    10.2.1 Print Review

    10.2.2 Part Workholding

    10.2.3 Tool Selection

10.3 Program Format

    10.3.1 Opening Statement

    10.3.2 Program Body

    10.3.3 Closing Statement

10.4 Address Codes

    10.4.1 G Codes

    10.4.2 M Codes

10.5 Tool Setting

    10.5.1 Length Offset

    10.5.2 Diameter Offset

10.6 Work Coordinate Offsets

10.7 Tool Nose Radius Compensation

10.8 Profile Cutting

    10.8.1 Programming Angles

    10.8.2 Programming Arcs

10.9 Drilling Canned Cycles

    10.9.1 G81 Drill Cycle

    10.9.2 G82 Spot Drill Cycle

    10.9.3 G83 Peck Drill Cycle (Deep Hole Drill)

    10.9.4 G84 Rigid Tapping Cycle

    10.9.5 Grooving

10.10 Threading Cycles

10.11 Parting Off

Learning Objectives

After reading this chapter, you should be able to:

- Explain how workholding affects lathe programming.
- Describe at least two different roughing and finishing inserts.
- Define the opening statement, body, and closing statement of a CNC lathe program.
- Explain the use of G codes in lathe programming.
- Describe when to use an M code in lathe programming.
- Enter lathe tool offsets on the appropriate offset page.
- Explain the use of tool nose radius compensation in lathe operations.
- Use G70, G71, and G72 canned cycles.
- Program angle moves in a lathe operation.
- Use G2 and G3 arc commands in lathe operations.
- Outline the types of drilling canned cycles and their functions for the CNC turning center.
- Explain the difference between the G32 and G76 threading cycles.
- Define *parting off*.

Key Terms

block-by-block threading

block threading method

cross threading

longhand threading

parting off

profile cutting

tool nose radius compensation

10.1 Introduction to CNC Lathe Programming

The process of CNC lathe programming is similar to CNC mill programming. Many of the canned cycles, G codes, and M codes are identical in the lathe and mill operations. Programming a cutting tool around a piece of material using the Cartesian coordinates is the same in the lathe and the mill. But there are many differences. This section will cover G codes and M codes that can be applied in lathe programming, including some new codes that are specific to lathe programming. We will discuss new canned cycles that are specific to lathe programs, and the workholding issues that are inherent to lathe parts.

This section will discuss programming specific to the Haas- and FANUC-controlled machines. These programs are G and M code programs and compatible with many different types of machines and controllers. The common opinion is that CNC lathe programming is "easier" than mill programming, but it does require a new set of skills to be successful.

10.2 Program Planning

Just like in programming a CNC machining center, the most important steps to creating an optimal CNC turning center program are the planning steps. Consider the machines available to create a specific part, how the material will be held, what tools will be needed, and what are the most effective cutting strategies.

The team approach to programming is just as important in lathe work as it is in mill work. Gathering input from operators, setup personnel, quality control, and the engineering staff can eliminate unnecessary steps and costly mistakes. Take time to study the print and identify critical features and machining challenges. Let your program address these concerns and make an acceptable part the first time, and as many more times as needed.

10.2.1 Print Review

As in Chapter 9, an integral part of the planning process before lathe programming can begin is to assess the design intent, interpreting the print dimensions to gain insight into the form, fit, and function of the finished product. In reviewing any print, it is critical to identify the part's key features and the locations of tolerances. As always, compliance with print specifications determines the acceptability of any part. **Figure 10-1** provides an example of the information that can be gathered by studying a lathe print. The long, dashed line represents the part centerline.

First, notice that all of the linear dimensions—2.125″, 4.125″, and 6.125″—are dimensioned from the left end or "back" of the part. The most critical of the three linear dimensions is 6.125″, with a tolerance of ± .002″. All of the diameters are ± .005″ except for the 5.500″ dimension, which is the closest tolerance at ±.002″. The two dimensions with a ± .002″ tolerance must be considered in workholding and finishing tool selection. The solid lines on the print represent

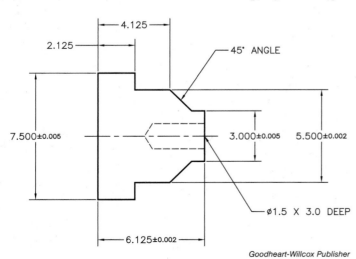

Goodheart-Willcox Publisher

Figure 10-1. A typical lathe part, cylindrical in shape, with multiple diameters, and a drilled center hole feature.

visible features, while the short, dashed line is a hidden feature in this view. **Figure 10-2** shows a solid, isometric view of the part, for better visualization.

At first glance, some linear dimensions appear to be missing. There is no linear dimension from the small diameter of the 45° angle. Because both of the adjoining diameters are defined and the end of the 45° angle is established, that length can be calculated. The hole in the center has a note that states, "Drill 1.5 diameter × 3.0 deep." A standard tolerance will apply to the drill size, and the 3″ depth refers to the depth of the full 1.5″ diameter, not to the tip of the drill.

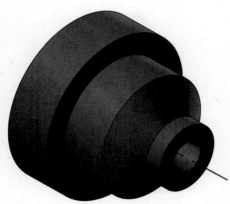

Figure 10-2. A solid model representation of Figure 10-1.

10.2.2 Part Workholding

Workholding is a critical decision that has to be made in the planning phase. If you don't know how you are going to hold the part, there is no way to determine how to machine it. Lathe parts can be held in multiple ways—3-jaw chucks, 4-jaw chucks, collet closers, face plates, and even custom fixtures. The ideal workholding solution provides the most rigidity, the most repeatability, and the most machining possible in each setup. The objective of workholding is to maximize machining and minimize setups or operations.

The part in **Figure 10-3** can be machined in two operations. By placing the solid material in a 3-jaw hydraulic chuck, the areas indicated by black arrows can be machined. Then the part is reversed 180° within the same chuck, and the features marked with green arrows can be finished in the second operation.

10.2.3 Tool Selection

The selection of lathe tooling is fairly straightforward. When cutting the outside diameter (OD) of a part, a roughing tool, finishing tool, groove tool, or threading tool is generally used. When cutting the inside diameter (ID) of a part, a drill, boring bar, groove tool, or ID thread tool should be used. The toolholders themselves are not as important as selecting the correct insert style, grade, and chipbreaker configuration. As discussed in Chapter 8, there are a variety of inserts used for machining, each with different applications.

OD and ID roughing tools are meant to take deeper cuts at high material removal rates. Although they give superior stability and resist fracturing, they will not provide high-quality finishes. **Figure 10-4A** depicts a

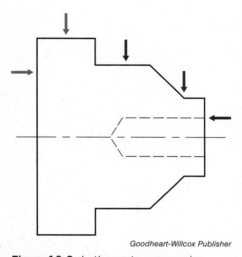

Figure 10-3. Lathe parts can require multiple operations. The colored arrows show one possible machining strategy.

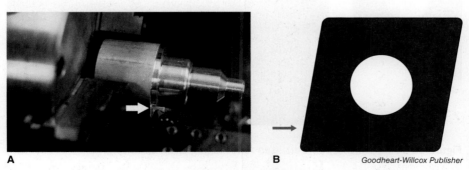

Figure 10-4. CNMG-style insert. A—Arrow shows the trailing support. B—Arrow shows the direction of cutting forces.

Goodheart-Willcox Publisher

Figure 10-5. WNMG-style insert. This insert has six cutting edges, making it more cost efficient.

CNMG-style roughing tool. The arrow shows cutting force direction, and the insert material (carbide) directly behind the cutting edge provides stability. This insert, **Figure 10-4B,** provides four cutting edges, two on the top and two on the bottom. **Figure 10-5** shows a WNMG-style insert used for roughing operations. This insert has six cutting edges, three on the top and three on the bottom. This tool also provides substantial support directly behind the cutting edge.

OD finishing tools do not provide the strength of a roughing tool, but they can take light cuts with superior finish and maintain higher tolerances. **Figure 10-6** shows a VNMG-style finishing insert and **Figure 10-7** shows a DNMG-style insert. Application will be determined by the part requirements and finish profile.

Goodheart-Willcox Publisher

Figure 10-6. VNMG-style insert used for finishing cuts. With little support behind the cutting edge, it will not produce large roughing cuts.

Goodheart-Willcox Publisher

Figure 10-7. A DNMG-style insert used for finishing or undercutting.

10.3 Program Format

After thorough planning, program writing can begin. Just like a mill program, a lathe program will have an opening statement, body, and closing statement, and the code for every tool used within the program will have these same components. This consistent format simplifies the coding process and helps the programmer spot any missing information. This section will cover codes specifically for Haas- and FANUC-controlled machines.

10.3.1 Opening Statement

Each program should have a consistent opening statement based on the machine for which it is written. The opening statement will contain the program number and information pertaining directly to the part being machined.

These are the first few lines of a sample lathe program:

```
O1234
(LATHE PART 1)
(DATE: 12-06-18)
(MATERIAL: STEEL, INCH: 1030, 200 BHN)
G20
G28 U0.W0.
(TOOL: 1 OFFSET: 1)
(OD ROUGH RIGHT: 80 DEG.  INSERT: CNMG-432)
```

In this example, **O1234** is the program number, which will always start with the letter "O." It will be the only letter O used in the program. Every program stored in a machine's memory must have a different program number.

The next lines in a program contain information regarding the part name or number, a program date, and possibly material type. Any information contained in parentheses is not read as code by the machine, rather it is a note intended for the operator or programmer.

After the first sets of parenthetical notes in our sample opening statement, there is a G20 line. The G20 designates that this program and its offsets will be in inches. If this program was created in metric units, the line would read G21. We will work with inch-based prints primarily. Remember that changing this code will also affect the machine coordinate and tool offset pages.

Our next line of code is a G28 U0. W0. This line sends the turning center to its machine home position, or the farthest point from the machine spindle centerline. Unlike machining center programming, no incremental, or G91 code, is necessary. In this G28 line, both the X (U) and Z (W) axes are moving simultaneously. Ensure that this is a safe function, as both axes will retract in an angular move directly to home. In some cases, the tool could collide with the material moving away in this direction. Many programmers prefer to return the X axis first, followed by the Z axis. In this case, a G28 is required for both lines, as follows:

```
G28 U0.
G28 W0.
```

The X and Z are replaced by axis substitution definitions of "U" and "W," so your home positioning move would be G28 U0. W0. The U and W are used because no incremental command is given.

10.3.2 Program Body

As in mill programming, the body of a lathe program will direct a machine through specific tools and precise movements to create a part according to the print. The code for each tool within the program will also have an opening statement, body, and closing statement.

The opening statement for a particular tool might look like this:

```
(TOOL: 1 OFFSET: 1)
(OD ROUGH: 80 DEG.INSERT: CNMG-432)
T0101
G97 S350 M03
G0 X7.3205 Z.2
```

T0101 tool changes to tool #1 and turns on the offset #1. This is different from the mill programming format. There is no separate M code for a tool change; simply calling that tool will force a change.

G97 S350 M03 is the direct revolutions per minute (rpm) input, the speed is 350, and M03 turns on the spindle in a clockwise rotation.

G0 X7.3205 Z.2 moves tool #1 at a rapid rate to the designated X and Z coordinates, 7.3205 and .2 in this example.

The body of this tool's program might look like this:

```
G99 G1 Z.1 F.01
Z-3.99
X7.4375
G2 X7.52 Z-4.0313 R-.0412
```

G99 G1 Z.1 F.01 places the machine in feed rate per revolution and makes a feed move to Z.1 at a rate of .010″ for every complete revolution the spindle makes.

Z–3.99, still in a feed rate, moves the machine to the designated Z coordinate.

X–4.0313, still in a feed rate, moves the machine to the designated X coordinate.

The closing statement for this tool might look like this:

```
G0 X7.57
G28 U0. W0. M05
T0100
```

G0 X7.57 is a rapid feed move of the machine to a safe position.

G28 U0. W0. M05 sends the machine home in X and Z and turns off the spindle.

T0100 cancels offsets for tool #1.

10.3.3 Closing Statement

The end-of-program closing statement contains the last lines of the program. It shuts the program off in a way that leaves the machine ready to run the next part in a safe manner, and in an optimum position to remove the part from the fixture and load new parts. For example:

```
M09
G28 U0. W0. M05
M30
```

M09 turns off coolant.

G28 U0. W0. M05 sends the machine to its home position in both X and Z, and turns the spindle off.

M30 ends the program and rewinds.

The story is now complete. This is the format this text will use going forward to create all programs.

10.4 Address Codes

Recall from Chapter 9 that address codes are single-letter characters at the beginning of each command that define what the computer should do with the numerical data that follows. They are used to designate commands or machine functions. **Figure 10-8** lists the available address codes and their functions.

 **Safety Note**

Know where the emergency stop buttons are located. One will always be located on the control panel, and others may be found in various locations on the machine. The emergency stop brings the program to an immediate halt.

Address Codes for Lathe Programming	
Address Code	**Function**
A	Angle of thread for G76
C	Chamfer or axis rotation of chuck
D	Depth of cut in G71 and G72; depth of first thread in G76
E	Precision feed rate in threading
F	Feed rate
G	Preparatory command, as in G01
L	Loop count repeat for canned cycles
N	Block number
M	Machine or miscellaneous function
O	Program number
P	Dwell time or start block # in G71 and G72
Q	End block # in G71 and G72
R	Arc radius
S	Spindle speed
T	Tool selection number
U	Alternate X axis
W	Alternate Z axis
X	X-axis motion
Z	Z-axis motion

Goodheart-Willcox Publisher

Figure 10-8. Available address code list for CNC lathe operations.

10.4.1 G Codes

Recall from Chapter 9 that the program address G, or preparatory command, puts the CNC control in a specific mode or state of operation. For example, G00 places the machine in a rapid mode, and G01 places the machine in a feed rate mode. And G codes can command canned cycles, or abbreviated multimovement cycles, such as G71, which commands a roughing turning cycle but does not produce any machine movement.

The table in **Figure 10-9** lists the most commonly used G codes in lathe programming, which can be used to produce most programs. There are several more G codes available, including machine-specific G codes for special functions. For specific codes for different controls, refer to the user's manual for that particular machine.

G Codes for Lathe Programming	
G Code	**Function**
G00	Rapid motion
G01	Linear motion; or Linear motion, chamfer, and corner rounding – modal
G02	CW circulation interpolation motion – modal
G03	CCW circular interpolation motion – modal
G04	Dwell (P) defined in milliseconds
G10	Programmable offset setting
G20	Inch coordinate system – modal
G21	Metric coordinate system – modal
G28	Rapid to machine zero point
G31	Feed until skip function – optional
G32	Thread cutting path – modal
G40	Tool nose compensation cancel
G41	Tool nose compensation, left – modal
G42	Tool nose compensation, right – modal
G50	Spindle speed maximum rpm limit (S) – modal
G52	Global work offset coordinate system shift – modal
G53	Machine zero positioning coordinate
G54	Work offset positioning coordinate #1 – modal
G55	Work offset positioning coordinate #2 – modal
G56	Work offset positioning coordinate #3 – modal
G57	Work offset positioning coordinate #4 – modal
G58	Work offset positioning coordinate #5 – modal
G59	Work offset positioning coordinate #6 – modal
G70	Finishing cycle
G71	OD/ID stock removal cycle
G72	End face stock removal cycle
G73	Irregular path stock removal cycle
G74	Face grooving OR high-speed peck drill cycle
G75	Peck grooving cycle OD or ID
G76	Threading cycle, multiple pass OD/ID

(continued)

Figure 10-9. A list of G codes used in lathe programming. There are more codes and some of these can be machine-controller specific.

G Codes for Lathe Programming *(continued)*	
G Code	**Function**
G81	Drill canned cycle – modal
G82	Spot drill/counterbore canned cycle – modal
G83	Peck drill deep hole canned cycle – modal
G84	Tapping canned cycle – modal
G92	Threading cycle – modal
G94	End facing cycle – modal
G96	Constant surface speed (CSS) – modal
G97	Constant nonvarying spindle speed, CSS off (S)
G98	Feed per minute (F) – modal
G99	Feed per revolution (F)

Goodheart-Willcox Publisher

Figure 10-9. *(Continued)*

The *Function* column of **Figure 10-9** can help determine when to use each G code. Categorizing G codes by function and using them to create some sample programs will help clarify the programming process.

- G00, G01, G02, and G03 are all movement commands: rapid, straight lines, and arcs.
- G20 and G21 define standard or metric inputs.
- G28 sends the machine home.
- G32, G76, and G92 are various OD or ID threading patterns.
- G40, G41, and G42 are commands for *tool nose radius compensation*. In milling we define the tool diameter or radius; in lathes we only compensate for the radius on the nose of the tool.
- G54–G59 are work coordinate offsets, used in a lathe to define multiple part locations.
- G70–G84 are canned cycle commands.
- G50 sets maximum allowable spindle speed.
- G96 sets constant surface speed (CSS).
- G97 sets fixed spindle speed.
- G98 sets feed distance per minute—how far the tool travels in one minute.
- G99 sets feed distance per every revolution of the spindle.

10.4.2 M Codes

The program address M, or machine function, acts as a switch. Although many M codes are the same for lathe and mill, some M codes are specific to machine type. As with mills, many lathe machines also come with "open M codes" that can be tapped into for optional equipment. Unlike G codes, only one M code can be used on each line of a program.

Figure 10-10 lists the most commonly used lathe programming M codes and their functions, which can be used to produce most programs. There are several more M codes available, including machine-specific M codes for special functions. For specific codes for different controls, refer to your user's manual.

M Codes for Lathe Programming	
M Code	**Function**
M00	Program stop
M01	Optional program stop
M03	Spindle on forward
M04	Spindle on reverse
M05	Spindle off
M08	Coolant on
M09	Coolant off
M10	Chuck open
M11	Chuck close
M21	Tailstock advance
M22	Tailstock retract
M30	Program end; return to start
M98	Subprogram call
M99	End of subprogram or program restart

Goodheart-Willcox Publisher

Figure 10-10. A list of M codes used in lathe programming. There are more codes and some of these can be machine-controller specific.

As with G codes, the list of M codes can be sorted by functional category to aid in programming.

- M00 is a mandatory program stop used for forcing a stop midcycle.
- M01 is an optional stop turned on and off by a switch.
- M03, M04, and M05 control spindle on and off; M03 will be used in most cases.
- M10 and M11 are used to clamp and unclamp chuck jaws. These will only work with the spindle stopped.
- M30 stops the program and rewinds to the beginning.
- M98 sends the machine to a subprogram in conjunction with a P address (see **Figure 10-8**).
- M99 rewinds and reruns the program; often called a loop.

10.5 Tool Setting

Tool setting in a turning center is much different than in a machining center. First, a turning center involves only two axes. Second, all lathe parts turn or spin around a common centerline, which can be used to the programmer's advantage. The most common tool setting method, and the one we will use to make our sample programs, is to set all tools to the spindle centerline in X and to use the face of the finished part as our Z-axis "0." **Figure 10-11** shows our sample part and the tool located at its X0., Z0. location.

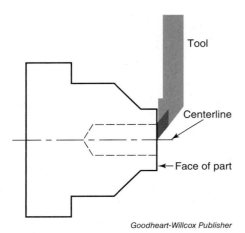

Figure 10-11. The most common position of the X0. is the centerline of the spindle, and Z0. is the finished face of the part.

10.5.1 Length Offset

The tool length offset measure, or Z-face measure, establishes a position for each tool from the machine home position to the intersection of the tool face and part face. The position in the Z axis as the tool is touching the material is the Z0. See **Figure 10-12**. The practical application for setting the Z offset is to make a light cut on the part material face to establish the part length.

Without changing the Z-axis position, press the Z-face measure key on the controller, **Figure 10-13**. This will record the current Z position for the active tool. Each subsequent tool will need to be touched to the same face to set its Z face.

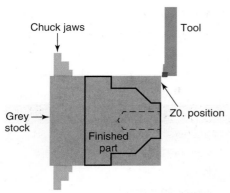

Figure 10-12. Making a small cut on the part face can be used to establish the Z0. tool position.

10.5.2 Diameter Offset

The diameter offset establishes the X-axis coordinate for each tool. In **Figure 10-11**, we showed that the X0. position for every tool will be the centerline of spindle rotation. To establish this position, we will make a light cut on the outside diameter. Without moving the tool position in the X axis, measure the diameter of that small cut, **Figure 10-14**.

In the Haas controller in **Figure 10-13**, push the X-diameter measure key. Other controllers may use a slightly different designation, but the process will be similar. When the *X Dia Mesur* button is depressed, a question appears on the screen asking for the diameter of the part, at which point the diameter measurement in **Figure 10-14** should be entered. The machine controller will calculate the current position to the spindle centerline and enter that data in the X-offset page for the active tool. This process can also be performed for inside diameter (ID) tools by drilling a hole in the part center, making a cut with the boring bar, and recording that diameter.

Figure 10-13. A Haas CNC controller has a *X Dia Mesur* button and *Z Face Mesur* button to simplify tool offsetting.

10.6 Work Coordinate Offsets

Early CNC turning centers did not have the same type of work coordinate system as CNC mills—the work coordinates were just the tool offsets from the home position. But creative programmers saw opportunities to make multiple parts in a single setup, and controller manufacturers implemented additional features to handle this, like a G10 work shift. The G10 allows you to make a single part and then shift all of the tools down the Z axis to make a second part.

Most current machines and controllers have implemented a simpler work coordinate system, similar to the machining center. In most cases this G54–G59 work coordinate system won't be used, unless making multiple parts—but it is possible to use it to solve other machining situations. For example, it is possible to set all tool offsets at the spindle face, or any known position, and then use a G54–G59 work coordinate to shift those coordinates to the desired Z-axis starting position. This is similar to milling tool offsets and can be a good technique if the lathe tools remain in the machine for multiple different parts. **Figure 10-15** shows how we can make multiple parts out of bar stock.

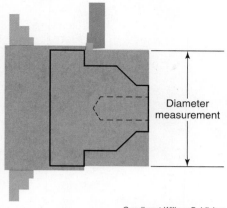

Goodheart-Willcox Publisher

Figure 10-14. A cut on the outside diameter can be used to establish the X-axis position. Enter the X diameter into the controller after the *X Dia Mesur* button is pushed.

10.7 Tool Nose Radius Compensation

The G41 and G42 codes command a compensation for the tool orientation and the radius on the tool tip. In milling, G41 and G42 were referred to as *cutter compensation* and *offset* for the tool diameter and cutter direction. It is similar in lathe programming, but cylindrical tools are not used in lathes. In lathe operations the controller needs to know on what side of the cut the tool is located as well as the tool nose radius. If tool nose radius is not used in a lathe program, both angles and radii will not be finished as programmed. It is similar to a milling operation, where if the tool diameter or radius is not entered, it alters the toolpath's finished cut dimensions.

G41 is tool nose compensation left. "Left" refers to the side of the machining process on which the toolholder is located, **Figure 10-16**. It is often used for ID features or when boring. Because of the multiple configurations of lathes—like dual spindle, dual chuck, vertical, or horizontal

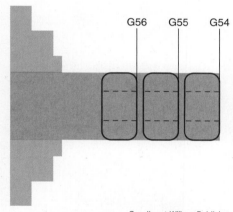

Goodheart-Willcox Publisher

Figure 10-15. Variable work coordinates can be used to machine multiple parts in one operation. The tool offset page will contain each Z-axis tool position and the X-axis offsets will remain the same.

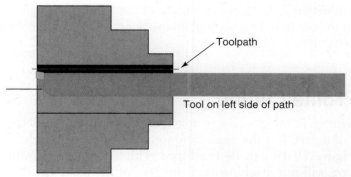

Goodheart-Willcox Publisher

Figure 10-16. A G41 tool nose radius compensation indicates the tool is on the left side of the machining direction.

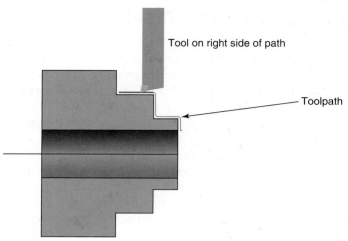

Tool on right side of path

Toolpath

Figure 10-17. A G42 tool nose radius compensation indicates the tool is on the right side of the machining direction.

machining centers—using the appropriate G41 or G42 code for the operation is necessary.

G42 is tool nose compensation right. In this case the tool is located on the right-hand side of the tool cutter path, **Figure 10-17.** This is often used for OD features, but machine variations or cutting directions can allow for tool nose compensation in either direction for OD operations. Visualize the cutting direction and determine if the toolholder is on the left- or right-hand side of the cut.

When setting tool offsets in a turning center, the tool offset page contains a column for tool nose radius. Simply insert the radius of the tool in that column and the machine controller will do the math calculations from there.

From the Shop

Walking the Path

Sometimes it's hard to visualize the right- or left-hand side of the path and choose the G41 or G42 code. A simple trick to determine the direction is to "walk the path." Imagine you are walking right on top of the material as it is being cut. You should be facing the same direction as the tool is traveling. Is the tool in your right hand, or is it in your left? That will determine right- or left-hand tool nose compensation.

10.8 Profile Cutting

Profile cutting is the process to machine the outside diameter or inside diameter of a part. It is the most common CNC lathe toolpath. The sample part in **Figure 10-18** can be machined with a simple program. For this example we will not machine the front face of the part, just the 1″ and 2″ outside diameters.

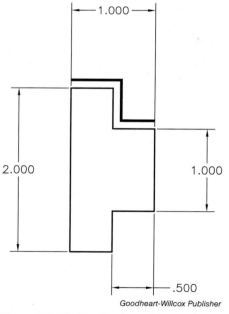

Figure 10-18. The final machine path will follow the part profile.

O3827	Program Number
(LATHE CONTOUR EX 1)	
(MATERIAL: 1030, 200 BHN)	
G20	Inch Mode
(TOOL 1: 80 DEG.INSERT: CNMG-432)	
G28 U0. W0.	Send Machine Home
G0 **T0101**	Call Tool 1 offset 1
G97 S402 M03	Spindle on forward
G0 X2.2 Z.5 M8	Safe position move
G42 X1.9 Z.2	TNR Comp on right
G50 S3000	Max Spindle Speed
G96 S200	CSS 200
G1 Z.1 F.01	
Z-.5	
X2.	Program Body next 50 lines
Z-1.05	
X2.2	
G0 Z.2	
X1.8	
G1 Z.1	
Z-.5	
X1.92	
X2.12	
G0 Z.2	
X1.7	
G1 Z.1	
Z-.5	
X1.82	
X2.02	
G0 Z.2	
X1.6	
G1 Z.1	
Z-.5	
X1.72	
X1.92	
G0 Z.2	
X1.5	
G1 Z.1	
Z-.5	
X1.62	
X1.82	
G0 Z.2	
X1.4	
G1 Z.1	
Z-.5	

X1.52	
X1.72	
G0 Z.2	
X1.3	
G1 Z.1	
Z-.5	
X1.42	
X1.62	
G0 Z.2	
X1.2	
G1 Z.1	
Z-.5	
X1.32	
X1.52	
G0 Z.2	
X1.1	
G1 Z.1	
Z-.5	
X1.22	
X1.42	
G0 Z.2	
X1.	
G1 Z.1	
Z-.5	
X1.12	
X1.32	
G40 X2.025	TNR Comp off
M9	
G0 G28 U0. W0. M05	Send Home spindle off
T0100	Tool Offset off
M30	Program Rewind

Even machining a simple part with two diameters and one step takes quite a few lines of code. Imagine the code for a complex part with multiple tools. But there is a way to make the program much shorter and reduce the chance for errors: the G70, G71, and G72 profiling canned cycles. G71 and G72 are the roughing cycles, and G70 is the finishing cycle. In the next example, the same part as **Figure 10-18** is programmed using G71. The G71 lines of code are broken out here:

G71 P100 Q110 U.02 **W.01** D.05 F.01
N100 G0 G42 X1. S200
G1 Z-.5
X2.
Z-1.05
N110 G40 X2.1

- **G71** turns on the roughing profile cycle.
- **P100** is the starting line number, or N, of the cycle. Notice the P100 and N100.
- **Q110** is the ending line number, or N, of the cycle. Notice the Q110 and N110.

 Tech Tip

The P and Q cannot be the same number, and all lines of code that define the profile are contained between those two line numbers. The line numbers in the program will start with the letter "N."

- **U.02** defines the amount of stock to leave in the X axis (.020″), or the radial amount.
- **W.01** defines the amount of stock to leave in the Z Axis (.010″).

 Tech Tip

As this is a roughing cycle, material is left for the finish pass.

- **D.05** is the depth of cut (.050″). This is radial, so this cuts a .100″ diameter.
- **F.01** is the feed rate in roughing (.010″ per revolution).

Here is the entire program:

```
O3287
(LATHE CONTOUR EX 1 G71)
(STEEL INCH: 1030, 200 BHN)
G20
G28 U0. W0.
(TOOL 1 80 DEG. INSERT: CNMG-432)
G0 T0101
G97 S402 M03
G0 X2.1 Z.15
G50 S3000
G96 S200
G71 P100 Q110 U.02 W.01 D.05 F.01
N100 G0 G42 X1. S200
G1 Z-.5
X2.
Z-1.05
N110 G40 X2.1
G0 Z.15
G28 U0. W0. M05
T0100
M30
```

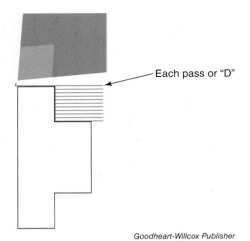

— Each pass or "D"

Goodheart-Willcox Publisher

Figure 10-19. Each cut or "pass" must be programmed as a single line of code. The illustration shows each pass the tool creates.

Using the roughing profile cycle, the program has been reduced from 76 lines to only 21 lines. Even better, to make an edit to depth of cut, simply change the "D.05" instead of searching the entire program and making many edits. **Figure 10-19** shows a visual of the cutting action of those 21 lines. Note that we only defined the finish profile, and the controller calculated all of the moves.

G71 greatly simplifies the programming of more complicated parts. For example, **Figure 10-1** (found earlier in the chapter) can be programmed using a G71 as follows:

```
O3288
(LATHE PART 1 G71)
(STEEL INCH: 1030, 200 BHN)
G20
G28 U0. W0.
(TOOL: 21 OFFSET: 21)
(35 DEG.  INSERT: VNMG-431)
G0 T2121
G97 S102 M03
G0 X7.5 Z0.
G50 S3000
G96 S200
G71 P100 Q110 U.02 W.01 D.05 F.005
N100 G0 G42 X-.2 S200
G1 X3.
Z-.7592
X5.5 Z-2.0092
Z-4.
X7.5
N110 G40 Z-6.125
G0 Z0.
G28 U0. W0. M05
T2100
M30
```

The line-by-line, long form of that program is about 300 lines, and the G71 version is 25. It is much easier to program this roughing cycle.

The G70 finishing profile cycle provides similar efficiency. It will follow the same part profile that was used in the G71 cycle. Needing only the line numbers (N) of those parameters, the G70 line is as simple as:

```
G70 P100 Q110
```

The G70 cycle will follow at the end of the G71 cycle as shown:

```
O3287
(LATHE CONTOUR EX 1 G71)
(STEEL INCH: 1030, 200 BHN)
G20
```

```
G28 U0. W0.
(TOOL 1 80 DEG.  INSERT: CNMG-432)
G0 T0101
G97 S402 M03
G0 X2.1 Z.15
G50 S3000
G96 S200
G71 P100 Q110 U.02 W.01 D.05 F.01
N100 G0 G42 X1. S200
G1 Z-.5
X2.
Z-1.05
N110 G40 X2.1
G70 P100 Q110        Finishing Cycle
G0 Z.15
G28 U0. W0. M05
T0100
M30
```

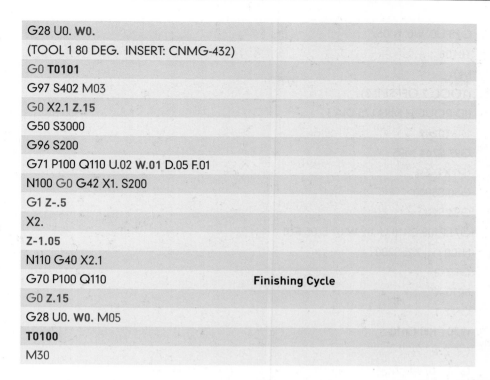

Figure 10-20. A sample lathe part with an outside diameter and a drilled and bored hole.

The G70 and G71 cycles are not just for OD cutting. They can also be used for ID boring operations. **Figure 10-20** shows the print for a simple bushing that will be made from solid stock and then bored to 2″ diameter.

The direct approach to machining this bushing is to drill a hole through the center and then use a boring bar with a G71 command to complete it. The only differences from the OD cut are the positioning of the tool before activating the cycle at the diameter of the drilled hole and that the "U" address in the cycle needs to be a negative number. This allows for excess stock to be left in the X– (negative) direction for a finish pass. See **Figure 10-21.**

Here is the program:

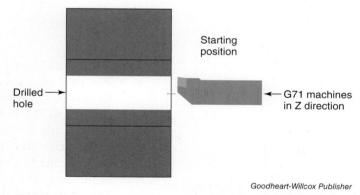

Figure 10-21. On a G71 bore cycle, the tool is positioned in the X axis below the hole diameter, and in a Z position in front of the part face.

```
O3289
(G71 BORE EXAMPLE)
(STEEL INCH: 1030, 200 BHN)
G20
G28 U0. W0.
(TOOL: 1 OFFSET: 1)
(DRILL 1. DIA.)
G0 T0101
G97 S200 M03
G0 X0. Z.25
Z.1
G99 G1 Z-3.5 F.01
G0 Z.25
```

```
G28 U0. W0. M05
T0100
M01
(TOOL: 2 OFFSET: 2)
(ID ROUGH MIN. .75 DIA.)
G0 T0202
G97 S764 M03
G0 X1. Z.1
G50 S3000
G96 S200
G71 P100 Q110 U-.02 W.01 D.05 F.01
N100 G0 X2. S200
G1 Z-3.
N110 X1.
G0 Z.1
G70 P100 Q110
G0 Z.1
G28 U0. W0. M05
T0200
M30
```

The G71 canned cycle specifically drives the tool in a Z– direction. The tool positions itself incrementally down in X and then makes all of its cuts in Z. Often it makes more sense to cut in the X direction, in which case the G72 canned cycle can be used. G72 operates exactly the same as a G71, only it cuts in the X direction. **Figure 10-22** shows a part where it makes sense to cut in the X-axis direction. Using G71, the tool would need to make many short passes in the Z– direction to take the 8″ stock down to the desired 1″ diameter. If the G72 cycle is used it will require far fewer passes, or travel distance, than using the G71. G72 will reduce cycle time significantly.

Here is the program using G72:

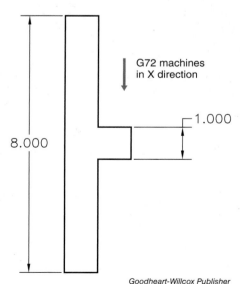

Goodheart-Willcox Publisher

Figure 10-22. The G72 roughing cycle is used to cut in the X-axis direction, as opposed to the G71's Z-axis direction.

```
O3290
(PROGRAM NAME: LATHE PART G72)
(MATERIAL: 1030, 200 BHN)
G20
G28 U0. W0.
(TOOL: 1 OFFSET: 1)
(OD ROUGH RIGHT: 80 DEG.  INSERT: CNMG-432)
G0 T0101
G97 S96 M03
G0 X8.1 Z.1 M8
```

```
G50 S3000
G96 S200
G72 P100 Q110 U.02 W.01 D.05 F.01
N100 G0 G42 X1.0625 S200
G1 Z-1.
N110 G40 X7.9375
G0 Z-.0313
M9
G70 P100 Q110
G0 Z.1
G28 U0. W0. M05
T0100
M30
```

10.8.1 Programming Angles

Programming angles in a CNC turning center is similar to programming them in a CNC machining center. Angles can be machined using the G71 and G72 canned cycles, or they can be created through direct point-to-point programming. The angle cut is programmed simply by positioning the tool at the start of the angle and then programming an X and Z positioning move to the end of that angle. **Figure 10-23** provides an example of point-to-point programming.

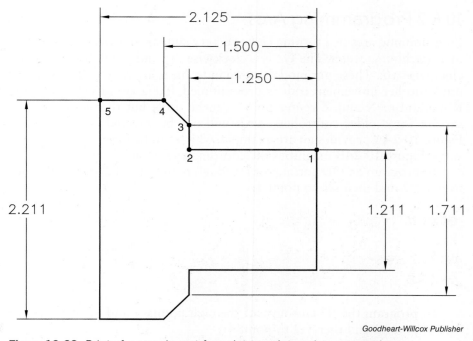

Goodheart-Willcox Publisher

Figure 10-23. Print of a sample part for point-to-point angle programming.

For our sample we will start at position 1, and move the tool to each position until we reach position 5.

O0921 (PROGRAM NAME: ANGLE SAMPLE)	
G20	
G28 U0. W0.	
(TOOL: 1 OFFSET: 1)	
(OD ROUGH RIGHT: 80 DEG. INSERT: CNMG-432)	
G0 T0101	
G97 S96 M03	
G0 X1.250 Z.1 M8	
G50 S3000	
G96 S200	
G01 X1.211 Z0. F.01	Point 1
Z-1.25	Point 2
X1.711	Point 3
X2.211 Z-1.5	Point 4 – Angle Move
Z-2.125	Point 5
G0 X2.3 Z.1	
G28 U0. W0. M05	
T0100	
M30	

Note that we have moved the tool to position 3 at the start of the angle and then programmed an X and Z move on the same line to create the angle.

10.8.2 Programming Arcs

Programming arcs in a turning center is very similar to programming them in a machining center. The G2 is a clockwise arc and the G3 is a counterclockwise arc. These are modal commands, meaning they will stay active until another movement code is programmed, like a G0 or G1. The I and K method of X and Z center points can be used, but this is only necessary in some older controllers, so it will not be covered in this section. **Figure 10-24A** provides an arc example, which can be programmed in two ways. **Figure 10-24B** magnifies the coordinates for each point.

To program an OD cutting tool through points 1 and 2, cut up the face to point 2 and then G3 to point 1:

G1 X0. Z0.
X4.5
G3 X5. Z-.5 R.5
G1 Z-3.05

To program the ID cutting tool through points 3 and 4, cut from the face at point 3 and then G2 to point 4:

G1 X2.5 Z0.
G2 X2. Z-.5 R.5
G1 Z-3.05

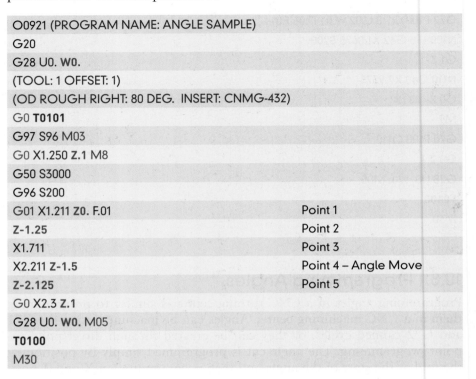

Arc example

A

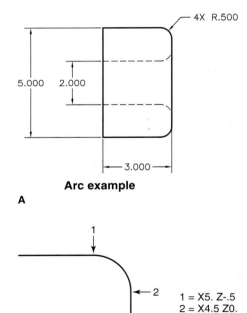

1 = X5. Z-.5
2 = X4.5 Z0.
3 = X2.5 Z0.
4 = X2. Z-.5

Arc points

B   *Goodheart-Willcox Publisher*

Figure 10-24. Programming arcs. A—The lathe part that requires arc programming in two positions. B—The X and Z locations of the start and end of each radii in the sample part.

10.9 Drilling Canned Cycles

Drilling a hole through the center of a part is a frequent operation on a lathe, made less complex with the advent of the CNC turning center and advancements in machining, **Figure 10-25**. Drilling canned cycles for the CNC turning center are very similar to the CNC mill canned cycles.

10.9.1 G81 Drill Cycle

The G81 drill cycle can be used for a spot drill or for another drilling operation. The drill simply travels to a Z depth and then returns. Remember, X0. is spindle centerline for all drill cycles, so all drill programs are completed at X0. The components of a G81 cycle are as follows:

- **G81**: calls drill cycle
- **F**: feed rate in inches per revolution
- **R**: rapid approach plane distance in Z
- **Z**: final Z depth of drill

Here is a short example of a G81 program:

Figure 10-25. A lathe performing a basic drilling operation.

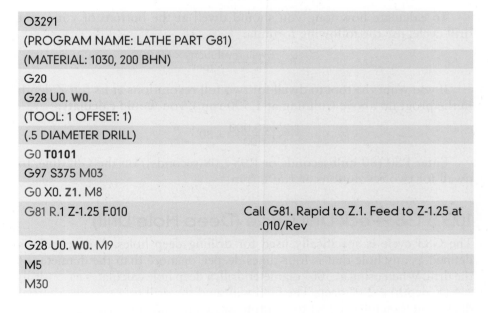

```
O3291
(PROGRAM NAME: LATHE PART G81)
(MATERIAL: 1030, 200 BHN)
G20
G28 U0. W0.
(TOOL: 1 OFFSET: 1)
(.5 DIAMETER DRILL)
G0 T0101
G97 S375 M03
G0 X0. Z1. M8
G81 R.1 Z-1.25 F.010          Call G81. Rapid to Z.1. Feed to Z-1.25 at
                              .010/Rev
G28 U0. W0. M9
M5
M30
```

10.9.2 G82 Spot Drill Cycle

The G82 spot drill cycle is very similar to the G81. The difference is an added dwell once the Z depth is achieved. This is helpful in a spot drill application or when using a chamfering tool, where you don't want the tool to hit the Z depth and instantly retract. This cycle will allow the tool to reach Z and then dwell for a short amount of time to allow the spindle to rotate around the tool, improving surface finish quality. The components of a G82 cycle are as follows:

- **G82**: calls spot drill cycle
- **F**: feed rate in inches per revolution
- **R**: rapid approach plane distance in Z
- **Z**: final Z depth of drill
- **P**: dwell time in milliseconds

Here is a short example of a G82 program:

```
O3291
(PROGRAM NAME: LATHE PART G82)
(MATERIAL: 1030, 200 BHN)
G20
G28 U0. W0.
(TOOL: 1 OFFSET: 1)
(1. DIAMETER CHAMFER TOOL)
G0 T0101
G97 S1500 M03
G0 X0. Z1. M8
G82 R.1 Z-.375 F.010 P80          Call G82. Rapid to Z.1. Feed to Z-.375 at
                                   .010/Rev. Dwell for 80 milliseconds
G28 U0. W0. M9
M5
M30
```

To calculate how long you should dwell at the bottom of your spot drill cycle, use the following formula:

$$P = \frac{\text{Dwell revolutions} \times 60{,}000}{\text{rpm}}$$

If you want the tool to dwell for two full revolutions at its full Z depth in the program above (running at 1,500 rpm), you would calculate:

$$\frac{2 \times 60{,}000}{1{,}500} = 80$$

Enter P80 (80 milliseconds or P.08 [.08 seconds]) on the G82 line, to dwell for two revolutions at 1,500 rpm.

10.9.3 G83 Peck Drill Cycle (Deep Hole Drill)

The G83 cycle is specifically used for drilling deep holes. A deep hole is defined as any hole that is four times deeper, or more than the diameter of the drill. When using a .500″ diameter drill, a deep hole calculates as .5 (drill) × 4 (X depth) = 2.0″ deep. The G83 will send the drill to a specified partial depth and then fully retract to clear out any chips and allow coolant to enter the hole. It will repeat that pecking cycle until the final Z depth is achieved.

 Tech Tip

Do not use this cycle with solid carbide drills. They can fracture when reentering the cut. The G81 cycle is preferred for carbide drills.

The components of a G83 cycle are as follows:

- **G83:** calls peck drill cycle
- **F:** feed rate in inches per revolution
- **R:** rapid approach plane distance in Z
- **Z:** final Z depth of drill
- **Q:** peck depth, incremental

Here is a short example of a G83 program:

O3292	
(PROGRAM NAME: LATHE PART G83)	
(MATERIAL: 1030, 200 BHN)	
G20	
G28 U0. W0.	
(TOOL: 1 OFFSET: 1)	
(.5 DIAMETER DRILL TOOL)	
G0 T0101	
G97 S500 M03	
G0 X0. Z1. M8	
G83 R.1 Z-2.25 F.010 Q.2	Call G83. Rapid to Z.1. Feed to Z-2.25 at .010/Rev. Retract at .200 deep)
G28 U0. W0. M9	
M5	
M30	

10.9.4 G84 Rigid Tapping Cycle

The G84 cycle allows the use of a conventional tap in a CNC turning center. This can be a fast and economical option, but tapping causes tremendous tool stress, and breaking a tap in an existing hole is always a concern. The use of a sulfur-based tapping fluid and a floating tool-holder can alleviate some of the stress. Calculating precise feed rates is critical in good thread formation. The components of the G84 cycle are as follows:

- **G84:** calls rigid tap cycle
- **F:** feed rate in inches per revolution
- **R:** rapid approach plane distance in Z
- **Z:** final Z depth of tap

Feed rate is calculated as 1/pitch or TPI of tap. In a 1/4-20 tap, that calculation is 1/20 = .050 feed. For a 5/16-18 tap the calculation is 1/18 = .0556 feed. This formula forces the tap to advance one tooth per every revolution of the spindle. Here is an example of a G84 cycle:

O3293
(PROGRAM NAME: LATHE PART G84)
(MATERIAL: 1030, 200 BHN)
G20
G28 U0. W0.
(TOOL: 1 OFFSET: 1)
(5/16-18 TAP)
G0 T0101
G97 S200 M03
G0 X0. Z1. M8

G84 R.1 Z-.5 F.0556	Call G84. Rapid to Z.1. Feed to Z-.5 at .0556/Rev)
G28 U0. **W0.** M9	
M5	
M30	

10.9.5 Grooving

Making OD grooves to accommodate seals, snap rings, or thread relief cuts is another common operation in the turning center. **Figure 10-26** shows an example of a print requiring an OD groove.

The groove is located .900″ from the front face, is .200″ wide, and the finish X dimension is 2.25″. Best practice advises against using a grooving tool that is the same size as the desired groove. It will leave a rough finish and jeopardize tolerance. Instead, use a narrower grooving tool and make multiple cuts. To create the groove in **Figure 10-26**, a grooving tool that is .125″ wide, **Figure 10-27**, can be used to make multiple cuts to achieve the .200″ width.

The issue is that the tool was zeroed out on the left-hand side as Z0., but the dimension is from the right-hand side with a width. **Figure 10-28** shows the tool being offset from the part face. Remember to consider the .125″ cutter width. When cutting at Z-1.025, the right-hand side of the groove is cut at .900″ from the part end.

Here is an example of a grooving cycle program:

(TOOL: 1 OFFSET: 1)	
(.125 WIDE GROOVE TOOL)	
G0 **T0101**	
G97 S500 M03	
G0 X3.1 **Z.1 M8**	
Z-1.065	This is the center of the groove
G1 X2.25 F.01	Feed down to X diameter
X3.1	Feed back out of part
Z-1.1	Move to finished Z position, left-hand side
X2.25	Feed down to X diameter

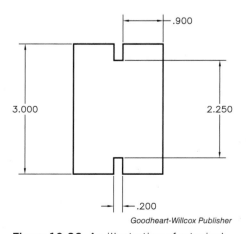

Figure 10-26. An illustration of a typical lathe part requiring a single grooving operation and program.

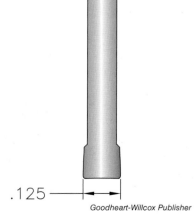

Figure 10-27. A .125″-wide grooving insert used to create part grooves or material part off.

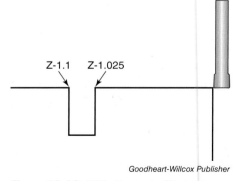

Figure 10-28. With the tool offset set on the part face, take into consideration the insert width when creating grooves or part off pieces.

X3.1	Feed back out of part
Z-1.025	Move to finished Z position, right-hand side
X2.25	Feed down to X diameter
X3.1	Feed back out of part

10.10 Threading Cycles

It is mandatory to use a threading canned cycle when trying to create either OD or ID threads. Programming straight moves will not allow the tool to synchronize with the rotary encoder signal and the tool will not reenter the threads at the same starting position. This is called *cross threading*.

One form of thread cutting is single point block programming, used when calculating each and every pass the threading tool will follow. This is called *longhand threading*, *block-by-block threading*, or simply the *block threading method*. This method takes many lines of code and substantially increases the opportunity for programmer error. On the plus side, it gives the programmer complete control of the programmed paths and depths of cut. It also gives the advantage of cutting tapered or straight threads with one easy cycle.

The longhand threading cycle uses the G32 canned cycle. The following sample program will cut 1/2-13 OD class 2A threads with a minor diameter of .404". A G97 for a fixed spindle rpm and a G99 for feed per revolution must be used. It is critical to have the same starting position in the Z axis for each cut so that the thread begins at the same position. The example will cut .020" deep with each cut for ease of visualization.

(TOOL: 1 OFFSET: 1)
(OD 60° THREADING TOOL)
G0 **T0101**
G97 S500 M03
G0 **X.6 Z.3** M8
X.48 (1ST DEPTH)
G32 Z-1. F.077 (1/TPI)
G0 X.6
Z.3
X.46 (1ST DEPTH)
G32 Z-1. F.077
G0 X.6
Z.3
X.44 (2ND DEPTH)
G32 Z-1. F.077
G0 X.6
Z.3
X.42 (3RD DEPTH)
G0 X.6
Z.3
X.404 (4TH DEPTH)
G32 Z-1. F.077

```
G0 X.6
Z.3
G28 U0. W0.
M30
```

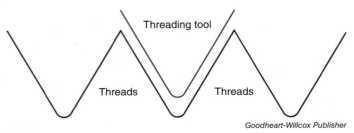

Figure 10-29. The V shape of a threading insert will engage more surface area as it travels down in the X axis.

Goodheart-Willcox Publisher

The G32 longhand threading cycle takes multiple lines of code, increasing the opportunity for error. To make these tapered threads, simply add an X diameter on each G32 line to signify the diameter along the taper.

A simpler cycle, requiring only a single line of code, is the G76 threading cycle, although it does reduce some flexibility for the programmer. Another benefit of the G76 cycle is that it has a decreasing depth of cut. Because of the threading tool's triangular shape, as it gets deeper into the threads it engages more surface area of the cutting tool, creating more tool pressure. See **Figure 10-29**.

The controller will calculate these changing depths of cuts based on the first cut depth and overall height of the thread. The components of the G76 cycle are as follows:

- **G76**: calls multiple repetitive thread cycle
- **X**: starting X diameter, **Figure 10-30**
- **Z**: final Z position of thread, **Figure 10-30**
- **K**: incremental or radial height of thread, **Figure 10-30**
- **D**: depth of 1st pass
- **F**: feed rate calculated at 1/TPI

The sample program below again cuts our 1/2-13 OD thread with a .404″ minor diameter. The starting X position can be .500. The final Z position will be Z-1.00. The K height will be .500 –.404/2, because this is a radial not diametric height, or .048″. The depth of the first pass will be .007″. The feed rate is calculated as 1/13 or .077.

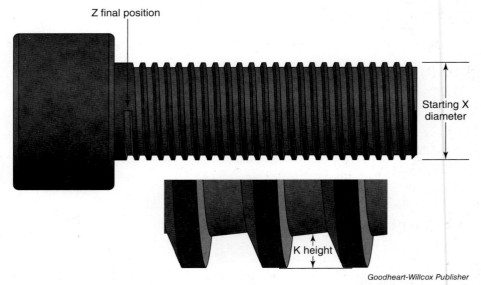

Goodheart-Willcox Publisher

Figure 10-30. The position for the programmed final Z position, the X starting diameter, and the K value needed to program a G76 thread operation.

(TOOL: 1 OFFSET: 1)	
(OD 60° THREADING TOOL)	
G0 **T0101**	
G97 S500 M03	Must be in feed per revolution
G0 X.6 **Z.3** M8	Positioning move
G76 X.5 Z-1.0 **K.048** D.007 F.077	
Z1.	
G28 U0. **W0.**	
M30	

10.11 Parting Off

As discussed earlier, there are occasions when it would be optimal to make multiple parts from bar stock in a single operation. This is usually done in the creation of smaller parts, where a length of bar stock is used to create the first piece, and then that piece is cut off or parted off. This operation is called *parting off*. Using either a work shift or a secondary work coordinate, the second piece can then be machined and also parted off. This process can be repeated as many times as possible, while bearing in mind that the further the bar is from the workholding, the less stable the bar will become.

Figure 10-31 provides an example, indicating that three parts that are each 1.000″ long should be made from a single bar stock. The parting tool is .125″ wide, so the distance from face to face of each part is set at 1.200″. This is enough distance from the first part to the second part to allow the .125″-wide cutting tool to fully cut off the part and for .075″ extra to face the second part. Remember, the tool must go to 1.125″ because the tool offset is on the left side of the tool. It is best practice to first make a depth cut beyond 1.125″, around 1.130″, before parting off the machined part. This will create a better surface finish and a flatter surface. Let's look at the code for this example:

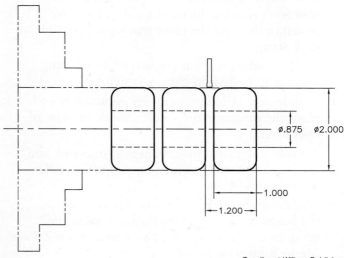

Goodheart-Willcox Publisher

Figure 10-31. An illustration of a possible part off program and operation. Remember to consider the tool width in the program calculations.

(TOOL: 6 OFFSET: 6)	
(.125″ PART OFF TOOL)	
G0 **T0606**	
G97 S500 M03	Must be in feed per rev
G0 G54 X2.1 **Z.3** M8	Safe move: use G54
Z-1.13	Move into position in Z
G1 X1.F.005	Make 1st depth cut
X2.1	Feed out in X
Z-1.125	Final Z position
X.850	Make cut off pass
G0 X2.1	
Z1.	
G28 U0. **W0.**	
M30	

Chapter Review

Summary

- Special review of workholding should be considered to maximize machining and minimize setups and operations. The ideal workholding solution provides the most rigidity, the most repeatability, and the most machining possible in each setup.

- A CNC lathe program consists of an opening statement, body, and closing statement.

- G codes are called preparatory commands and place the machine in a specific mode or state of operation.

- M codes are called miscellaneous functions and serve as "switches" to turn on and off specific machine functions.

- The Z-axis tool offsets can be entered into the offset page by positioning the tool in the programmed Z0. position and pressing the *Z Face Mesur* button on the control.

- The X-axis tool offset is established by cutting the material to a known diameter and then pressing the *X Dia Mesur* button on the control. You must then enter the known diameter size into the control at the prompt.

- Tool nose radius compensation is used to program the correct orientation of the cutting tool and compensate for the radius on the cutting tool nose.

- The G70, G71, and G72 are profiling canned cycles used to efficiently program a lathe operation. The G71 and G72 are roughing cycles that eliminate extended program lengths and simplify editing. The G70 cycle is a finishing cycle that creates the finish cut in the G71 or G72 cycle.

- Angle moves are programmed in a lathe by placing an X- and Z-axis move on the same line of programming.

- The G02 and G03 commands are used to program arc or radius moves in a lathe program. G02 creates a clockwise arc and G03 creates a counterclockwise arc.

- G81, G82, G83, and G84 are canned drilling cycles for the CNC turning center. The G81 drill cycle, in which the drill simply travels to a Z depth and returns, can be used for spot drilling or other drilling operations. G82 adds a dwell once the Z depth is achieved, helpful in a spot drill application or when using a chamfering tool. G83 is used for drilling deep holes, sending the drill to a specified partial depth and then fully retracting to clear out any chips and allow coolant to enter the hole. It will repeat that pecking cycle until the final Z depth is achieved. G84 is a rigid tapping cycle. Making OD grooves to accommodate seals, snap rings, or thread relief cuts is another common operation in the turning center.

- G32 is the longhand threading cycle. While requiring many lines of code and increasing the opportunity for programmer error, G32 gives the programmer complete control of the programmed paths and depths of cut and gives the advantage of cutting tapered or straight threads with one easy cycle. G76 reduces some flexibility for the programmer, but it is a simpler threading cycle, requiring only a single line of code. Another benefit of the G76 cycle is that it has a decreasing depth of cut.

- Parting off is the creation of multiple parts from bar stock in a single operation by cutting off each part from the stock as it is completed.

Review Questions

Answer the following questions using the information provided in this chapter.

Know and Understand

1. It is always a good idea to assemble a _____ when preparing to program a part and move into production.

   A. fixture
   B. toolholder
   C. team
   D. set of plans

2. The ideal workholding solution provides the most _____ and the most _____.

   A. efficiency, cost saving
   B. rigidity, repeatability
   C. speed, accuracy
   D. ease of use, speed

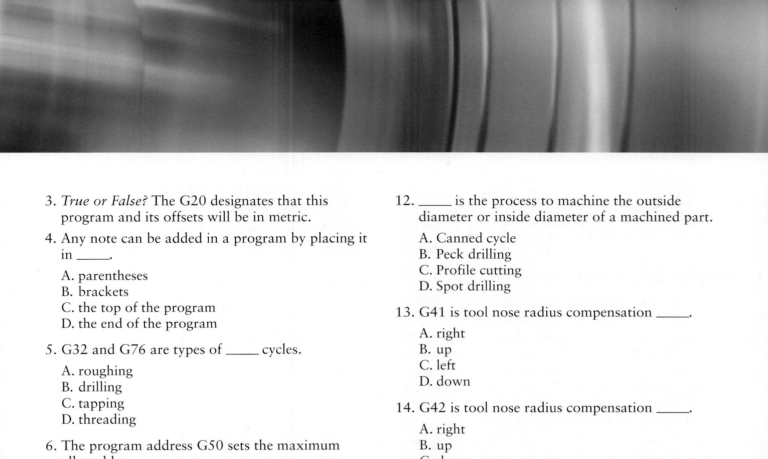

3. *True or False?* The G20 designates that this program and its offsets will be in metric.

4. Any note can be added in a program by placing it in _____.

A. parentheses
B. brackets
C. the top of the program
D. the end of the program

5. G32 and G76 are types of _____ cycles.

A. roughing
B. drilling
C. tapping
D. threading

6. The program address G50 sets the maximum allowable _____.

A. drilling depth
B. Z depth
C. spindle speed
D. thread height

7. *True or False?* The diameter offset is the establishing of the X-axis coordinate for each tool.

8. The _____ is a canned cycle used to make a finishing contour path.

A. G71
B. G70
C. G40
D. G82

9. The _____ canned cycle is a roughing profile contour that machines in the X-axis direction.

A. G71   C. G40
B. G70   D. G72

10. The _____ canned cycle is a roughing profile contour that machines in the Z-axis direction.

A. G71   C. G40
B. G70   D. G72

11. _____ stops the program and rewinds to the beginning.

A. M98   C. M05
B. M99   D. M30

12. _____ is the process to machine the outside diameter or inside diameter of a machined part.

A. Canned cycle
B. Peck drilling
C. Profile cutting
D. Spot drilling

13. G41 is tool nose radius compensation _____.

A. right
B. up
C. left
D. down

14. G42 is tool nose radius compensation _____.

A. right
B. up
C. down
D. left

15. The _____ and _____ codes are the simplest method to program arcs.

A. G28, G31
B. G70, G71
C. M03, M04
D. G02, G03

16. In the G76 threading cycle, what does the letter "K" represent?

A. Dwell time
B. Incremental peck amount
C. Incremental thread height
D. Final drill depth

17. In a G83 peck drilling cycle, what does the letter "Q" represent?

A. Dwell time
B. Incremental peck amount
C. Incremental thread height
D. Final drill depth

18. When machining 1/2″-13 threads on a machining center, what is the feed rate in inches per revolution? (Round to 3 decimal places.)

A. .500 ipr   C. .013 ipr
B. .770 ipr   D. .077 ipr

19. When machining 1″-8 inside diameter threads on a machining center with a spindle speed of 500 rpm, what is the feed rate in inches per revolution? (Round to 3 decimal places.)

A. .125 ipr
B. .800 ipr
C. .013 ipr
D. .077 ipr

20. *True or False?* In a G32 threading cycle all cut diameters must be programmed.

Apply and Analyze

1. In CNC lathe operations, what are some different types of workholding?

2. Using program #1212, a 1/2″ diameter drill as tool #1, and drilling 1.25″ deep, write a program to spot drill a hole through the center of a lathe part. Spindle speed is 1,500 rpm and feed rate is .010 ipr.

3. Using program #2200, a 3/4″ diameter drill as tool #2, and drilling 3.00″ deep, write a program to peck drill a hole through the center of a lathe part. Spindle speed is 1,000 rpm and feed rate is .008 ipr.

4. For the supplied print create a G70–G71 program to rough and finish the part. Program #3223, tool #1 (CNMG 432) for contouring, constant surface speed of 200 with a maximum spindle speed of 3,000 rpm. The depth of cut is .050″, and leave .010″ of material in X and Z for the finish pass.

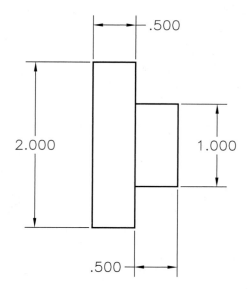

5. For the supplied print create a G70–G72 program to rough and finish the part. Program #1450, tool #1 (CNMG 432) for contouring, constant surface speed of 200 with a maximum spindle speed of 3,000 rpm. The depth of cut is .050″ and leave .010″ of material in X and Z for the finish pass.

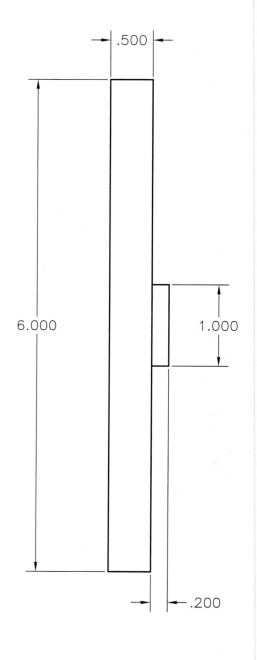

6. For the supplied print create a grooving program. Program #0543, tool #1 (.125″-wide groove tool), constant surface speed of 150 with a maximum spindle speed of 1,500 rpm and a .010 feed rate.

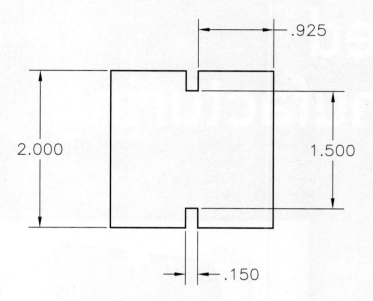

7. Create a program #3098 using a G76 threading canned cycle to create 1/2″-13 threads. The minor diameter is .404″ and the first depth cut is .010″ deep. The length of threads are .750″. Tool #3 is a 60° threading insert.

Critical Thinking

1. Lathe part programming creates unique challenges in workholding. What are some of the ways the program can address these challenges and minimize those effects?

2. Lathe part programming and production can provide an excellent career path and opportunities. Take some time to find some typical items that you see every day that were produced on a CNC turning center. How would you determine how that part was machined and what type of industry produces those parts?

3. Write a short narrative on the differences between mill and lathe programming. Describe what career industry you would like to explore and analyze if that field uses mostly mill or lathe programming, or if it requires skill and expertise in both.

11 Computer-Aided Manufacturing

Chapter Outline

11.1 Introduction to Computer-Aided Manufacturing

11.2 CAD versus CAM

    11.2.1 Computer-Aided Drafting

    11.2.2 Computer-Aided Manufacturing

11.3 Geometry Creation

    11.3.1 2-Dimensional Geometry

    11.3.2 Points

    11.3.3 Lines

    11.3.4 Arcs or Circles

    11.3.5 3-Dimensional Surfaces

    11.3.6 Solid Creation

11.4 Geometry Import

11.5 Toolpath Generation

    11.5.1 Drilling Toolpath Selection

    11.5.2 Contour Toolpath

    11.5.3 Creating Pockets

    11.5.4 Solids Machining

    11.5.5 Surface Machining

11.6 Associated Geometry

11.7 Job Setup

11.8 Post-Processing: From Screen to G Code

Learning Objectives

After reading this chapter, you should be able to:

- Define the term *CAD*.
- Outline the benefits of computer-aided drafting (CAD) over conventional drafting.
- Explain the use of a CAM system in the modern manufacturing environment.
- Summarize the relationship between CAD and CAM systems and how they function together.
- Describe cutting technique.
- Discuss what is meant by *geometry creation*.
- Explain the difference between 2-dimensional and 3-dimensional geometry.
- Describe the technique of using surface geometry and how this differs from solid geometry.
- Create basic toolpath operations in a CAM system.
- Explain what is meant by *associated geometry*.
- Describe the process of importing files between CAM and CAD systems.
- Explain the benefits of information transfer in setup sheets.
- Explain post-processing.

Key Terms

associated geometry

computer-aided drafting (CAD)

computer-aided manufacturing (CAM)

data control

Direct Exchange Format (DXF)

feature-based machining

geometry creation

high-speed machining (HSM)

information transfer

Initial Graphics Exchange Specification (IGES)

layers

levels

post-processing

setup sheet

solid model

surfaces

toolpath

11.1 Introduction to Computer-Aided Manufacturing

The last two chapters have discussed "manually" writing programs and the codes associated with those programs. Alongside advancements in CNC machining technology, there have also been advancements in CNC program creation. As production parts have become increasingly complicated, the use of computers to produce difficult programs and simulate the toolpaths has become more prevalent. The ability to simulate manufacturing in a computer software environment is known as computer-aided manufacturing, or CAM.

11.2 CAD versus CAM

Several times in this text the terms *CAD* and *CAM* have been used. It is important that we take a look at these two tools at our disposal and identify the differences between them. They are each distinct in their use and purpose, but they are also uniquely twined together. This chapter examines the differences and then focuses primarily on the use of CAM systems in manufacturing.

11.2.1 Computer-Aided Drafting

Drafting, or print creation, has rapidly evolved over the past few decades. Drafting has long been its own unique field of study and expertise. For decades drafting was done with a pencil and paper, and even simple drawings took hours to create. More problematic, when a drawing needed a revision, it often had to be totally recreated. This created the dilemma of drawing storage and print revision history, as cabinets full of large-scaled drawings were filed for later use. With the onset of *computer-aided drafting (CAD)*, all of this changed.

CAD is the graphical process of creating true-to-form and scale parts, assemblies, or renderings. It uses lines, vectors, and arcs to create a digital recreation of a concept or part. Up until just a few years ago, this was almost strictly a 2-dimensional representation, but every new system now fully utilizes 3-dimensional features with solids or surfaces. CAD is now used almost exclusively in part creation for the manufacturing field.

Drawings in CAD are to actual size, so no scaling is required. For example, in **Figure 11-1** there is a simple drawing of a quarter and a dollar bill. Regardless of how these are printed or appear on paper, in the CAD program they are exact representations of the precise size of a quarter and a dollar. This is very important in drafting objects and displaying them in the correct scale or proportionate size. In a CAD program you can draw an assembly, a house, or even an entire city exactly to its intended size. There is no need to create all of the objects to half-size so they fit on the drawing paper; the computer does not have a limit to the geographic size of drafted objects.

Another beneficial aspect of CAD drawings is the perfection of geometric shapes. If an arc is drawn, it

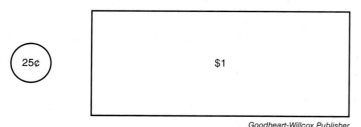

Goodheart-Willcox Publisher

Figure 11-1. A 2-dimensional CAD drawing. The model is an exact representation of the size and shape of the bill and quarter.

is a geometrically perfect arc, a set of parallel lines are exactly parallel, and a set of perpendicular lines are exactly 90° apart. Hand-drawn prints often suffer from inconsistencies in both shape and scale.

Originally, CAD was only designed for 2-dimensional multiview drawings—basically what was achievable with pencil and paper. It did not take long for CAD developers to move into three dimensions, which makes it easy to create multiview prints by simply rotating the part through the correct projection. For example, **Figure 11-2** shows a fairly complex solid modeled part that needs a multiview print. Most CAD operators can draw this solid easily enough, and then with special features built into the software, they can display this as a multiview drawing as in **Figure 11-3**. The last step is to dimension and tolerance as required.

CAD programs have developed rapidly, and the ease of added tools and displays make massive print projects a manageable task. Developers have also created *layers*, or *levels* that can be visible or invisible, allowing the operator to show or hide the necessary items depending on their desired print. For example, an architect can draw an entire building, with all of the plumbing, framing, and electrical in a single drawing, and then isolate only the electrical for the electricians and the plumbing for the plumber, with a single click of a mouse. In a manufacturing environment, large assemblies can be created and single parts can be isolated for machining purposes.

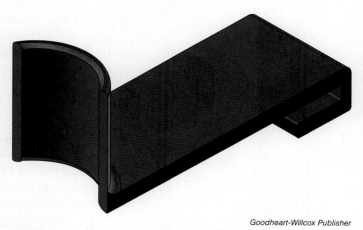

Goodheart-Willcox Publisher

Figure 11-2. A 3-dimensional CAD drawing of a spool holder from a 3D printer.

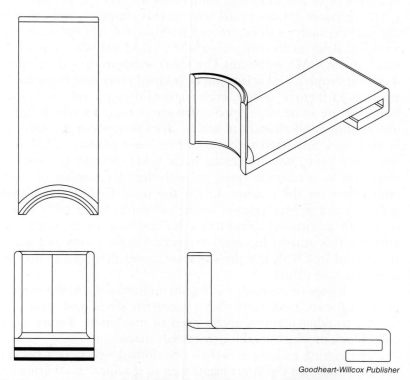

Goodheart-Willcox Publisher

Figure 11-3. This is a third-angle projection print with (counterclockwise from top left) top, front, right, and isometric views.

In the modern machine shop, machinists are often given already-created CAD drawings to machine. It is possible that a machinist would be required to draw a part in a CAD program, but most machine shop tasks will be completed using only computer-aided manufacturing (CAM) software.

Thinking Green

Government Programs as Green Catalysts

Many government programs have established green initiatives to encourage more sustainable practices in manufacturing. These initiatives aim to reduce energy use, operating costs, and utility bills to promote a cleaner future. For example, the Environmental Protection Agency (EPA) created a lean manufacturing plan that streamlines the mass production of products. The EPA's plan rearranges production activities to create a more orderly and efficient workflow. Another government effort, the Save Energy Now initiative by the U.S. Department of Energy (DOE), provides energy savings assessments to large U.S. manufacturing plants. These assessments are conducted by DOE qualified energy-efficiency experts and help identify immediate and long-term opportunities for sustainability.

11.2.2 Computer-Aided Manufacturing

Computer-aided manufacturing (CAM) is the process in which machining parameters are defined using computer software. Depending on the complexity of the drawing, sometimes a CAD file is imported into the CAM software (as discussed later in this chapter), and other times simple part geometry can be created within the CAM software itself. In the sophistication of drafting capability, CAM software has evolved separately from CAD software. The CAD software has developed to make drawing simpler, and it is being developed more and more for large-scale projects in factory design, architectural design, and even city planning. Some of the most developed software is used to create full models of automobiles and aircraft, in which all of the testing is resolved and there is no longer any need to build test or scale models. CAM software, on the other hand, has integrated basic CAD technology, and its development has focused on keeping up with the vast number of machines and controllers on the market. Often the term *CAD/CAM* is used, which means it is a hybrid version of the software, but it truly is a CAM system with additional capabilities for drafting or drawing any part or additional features. In simpler terms, simple parts and geometry can be created in CAM, but the software is not designed to do complicated assemblies or prints.

The basics of computer-aided manufacturing are the same regardless of the software used. Start with a geometric shape and make a determination as to what features are in need of machining. **Figure 11-4** shows a solid block in which the blue slot needs machining.

The cutting technique is then determined—such as a pocket, drill, or contour, and then the appropriate tool to machine this feature can be chosen. A contour toolpath with a 3/8″ end mill will be used to create this slot.

Goodheart-Willcox Publisher

Figure 11-4. A typical part that requires a machining process to be performed.

The CAM software will generate a path based on the given parameters and display that on the screen, **Figure 11-5**. When the part machining functions are complete, the CAM software will generate the G- and M-code program for the programmer.

 Tech Tip

In the following sections, we will discuss creating geometry and toolpaths. The software displayed is Mastercam™, the leading retailer of CAM software on the market. Although there are many other brands of software, the nomenclature of geometry and tool creations is extremely similar. The use of Mastercam™ is for demonstration purposes only.

11.3 Geometry Creation

The creation of geometry is a vital step in working with a CAM system. Even when working with an existing part file it is often necessary to create some additional geometry for the tool to follow. *Geometry creation* simply means drawing lines, arcs, or solids that define a part's shape. The tool follows existing geometry, so it is important that the geometry is created accurately. The solid we see in **Figure 11-5** started with just a few simple lines and arcs, **Figure 11-6**. The solid was created from those simple lines and arcs using CAM solid modeling. Although the software can use the 2-dimensional geometry for toolpaths, the benefit of 3-dimensional CAM will be explored later in this chapter.

11.3.1 2-Dimensional Geometry

Most toolpaths can be generated from basic 2-dimensional (2D) geometry. The objective of creating geometry is to designate a line, arc, or point for the tool to enter or follow. With that in mind, be very careful when drafting entities that do not connect or that lay on top of each other. The definition of 2D geometry is geometry that only lies in two planes, usually X and Y, but it can be any two planes. This is the simplest form of toolpath creation, but it relies heavily on the operator to manage more of the variables.

Figure 11-7 shows some of the most commonly used icons in the Mastercam™ geometry toolbar used to create standard 2D geometry. In this toolbar we have points, lines, and arcs. Within each of these categories are multiple options for creating these three types of 2D geometry.

Goodheart-Willcox Publisher

Figure 11-5. Virtual simulation of a standard toolholder and end mill machining a slot.

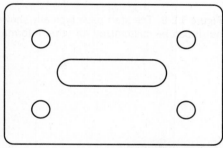

Goodheart-Willcox Publisher

Figure 11-6. The 2-dimensional geometry that was used to create the 3-dimensional model.

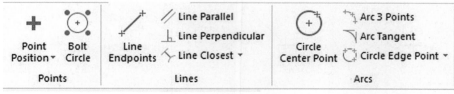

Goodheart-Willcox Publisher

Figure 11-7. The wireframe toolbar used to create simple 2-dimensional geometry.

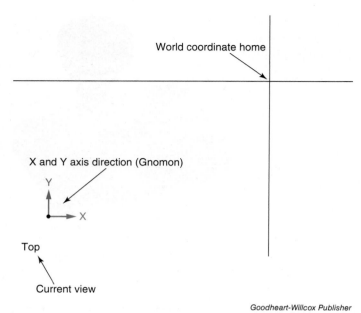

Figure 11-8. The start page typically shows orientation and view, but it can be customized for other information.

Figure 11-8 shows three features on the opening screen of Mastercam™: the World Coordinate Home; the X and Y axis directions, called a Gnomon; and the current view name. The World Coordinate Home is the starting location for each CNC program within CAM. If your Work Coordinate System in your setup is the top-left corner of your part, your geometry should have the same starting location in the World Coordinate Home. The X and Y Gnomon shows the direction of the axes. If we tilted this view somewhat, we could see that there is also a Z arrow. The arrow direction denotes the positive direction for that axis. The current view, especially in 2D work, should be set to *top*. This top view shows the part from a perspective as if you were looking down the cutting tool.

11.3.2 Points

A point is just a position in the X and Y axes. By itself, a point cannot be toolpathed, but it can be used as a tool's starting position or as a drill location. When creating a drill toolpath, drawing the hole diameter is not required; only a position to locate the drill tip center point is needed. Another icon on the point toolbar is a bolt circle. This is a shortcut to creating a series of points with a common radius, evenly spaced (equal angles) from a center point. **Figure 11-9** shows the result of a 10″ bolt circle with eight points.

Because this is a bolt circle, all points are 5″ from the center point and 45° apart. Eight holes divided equally in a 360° circle equal 45° apart, **Figure 11-10**.

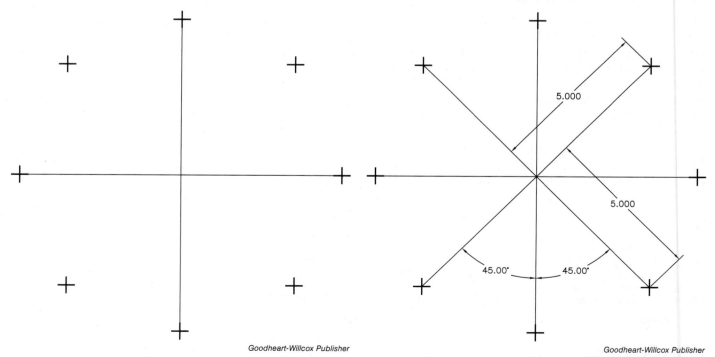

Figure 11-9. The bolt circle shortcut creates points at equal distance and angle from a given center point.

Figure 11-10. The analysis of the point locations verify distance and direction.

11.3.3 Lines

Lines are the most common form of 2D geometry. Lines can be drawn horizontally, vertically, or at any angle. As discussed in Chapter 4, the mathematical definition of a line is a straight geometric element with no thickness and no ends. What we are drawing are actually line segments, or parts of lines. Our "lines" have end points as we define them in our geometry, **Figure 11-11**.

11.3.4 Arcs or Circles

Arcs or circles can be internal or external features. An arc is usually considered a non-closed circular shape or less than 360°. Arcs can be used where two lines intersect to create a fillet, or rounding, at a corner, inside a pocket, or as part of the defined part shape. **Figure 11-12** shows two uses of fillets. Circles are primarily used to define hole features on mills or bores on lathes. **Figure 11-13** shows a typical milling part from a previous example. It displays multiple applications of circles and arcs.

11.3.5 3-Dimensional Surfaces

Not all CAD/CAM creation is as simple as squares, rectangles, or circles. Machining often deals with extremely complicated geometric shapes that must be defined in more than just a 2-axes orientation. Solids are examined in the next section, but there is another way to define 3-dimensional features, called *surfaces*. Surfaces are best compared to stretching a skin over a frame. Surfaces have no thickness or mass, and many surfaces are often "stitched" together to create complicated forms. **Figure 11-14** shows a surfaced part. Notice the complexity of the shape and the ever-changing contours of the surface. Surfaces create a unique set of machining challenges, discussed in Section 11.5, Toolpath Generation.

11.3.6 Solid Creation

The greatest advances in both CAD and CAM have come in the implementation of solid models. *Solid model* is a term for a 3-dimensional CAD model that has mass and volume. What is represented on the screen is a

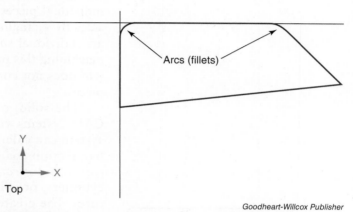

Goodheart-Willcox Publisher

Figure 11-11. Lines can be classified as horizontal, vertical, or angular in relation to the current view.

Goodheart-Willcox Publisher

Figure 11-12. Fillets are used to blend two lines together without sharp corners.

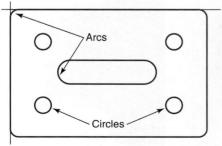

Goodheart-Willcox Publisher

Figure 11-13. Circles and arcs can be utilized to create desired part features.

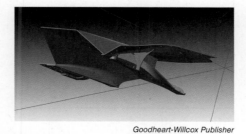

Goodheart-Willcox Publisher

Figure 11-14. Complicated shapes and curves can be achieved by blending surfaces together.

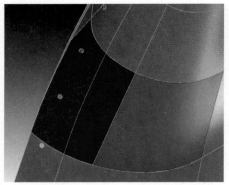

Goodheart-Willcox Publisher

Figure 11-15. This shape is created by stitching together multiple surfaces.

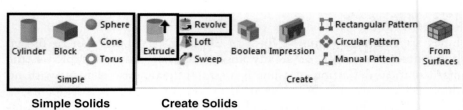

Simple Solids   **Create Solids**

Goodheart-Willcox Publisher

Figure 11-16. The Solids toolbar can be used for simple solids or created solids.

solid piece of material. Solid models have multiple advantages in computer-aided manufacturing. The problem that solid models solve can be seen by taking a second look at surfaces. Recall that surfaces are stitched together, individual pieces of geometry. **Figure 11-15** shows three surfaces side by side. By changing the colors, you can see that the blue, red, and green are individual surfaces that are not truly connected. The disadvantage in machining this part is that if you machine the green surface, the CAM system does not consider the surfaces next to it because they are individual pieces.

The solid, on the other hand, is truly one piece. In fact, in many CAD systems you can designate the material type of a solid, and the system can calculate the weight and volume. To create a solid, the toolbar section for solids, **Figure 11-16**, has two main categories, *simple solids* and *created solids*. The simple solids can create simple blocks, cylinders, or spheres. These can be useful in creating simple solid features. The created solids give you more diversity and use 2D geometry to define the solid shape. *Extrude* and *Revolve* are the most used tools to create solids.

Figure 11-17A shows a 2D drawing that can be used to create the solid model in **Figure 11-17B**. This model was created using the *Extrude* tool from the solids toolbar.

Figure 11-18A uses the *Revolve Solid* tool to form **Figure 11-18B**. To use the *Revolve Solid* tool, two entities must be defined, the exterior shape and the axis of rotation. The exterior shape will be rotated around the axis of rotation, and a solid will be formed inside that boundary.

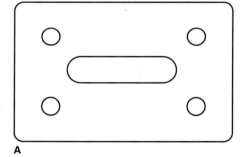

A

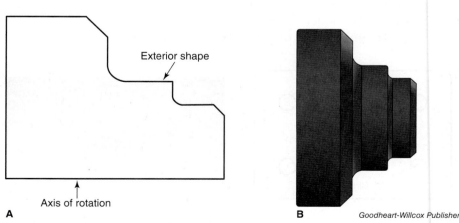

B   *Goodheart-Willcox Publisher*

Figure 11-17. A—2D drawing of the desired part features. B—The extrude solid function was used to create a solid from the 2D drawing.

Figure 11-18. A—The axis of rotation and exterior shape must be defined in the revolve solid function. B—The resulting solid body from the 2D drawing in Figure 11-18.

11.4 Geometry Import

As has been shown, 2-dimensional geometry can be created from a print or sketch, or full 3-dimensional solids can be created from that 2-dimensional drawing. Creating correct geometry in any CAM system is going to take some practice, but it is imperative that it is done correctly. You can never make the right program if you start with the wrong geometry.

The most common practice in the modern manufacturing environment is for the manufacturer to receive a CAD file from the customer. There are a couple of reasons for this. First, the customer's CAD file has been checked and approved by them and their engineering staff. They have already ensured that the provided file is the finished part they want. Most aerospace and automobile prints now have a note that dictates that the CAD model is the determining factor in part creation—so, if there is a discrepancy, the CAD file is the referenced document. Second, the customer wants to control its data and maintain file integrity. Once the file leaves its native system or is recreated by another system, they have lost control of the data and its integrity.

This practice creates an issue with importing files because the engineering staff is not using a CAM system to create models; they are using a CAD system, and there are a variety of different systems. Therefore, every CAM system includes an option to translate or import CAD files. When purchasing a CAM system, it is important to find out what types of files can be imported. One of the oldest file types that can be imported is an *Initial Graphics Exchange Specification (IGES)* file. The CAD system can "save as" an IGES file, and most CAM systems can import these files. IGES files will import all 2D geometry, but when importing solids, it will import them as surfaces. Another file type is the *Direct Exchange Format (DXF)*. It is a newer file format that was developed specifically by a CAD company to convert its files out to other systems.

 From the Shop

Is My Surface Solid?
Exchanging files from software to software can be a real issue. Many CAM systems have an analysis tool that make it possible to check imported surfaces and solids for connectivity. Even an imported surface can be reconstructed as a solid in most cases. Just remember that this conversion can have minor inconsistencies and give your CAM software some nightmares.

Issues importing and exporting massive files between systems are common. CAD/CAM developers are making progress into integrated systems, and fully integrated CAD/CAM systems with full CAD and full CAM capabilities are on the horizon.

11.5 Toolpath Generation

Once a CAD file is imported or geometry created, toolpaths can be applied. A *toolpath* is the cutting strategy used to create a part. First, determine what can be cut, based on fixturing and print requirements. Second, determine the type of toolpath required to machine that feature as fast and

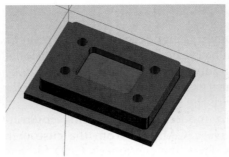

Figure 11-19. A sample part that will require four drilled holes.

Figure 11-20. When selecting a hole to be drilled, a point will appear in the center for graphic display.

accurately as possible. Third (and last), decide on the tool required for the cut. In Mastercam™ there are 63 different types of toolpaths and countless numbers of variables in every toolpath. This chapter will examine three of these toolpaths that will be used the majority of the time.

Figure 11-19 provides a sample part for which three different toolpaths are needed: drill, contour, and pocket. The following sections break these toolpaths down individually to make them easier to see. (At this point the dimensions, tolerances, and other details to complete this part are not relevant, as this example focuses on picking toolpath strategies and the operation of CAM systems.)

11.5.1 Drilling Toolpath Selection

In the sample part from **Figure 11-19**, it is pretty easy to see that four holes need to be drilled. It is always best practice to spot drill or center drill a hole before drilling. This helps start the drill in a straight path as it enters its cut. Start with a 90° spot drill. From the toolpath selection screen, select a drill toolpath. Then, back on the geometry, select the point at which to drill, a point in the center of the drilled hole location, **Figure 11-20**.

Once the points are selected, the tool is selected and then the cutting parameters are set. **Figure 11-21** shows the *Drill Cycle Selection* screen. The selection of different cycles will create different outputs in your machine code. For example, *Drill/Counterbore* outputs a G81, *Peck Drill* outputs a G83, and *Tap* outputs a G84.

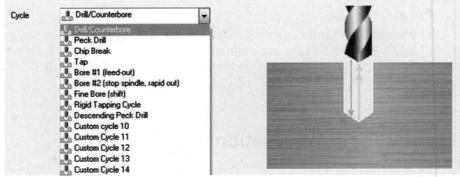

Figure 11-21. The drilling cycle drop-down menu will provide options for the desired drill path.

Finally, look at the *Parameters* page. The *Parameters* page sets the cutting clearance planes (R Planes) and cutting depths, **Figure 11-22**. It is important to understand whether the parameter should be set in Absolute or Incremental values, and values are set as positive or negative numbers. The absolute setting places that exact value in the Z height, and the incremental places that starting position incrementally above the geometry selected.

The next tool is the drill required to drill the through holes to print specifications. Select the appropriate drill cycle and parameters for this tool in the same way as the spot drill.

11.5.2 Contour Toolpath

A contour toolpath is used to direct a tool along a specific set of lines or arcs. When a tool should follow a set of connected geometry, the contour toolpath is the most commonly used cycle. **Figure 11-23A** shows our sample part being machined by a 1″ end mill with a contour toolpath. Simply pick the contour toolpath from the toolpath menu, **Figure 11-23B**. Then, select the geometry the tool should follow, **Figure 11-23C**. Finally, select or create the tool, and a contour toolpath can be

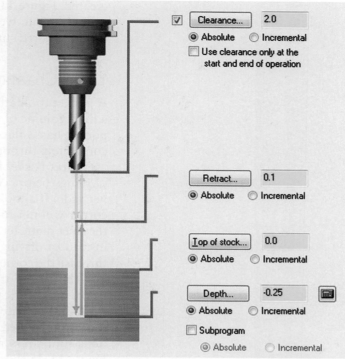

Goodheart-Willcox Publisher

Figure 11-22. The drilling parameters page will prompt for information to complete the drilling cycle. The selections will determine the programmed path.

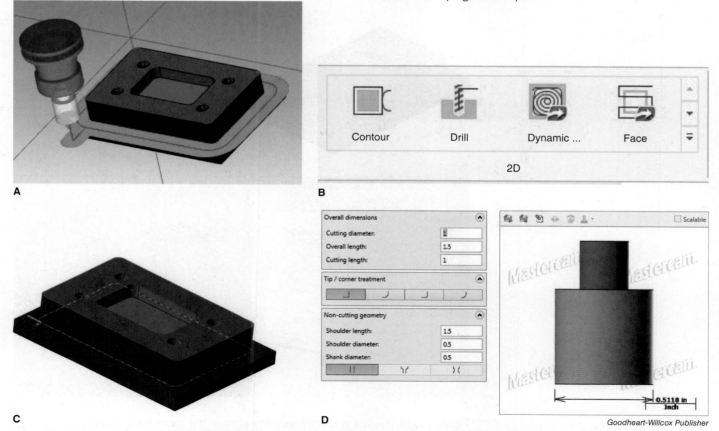

Goodheart-Willcox Publisher

Figure 11-23. A—A simulation of a 1″ end mill performing a contour toolpath. B—The contour toolpath is located in the toolpath toolbar. C—The graphical representation of the geometry selected. D—Tools can be customized or created in a CAM system.

created, **Figure 11-23D**. All CAM systems will have the ability to automatically calculate feeds and speeds for your tool. By designating the material type or desired surface feet per minute (sfm), calculations will be completed.

11.5.3 Creating Pockets

A pocket toolpath is used with a closed, or partially closed, boundary with a high volume of material removal, **Figure 11-24A**. In our sample part, the pocket has a flat bottom, but a pocket can be used for material removal completely through the part as well.

Select *Pocket* from the toolpath menu, **Figure 11-24B**. Consider a couple of things before cutting the pocket. **Figure 11-24C** is an analysis of the corner radii. If a tool with a larger diameter than the fillet diameter is used, the corner will not fully cut. It is best practice to use a tool that is smaller than the fillet diameter so that the machine can cut a radius as it turns the corner instead of driving directly into the corner and then changing directions. In this case the corner has a 1/2″ diameter, which can be cut with a 3/8″ end mill.

The second thing to consider is entry strategy. How to enter a cut may be the most critical decision. Because this is a closed pocket, we cannot enter from off the material; instead the tool must dive directly into the part. If the tool is positioned to go straight down in the Z axis, this will cause tremendous tool stress and load and most likely cause catastrophic tool failure. **Figure 11-25** shows the *Entry Motion* page. These parameters will allow the tool to either Helix Entry or Ramp Entry into the part, both of which are

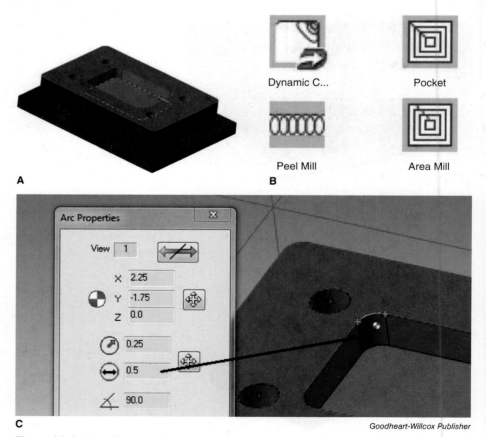

Dynamic C...

Pocket

Peel Mill

Area Mill

A

B

C

Goodheart-Willcox Publisher

Figure 11-24. A—The graphical representation of the geometry selected for the pocket. B—The pocket toolpath is located in the toolpath toolbar. C—Analysis of part features can be used to determine tooling or machining strategies.

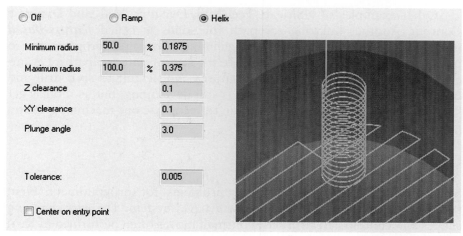

Goodheart-Willcox Publisher

Figure 11-25. Entry motion variables can be controlled in the CAM system for best results.

viable options. **Figure 11-26** shows the result of a pocket toolpath for our sample part.

11.5.4 Solids Machining

Although our toolpath sample part is a solid, we are missing out on some of the advantages of machining solid models using the previous methods. The advantage of a solid over surfaces or 2D geometry is that the CAM system "sees" all of the entities of a solid. The inclusion of solid modeling in CAM systems has created some additional toolpaths that have changed the way we machine. These toolpaths are known as *high-speed machining (HSM)*. Now that the CAM system sees the part as one entity, the machining process can be sped up without concern for violating any adjoining part feature. Hand in hand with advanced toolpaths are machine advances in speeds, feeds, and rigidity and tool cutter advances in carbide and cutter insert design.

The theory behind high-speed machining is to use the entire flute length at very high spindle speeds and high feed rates, while staying engaged in the part. The most volatile moments in machining are entering and exiting. So by keeping the tool in contact with the material constantly and removing only small amounts of material with the flutes, the machine can travel at 600–800 ipm, depending on material type. Looking back at the feed and speed calculator from Chapter 4, notice that this high feed rate will require a machine that can turn 10,000–12,000 rpm and that five or six flute cutters might be needed to reduce the chip load on the tool.

Figure 11-27 shows a 2D Area Mill HSM toolpath for our pocket. Notice that the toolpath is making smaller step-over moves, but it can be programmed much faster than the surface toolpath methods and with maximum material removal rates.

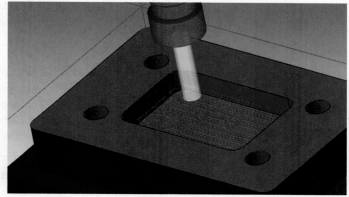

Goodheart-Willcox Publisher

Figure 11-26. Graphical lines display the entry motion, path, and finish machining of the pocket toolpath.

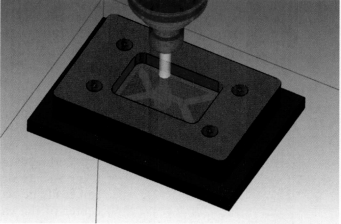

Goodheart-Willcox Publisher

Figure 11-27. A visual display of the toolpath in a high-speed machining operation. These toolpaths reduce tool engagement and are performed at high feed rates.

Other examples of HSM are Peel Mill, Dynamic Mill, and Dynamic Contour. Another amazing toolpath for solids is called *feature-based machining (FBM)*. Feature-based machining allows the programmer to pick a solid and let the software decide the appropriate toolpaths. Parameters can be set to recognize hole features and default to spot drill and drill paths. Feature-based machining is still developing, but as models get "smarter" and our CAM systems begin to recognize our machining strategies, FBM will become commonplace.

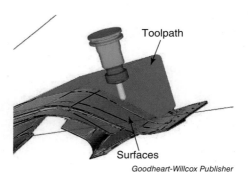

Goodheart-Willcox Publisher

Figure 11-28. A surface toolpath used to create complex geometric shapes.

11.5.5 Surface Machining

Surface machining presents distinct problems for programmers. First, surfaces cannot be machined without a CAM system. The ever-changing shapes and contours create a series of small moves often performed in X, Y, and Z axes simultaneously. The sheer math involved is staggering. Because the surface is ever changing, a ball end mill will be required to complete the surface toolpath. **Figure 11-28** shows an example of a surface parallel path.

Once again, there are too many surface toolpath options to explain in this text. Practice different styles and techniques in the CAM system to understand the variances between different paths.

Safety Note

CNC machines are equipped with door interlock latches that "lock" the door closed while the machine is in operation. The doors must remain closed and the operator protected at all times when operating or setting up a CNC machine. Disabling or altering the door interlocks can cause devastating injuries and is a violation of federal OSHA regulations.

11.6 Associated Geometry

Computer-aided manufacturing systems are very intelligent. Most good CAM systems were developed by machinists or programmers *for* machinists, so they take into account the manufacturing environment and the ever-changing atmosphere in which machinists work. It is possible to receive a print or CAD model that is "oriented" to one location or position, like the part center or the top-left corner, but the machine setup dictates a different location for the work coordinate system (WCS), or part zero. In most aerospace CAD files, the part is actually drawn exactly where it sits in the fully assembled aircraft, which means it is oriented in a position nowhere near where machine positions could be set. When longhand programming was the only option, it was a massive undertaking to move the machine program to a different WCS. All of the X+s became X−s, or the Y or Z values changed. It is easy to see the possibility for a catastrophic mistake.

With a CAM system, the toolpath is associated with the geometry. If it is easier to draw and toolpath to the top left-hand edge of a part, that can easily be accomplished. But if it is easier to machine the part from the center of the workpiece, the part can be moved and all of the toolpaths move with the geometry. With all of the toolpaths created in the preceding examples, we selected the drill points, contour lines, and pockets. If we move that geometry, the toolpath also moves. This *associated geometry* eliminates programming and set-up errors.

11.7 Job Setup

After a programmer receives a CAD file, decides on the correct machining approach, and creates the necessary toolpaths, the job is not yet complete. This information still needs to be conveyed to the setup person and operator, and the data needs to be controlled. Part programs need revision numbers, which must be controlled just as prints need document control. Too many machine shops have quoted a job on a Revision "C" print and then used an old Revision "B" print to manufacture the part. This can be crippling to the bottom line. In the same way, many setup machinists have machined a part using an old program that does not recognize the newest part revision.

There needs to be a system of *information transfer*. Simply, all of the information gathered in the programming stages needs to be conveyed to the shop floor and input in a CNC machine controller. Where is the part's WCS zero point? What type of fixturing is used? What are the part number and print revision? What tools are being used? The programmer's job is to make the setup and the operator's job as easy and error-free as possible by communicating all of this information in a clear, comprehensive manner.

Most CAM systems have an option for *setup sheets*. Setup sheets are a series of pictures and diagrams with all of the setup information embedded. The trick to these sheets is that all of the information must be filled in within the CAM system. Mastercam™ has the ability to show tool lengths and toolholders in its display. By inputting these parameters while programming, this information will be transferred into the setup sheets. The following figures demonstrate the setup sheet for the sample part used throughout this chapter and some of the benefits of supplying this to the setup person along with the program.

Figure 11-29 shows the general information page. It lists part name, customer, programmer, program or part revision, and a snapshot of the

GENERAL INFORMATION

PROJECT NAME:	CAM Sample Part	
CUSTOMER NAME:	CNC Manufacturing Technology	
PROGRAMMER:	R. Calverley	
DRAWING:	CAM Sample	REVISION: C
DATE:		
TIME:	4:12 PM	

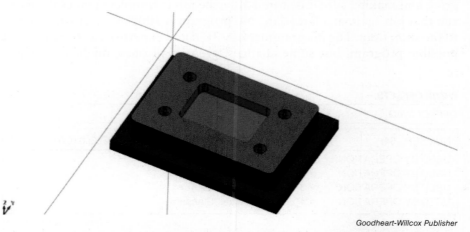

Goodheart-Willcox Publisher

Figure 11-29. The general information page of a setup sheet with detailed information on a part.

PROGRAM NUMBER:	0
SPINDLE SPEED:	2674 RPM
FEEDRATE:	16.044 inch/min
CLEARANCE PLANE:	2.0
RETRACT PLANE:	0.0
FEED PLANE:	0.1
DEPTH:	-0.25
STOCK TO LEAVE:	0.0
COMP TO TIP:	NO
WORK OFFSET:	0

TOOL INFO	*1/2 SPOTDRILL*

TYPE:	Spot Drill	
NUMBER:	1	
DIAMETER:	0.5	
CORNER RADIUS:	0.0	
LENGTH OFFSET:	1	
DIAMETER OFFSET:	1	
MATERIAL:	HSS	
NUMBER OF FLUTES:	2	
FPT:	SFM:	
MFG CODE:		
ASSEMBLY:		
HOLDER:	C4C3-0032	C:
TIME:	00:00:07	

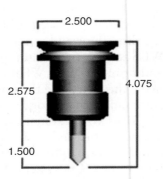

Goodheart-Willcox Publisher

Figure 11-30. Setup sheets should contain details on operation type and tool information.

part with WCS home position. **Figure 11-30** shows each individual toolpath page. Displayed here is the toolpath for our spot drill operation. It contains information on speeds, feeds, final depth, tool number, and toolholder number. Every individual operation will have a separate page to describe tool and operation type. **Figure 11-31** shows all of the tools used and the WCS number—G54 in this case.

Many machine shops will have some variation of a setup sheet. The format is not important as long as all of the information is transmitted to the person running the machine. Do not make the operator guess what the programmer might have been thinking; it is a waste of time that delays machining.

The programmer must also oversee *data control*. Controlling the program and making sure it is stored after the job is complete makes it easy to run that job again at a later date. No program is 100% perfect when it gets to the shop floor. The programmer uses skill and expertise to create the best possible program, but some changes, even small ones, might be needed.

WORK OFFSETS

OFFSET INFO

NUMBER: 54	PLANE: Top	ORIGIN: 0.0, 0.0, 0.0
USED BY OPERATION:	# 1	1 - Drill/Counterbore
USED BY OPERATION:	# 2	2 - Peck Drill
USED BY OPERATION:	# 3	3 - Contour (2D)
USED BY OPERATION:	# 4	4 - Pocket (Standard)

Goodheart-Willcox Publisher

Figure 11-31. Include as much information as possible on the setup sheet to reduce errors and setup time.

Feeds and speeds may be optimized, or clearance planes reduced when safe. Whatever those small changes are, go back into the CAM system, update the toolpaths to reflect any changes, and save that program appropriately.

11.8 Post-Processing: From Screen to G Code

Post-processing is the term for turning toolpaths within the CAM system into actual G- and M-code programs that a machine can run. For most CAM systems, this is as simple as selecting the machine file for your machine and then telling the CAM system to create code for your toolpaths. Programs from 100 to 100,000 lines long can be processed in a matter of seconds. How does this process of converting the information from the computer screen into an actual program that can be run on a machine work? In the CAD, toolpath, and processing workflow, multiple files are created for the CAM system to use. The first one is the "CAD" file, or workpiece geometry file. The second is one the user never really sees or deals with, but it is key. It is the *generic machine code file*. It stores all of the different tool selections and parameters in a hidden file for the post-processor to use. The third file is the actual program file, or the G- and M-code file. Program files can be created as a variety of different file types, though they are nothing more than very basic text files. In fact, in the past, fax programs were used to send these back and forth to the machines because program files are just letters and numbers.

Remarkably, the generic machine code file gets "filtered" through whatever machine is chosen. So, if a program needs to run on a Haas VF2 machining center, post to that file. If that same job needs to be run again on a FANUC-controlled machine, simply change the machine selected from the *Machine Group: Properties* of your toolpaths (**Figure 11-32**) and repost the program. The program will change to run on that controller. These CAM post-processors can also be altered to output in different formats. The post-processor is really a translator from the generic machine code to the actual machine controller.

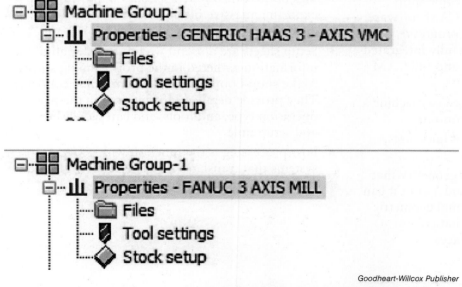

Goodheart-Willcox Publisher

Figure 11-32. The specific machine can be selected or modified in a CAM system to alter the final machine program.

Chapter Review

Summary

- Computer-aided drafting (CAD) is the graphical process of creating true-to-form and scale parts, assemblies, or renderings, using lines, vectors, and arcs to create a digital recreation of a concept or part.

- The benefits of CAD over conventional drafting include: reduced drafting time; elimination of the need for physical print storage and revision history; CAD drawings are to actual size, so no scaling is required; perfection of geometric shapes; the ability to create 3-dimensional drawings with multiview prints by simply rotating the part through the correct projection; and layers or levels that can be hidden or displayed depending on the desired print.

- Computer-aided manufacturing (CAM) systems provide the ability to quickly produce programs and simulate toolpaths for the manufacturing of parts, which can often be more complicated in the modern manufacturing environment.

- Simple parts and geometry can be created in CAM, but the software is not designed to do complicated assemblies or prints, as in CAD. Machinists are often given CAD drawings by the customer to machine, but most machine shop tasks will be completed using only CAM software. CAD/CAM developers are making progress developing integrated systems, and fully integrated CAD/CAM systems with full CAD and full CAM capabilities are on the horizon.

- Cutting technique is the method used to machine a feature, such as a pocket, drill, or contour.

- Geometry creation is the drawing of lines, arcs, or solids that define a part's shape.

- Two-dimensional (2D) geometry is geometry that only lies in two planes, usually X and Y (but it can be any two planes). Three-dimensional geometric shapes are defined in three planes that create depth. They include solids and surfaces.

- Surfaces, which have no thickness or mass and are often many surfaces "stitched" together to create complicated forms, cannot be machined without a CAM system. The ever-changing shapes and contours create a series of small moves often performed in X, Y, and Z axes simultaneously. Because the surface is ever changing, a ball end mill will be required to complete the surface toolpath. Solids, on the other hand, are truly one piece. CAM systems "see" all of the entities of a solid and include solid modeling and high-speed machining (HSM) toolpaths, which allow for the machining process to be sped up without concern for violating any adjoining part feature.

- Associated geometry is the ability within a CAM system to move the part and have all of the toolpaths associated with that part move with the geometry, eliminating programming and setup errors.

- Every CAM system includes an option to translate or import CAD files, either as IGES or DXF files. IGES files will import all 2D geometry, but when importing solids, it will import them as surfaces. DXF is a newer file type that was developed specifically by a CAD company to convert its files out to other systems. Issues importing and exporting massive files between systems are common.

- Setup sheets are a useful way to transfer all of the information gathered in the programming stages to the shop floor and the CNC machine controller. They provide detailed information on the part, operation type, and tools, and can reduce errors and setup time.

- Post-processing is the function within CAM systems that translates the generic machine code to the actual machine controller.

Review Questions

Answer the following questions using the information provided in this chapter.

Know and Understand

1. _____ is the graphical process of creating true-to-form and scale parts, assemblies, or renderings.
   A. CAM
   B. CAD
   C. Post-processing
   D. Toolpath generation

2. *True or False?* One benefit of CAD drawings is the perfection of geometric shapes.

3. Computer-aided manufacturing, or CAM, is the process in which we define the _____ using computer software.
   A. machine controller
   B. tool offsets
   C. machining parameters
   D. machining limits

4. The creation of _____ is a vital tool in working with a CAM system.
   A. geometry
   B. toolholders
   C. post-processors
   D. toolbars

5. The object of creating geometry is to designate a _____, _____, or _____ for the tool to enter or follow.
   A. surface, solid, path
   B. path, arc, circle
   C. point, hole, radius
   D. line, arc, point

6. *True or False?* The Cartesian coordinate system is the starting location for your CNC program.

7. A(n) _____ is just a position in the X and Y axes.
   A. tool
   B. point
   C. arc
   D. line

8. Lines can be drawn _____, _____, or at any angle.
   A. vertically, horizontally
   B. left, right
   C. up, down
   D. long, short

9. An arc is considered a non-closed circular shape, or less than _____.
   A. 90°
   B. 180°
   C. 540°
   D. 360°

10. *True or False?* Solids have no thickness or mass, and many solids are often "stitched" together to create complicated forms.

11. _____ is a term for a 3-dimensional CAD model that has mass and volume.
    A. Surface
    B. Solid block
    C. Solid model
    D. Weighted model

12. The toolbar to create solids has two main categories: _____ and _____.
    A. toolpath, geometry
    B. simple solid, create solid
    C. extrude, revolve
    D. contour, pocket

13. When using the *Revolve Solid* tool you must define two entities, the _____ and the _____.
    A. toolpath, geometry
    B. sweep, angle
    C. entry motion, exit motion
    D. exterior shape, axis of rotation

14. A(n) _____ is the cutting strategy used to create a part.
    A. toolpath
    B. contour
    C. toolholder
    D. end mill

15. *True or False?* The contour toolpath is used to direct a tool along a specific set of lines or arcs.

16. The _____ is used with a closed, or partially closed, boundary with a high volume of material removal.

   A. contour toolpath
   B. surface toolpath
   C. pocket toolpath
   D. solid toolpath

17. *True or False?* How we enter a cut may be the most critical decision.

18. The theory behind _____ is to use the entire flute length at very high spindle speeds and high feed rates.

   A. high-speed machining
   B. surface toolpath
   C. deep hole drilling
   D. peck drilling

19. With a CAM system, the toolpath is associated with the _____.

   A. toolholder
   B. chain
   C. geometry
   D. post-processor

20. _____ is the term for turning toolpaths from the CAM system into actual G- and M-code programs that a machine can run.

   A. Manipulating
   B. Chaining
   C. Selection
   D. Post-processing

Apply and Analyze

1. What are some advantages of computer-aided manufacturing (CAM)?

2. What is the basic difference between CAM software and CAD software?

3. What is feature-based machining and how will it streamline the programming process?

4. What are 2-dimensional geometry and 3-dimensional geometry? Are there advantages to creating toolpaths from different geometry?

5. Name some different types of 2-dimensional geometry.

6. Why is the contour toolpath one of the most commonly used cycles?

7. What is the difficulty in creating a pocketing toolpath?

8. What is the advantage to using solids to create toolpaths?

9. When might the machining of surfaces need to be used? Why would a CAD system create a surfaced part?

10. The term *associated geometry* refers to the toolpath being directly associated with the geometry from which it was created. Why is this important in CAM software?

11. Job setup sheets should always be provided to the operator and/or setup person. Why is it critical that a job setup sheet be included and thorough?

12. Post-processing of the toolpath converts the toolpath strategy into a usable G- and M-code program the machine can execute. How does this process streamline the manufacturing process if the parts are machined on different machines or different controllers?

Critical Thinking

1. We have discussed CAD software and CAM software. The industry is progressing to full integration, where all the "drawing" or CAD features and design are completed in the same platform or software as the CAM toolpaths are created. This integration eliminates issues with exchanging file types and with design changes affecting machining strategies. Why is this important to today's engineering and manufacturing teams that cooperate from all parts of the world?

2. Complex engineering designs and industry competition have challenged the manufacturing sector to keep pace with machinery, accuracy, and speed of production. Today's modern machine shops are becoming dependent on CAM software to program and produce these challenging demands. What new career paths are opened to this demand? What additional skills does today's machinist need to compete in the workforce?

3. Write a short summary of what computer-aided manufacturing might look like in 30 years. Consider the future of the current technology, like artificial intelligence, cloud-based computing, and augmented reality. How will this technology influence CAM systems and user interfaces?

A model of a car being created by two computer-controlled robot arms.

12 Multiaxis Programming

Chapter Outline

12.1 Multiaxis Machining

12.2 Multiaxis Turning

    12.2.1 2 + 1 Turning

    12.2.2 3-Axis Turning

12.3 Multiaxis Milling

    12.3.1 3 + 1 Milling

    12.3.2 Full 4-Axis Milling

    12.3.3 3 + 2 Milling

    12.3.4 Trunnions

    12.3.5 Fully Integrated 5-Axis Milling

    12.3.6 Is This 5-Axis Work?

    12.3.7 Full-Axis Engagement

12.4 Mill-Turn Machines

Learning Objectives

After reading this chapter, you should be able to:

- Discuss the benefit of multiaxis machining.
- Identify additional axes of movement in a CNC turning center that makes it a multiaxis machine.
- Explain what is meant by live tooling lathe.
- Explain the limitations of live tooling lathe machines.
- Define 2 + 1 turning.
- Define the term full 3-axis turning.
- Explain the different G and M codes required to perform live tool turning.
- Define 3 + 1 milling.
- Outline the multiple types of add-on equipment used for multiaxis milling.
- Describe full 4-axis milling operations and differentiate them from 3 + 1 milling operations.
- Discuss the value of 3 + 2 milling capabilities.
- Explain the term full 5-axis machining.
- Identify the various nomenclature of machine parts that comprise a 5-axis machine.
- Define the term full-axis engagement.
- Explain the use of the mill-turn machine.

Key Terms

2 + 1 turning	full-axis engagement	multiaxis
3 + 2 milling	harmonics	rotary
center of rotation	head-table design	surfacing toolpath
full 3-axis turning	indexer	tool engagement
full 4-axis milling	live tool	trunnion
full 5-axis milling	mill-turn machine	

12.1 Multiaxis Machining

The term *multiaxis* has been used throughout this text. As seen in Chapter 7, many different machine types fall under the category "multiaxis" machinery and various components can be added to machines to make them multiaxis. The two machines used most in the manufacturing world—the lathe and the mill—have either two or three axes of movement. In the lathe there are normally two axes of movement—the X and the Z, and in the mill there are 3 axes of movement—the X, Y, and Z. Any time there are more axes of movement than standard, it is considered ***multiaxis***.

So what is the benefit of multiaxis machining? Previously in machining, parts like those in **Figure 12-1** would be machined one side at a time, one setup at a time, and they might require lathe *and* milling operations. Many machinists and machine shop owners will say they make parts that way *"because that's the way we've always made them."* Those machine shops will not survive in today's rapid-paced, ever-changing manufacturing environment. The innovative, forward-thinking machinists are always looking for a faster, more cost-effective, and better-quality machining process. Many shops have turned to multiaxis equipment like rotaries, trunnions, and even fully integrated mill-turn machines to solve complex machining issues. It cannot be stressed enough that these are machine solutions being developed by machinists and machine shop owners, and all of these machine parts are manufactured by machinists. There are not many fields in which you can identify a need, design and CAD the solution, and then go out to your machines and create the pieces.

Someone who did just that was Gene Haas, who in 1978 started a small machine shop with two employees. In 1980 he noticed one of his employees was taking a long time using a manual indexer on a mill. Haas thought he could speed up the process by using a stepper motor to drive the indexer and then programming its movement. In 1983 Haas Automation debuted its programmable 5C colleted ***indexer*** at WESTEC, the largest machine tool show on the West Coast. By 1984 he had a patent, and in 1986 he started mass production of the HBI-5C programmable indexer, similar to the one in **Figure 12-2**. Gene Haas grew that idea into making fully enclosed CNC machines, and he is now the largest manufacturer of CNC equipment in North America. It all started in a machine shop with a need to do multiaxis work smarter and faster.

This chapter discusses the benefits and programming of various multiaxis equipment. Up until a few years ago, there was only one real downside to multiaxis equipment, and that was the cost. Since 2010 the cost has come down significantly, and the use of multiaxis equipment has multiplied in the field. Whereas multiaxis machining used to be considered high-level programming and high-level equipment, today it is becoming the norm around the globe. In most cases, a CAM system is required to perform multiaxis programming.

CNC Software, Inc.

Figure 12-1. Multiaxis machined parts made in a single operation.

Goodheart-Willcox Publisher

Figure 12-2. The Haas HBI-5C programmable indexer, as mounted on a machining center table.

Tech Tip

In multiaxis machining, there is a great variety of G and M codes between different controls and machine types. This text displays codes primarily used for Haas machines, due to their wide-scale availability. Most FANUC-controlled machines will accept the same code and add-on equipment. Although the processes outlined here are comparable for other machines and controllers, consult with the machine manufacturer's programming manuals for specifics.

12.2 Multiaxis Turning

The standard CNC turning center has two axes of movement, the X and the Z. There are a couple ways to make a turning center into a multiaxis machine. The spindle can be oriented and locked into place; this is called the C axis. With the spindle oriented in place, a "live tool," such as a drill or end mill, can be used to drill or slot the turned piece. There can also be a Y axis of movement where the tool can be moved off the part centerline to machine off center. **Figure 12-3** shows an example of a part made through multiaxis turning.

12.2.1 2 + 1 Turning

The term *2 + 1 turning* refers to the use of a 2-axis turning center in which the spindle can be programmed to any rotational angle and locked into place to provide a third axis. A *live tool* is used to create features on the face or OD of a part. A lathe live tool is one that is driven axially or radially to perform such secondary operations as drilling, milling, or tapping—both on the face of the part and around the diameter. The sample part in **Figure 12-4** provides an example.

This is a typical 2 + 1 application. The picture displays a coupling with two through holes. Also shown is the stock representation and the chuck jaw orientation. The order of operations would be to drill and bore the ID, then face and turn the OD, creating the two diameters shown. Without the use of multiaxis turning, this part would then have to be removed from the lathe and placed in a mill, most likely an indexer, to create the two through holes. That means a second machine would be tied up machining, with an entirely different setup and cycle time. A great advantage to multiaxis turning is that parts can be completed in one operation, on one machine. Adding a second operation magnifies the chance for human error or fixture repeatability errors. The following program is the hole-making operation for Hole #1 using 2 + 1 turning:

CNC Software, Inc. Mastercam

Figure 12-3. A typical turning center part in which cross hole drilling is performed in a single operation.

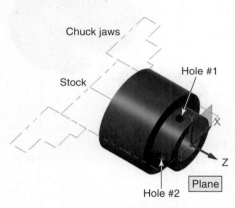

Goodheart-Willcox Publisher

Figure 12-4. A 3-dimensional CAD model of a sample lathe part that requires turning and cross hole drilling.

O3295	
(MULTI SAMPLE 1)	
(MATERIAL - STEEL INCH - 1030)	
G20	
(TOOL - 6 OFFSET - 0)	
(LTR. Q DRILL)	
T0600	
M154	C Axis Engage

G97 P1610 M133	Turn on live tooling to 1610 rpm
G98	Feed Per Minute
G17	
G00 G54 **Z1.** C90.	Rotate to 90° or C90.
X2.	
G81 R1.5 F4.12 X.5	Drill to X.5
G00 G80 X2.	
M155	C Axis Disengage
M135	Live Tooling Stop
G28 U0.	
G28 **W0.**	
M30	

Goodheart-Willcox Publisher

Figure 12-5. Computer-aided manufacturing simulation of the first hole drilled in the 3-dimensional model.

In the program there are notes in parentheses indicating G or M codes not found in 2-axis turning that are specific to 2 + 1 multiaxis machining. This program creates a drilled hole as shown in **Figure 12-5.**

By simply adding another C position and the Z coordinate of the second hole, Hole #2 can be created, as seen in **Figure 12-6.**

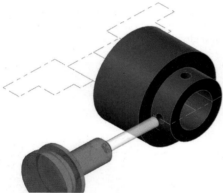

Goodheart-Willcox Publisher

Figure 12-6. Simulation of the second drilled hole. Notice the tool orientation to the second drilled hole.

O3295	
(MULTI SAMPLE 1)	
(MATERIAL - STEEL INCH - 1030)	
G20	
(TOOL - 6 OFFSET - 0)	
(LTR. Q DRILL)	
T0600	
M154	C Axis Engage
G97 P1610 M133	Turn on live tooling to 1610 rpm
G98	Feed Per Minute
G17	
G00 G54 **Z-.4** C90.	Rotate to 90° or C90.
X2.	
G81 R1.5 F4.12 X.5	Drill to X.5
C180.	Additional Hole Rotation for Hole #2
G00 G80 X2.	
M155	C Axis Disengage
M135	Live Tooling Stop
G28 U0.	
G28 **W0.**	
M30	

12.2.2 3-Axis Turning

Full 3-axis turning is the process of moving a primary axis simultaneously with a secondary axis. **Figure 12-7** adds a slot to our existing part file. Notice that this slot is angled slightly across the part. To perform this operation, rotate the C axis in time with a Z-axis movement.

Rotating slot

Goodheart-Willcox Publisher

Figure 12-7. A rotating slot added to the sample part. This part requires full 3-axis machining.

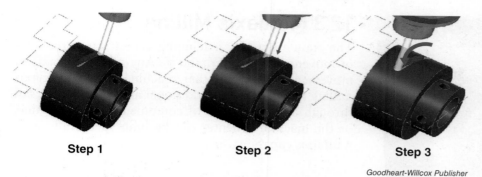

Step 1　　　　**Step 2**　　　　**Step 3**

Figure 12-8. Step-by-step view as the slot is created.

The operational sequence, or programming strategy, would be to rotate the C axis to the starting angular position of the slot, position the end mill, move it down in X to the correct depth, and then rotate the C while simultaneously moving Z in a negative direction. **Figure 12-8** shows the steps involved.

Although this type of programming can be programmed manually, it is most often programmed through a CAM system. In the CAM software lathe operations, there are options to create 3-axis milling or drilling functions. The sample in **Figure 12-8** would require selecting *C Axis Contour* and then selecting the appropriate tool and toolpath. This will create a type of "milling" operation in your lathe toolpath, **Figure 12-9**. You must be using a lathe that is equipped with C-axis capability to perform this operation.

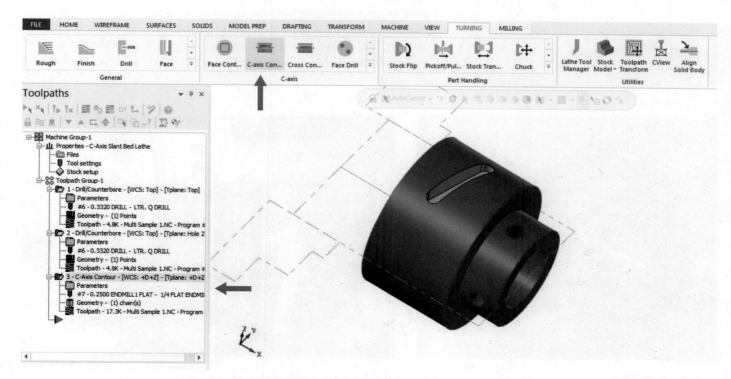

Figure 12-9. The *C-axis contour* toolpath is selected in the *C-axis* toolbar.

Haas Automation Inc.

Figure 12-10. A platter-style, 4th-axis rotary. Material or fixtures can be attached to this fixture.

12.3 Multiaxis Milling

The standard CNC machining center has three axes of movement, the X, Y, and Z. Any additional axis of movement transforms that machine into a multiaxis milling machine. These additional axes can be achieved through adding a piece of equipment to the machine, or the machining center can be built specifically with multiaxis capability.

12.3.1 3 + 1 Milling

3 + 1 milling is the process of adding one additional axis to a 3-axis machining center. Usually an additional axis is called a *rotary* because it rotates around a single axis. **Figure 12-10** and **Figure 12-11** show two different styles of rotary axis components. Both of these units can be mounted in a vertical or horizontal position, depending on the type of operation being performed.

Depending on the machine type and setup, these axes are commonly referred to as the A axis. The correct nomenclature would be *A* if it rotated around the X axis, *B* if it rotated around the Y axis, and *C* if it rotated around the Z axis. But for standardization and ease of programming in a multiple setup piece of equipment, standard practice is to call them all the A axis. 3 + 1 milling can be very effective for multiple part setups (**Figure 12-12** and **Figure 12-13**) or multisided work where it would require multiple setups to complete a part.

Figure 12-14 shows an example of 3 + 1 machining: a rectangular-shaped part with machining on the flat faces of two sides. For clarity, there is no clamping or workholding shown on the 3D model. To program this part, decide where program zero or work coordinate system (WCS) will be located. Many programmers prefer to place the center of

Goodheart-Willcox Publisher

Figure 12-11. A 5C collet indexer. Round stock material can be secured in a collet closer for 4th-axis machining.

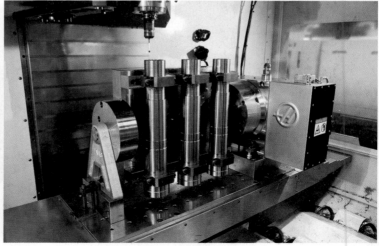

Haas Automation Inc.

Figure 12-12. A part fixture can be mounted on a 4th-axis rotary for multipart machining.

the block concentric with the center of rotation of the rotary axis, in which case just one work coordinate (G54) is needed to program both faces. As the rotary rotates the block, the block spins on the same centerline. This requires additional setup time to make sure the block has a common centerline with the rotary. This also requires a more stable and accurate fixture or workholding to allow multiple parts to be run and maintain that position without setting up each block individually.

The second option is to set a WCS on each rotation, probably 0° and 90° on this part. The WCS (G54 and G55) can be set on any corner, or on the centerline, of any side. That allows for an easier setup, but it still requires accurate repeatability with workholding.

To actually program the part features, simply position the rotary at 0° and then run the 3-axis program as if it were in a vise or any other normal setup. Then rotate the part to 90° and run the toolpath for the second side. It is often a good practice to test cycle times by trying to machine one side completely, then the second side completely. Then try running each tool on both sides completely, rotating the piece back and forth as needed, to compare methods and decide which is most efficient. Recall that CNC machining is all about speed and accuracy.

Haas Automation Inc.

Figure 12-13. Multiple part and fixture designs can be assembled for multiaxis machining.

12.3.2 Full 4-Axis Milling

Full 4-axis milling is similar to 3-axis turning. In *full 4-axis milling*, one or all of the primary axes are moving while the additional axis is utilized simultaneously. It is a little more simplistic, given that the milling machine is already moving in an inches per minute (ipm) feed rate. Parts milled using this method can become extremely complex, but **Figure 12-15** provides a part with a fairly straightforward programming strategy and code to complete it.

This part has a 10° rotated slot, starting at X–1. and ending at X–2. Because of the rotation of the slot, a simple X- and Y-axis movement will not suffice. The programming code will look like this:

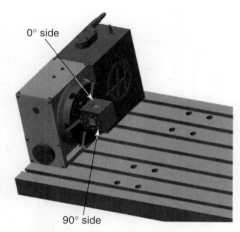

Haas Automation Inc.

Figure 12-14. A rectangular cube with machining on multiple sides.

```
O3312 (10° SLOT)
M6 T1 (1/4" END MILL)
G0 G90 G54 X-1.5 Y0. A0. M3 S7640
G43 H1 Z1.5
G1 Z.85 F10.
X-3. A10. F76.4
Z1.5
G0 X-1.5 A0.
G1 Z.75 F10.
X-3. A10. F76.4
Z1.5
G0 G91 G28 Z0.
M30
```

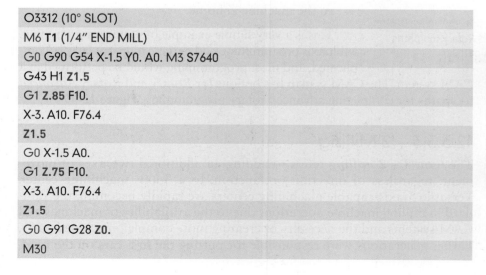

Goodheart-Willcox Publisher

Figure 12-15. A computer-aided drafting model of a part requiring rotational slot machining.

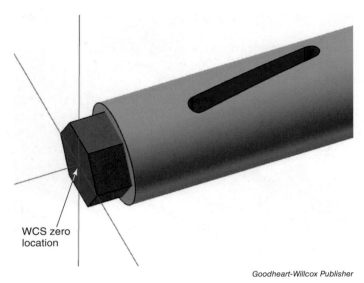

Goodheart-Willcox Publisher

Figure 12-16. The WCS zero is set at the center of rotation and part face in this example.

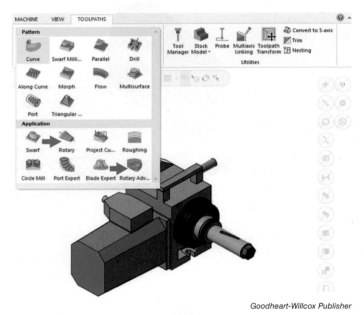

Goodheart-Willcox Publisher

Figure 12-17. Either the *Rotary* or *Advanced Rotary* toolpaths can be selected to create 4th-axis rotary toolpaths.

Look at some individual lines to fully understand the code and the machining strategy involved. First establish "where" on the part to create the WCS zero point. With a single slot or single rotation part, it might be simpler to set the X and Y WCS at the face or end of the part and the Z coordinate or tool height offset at the top of the part. But most programmers doing multiaxis machining will set the Z or tool height offset at the center of rotation. Both strategies will work as long as the programmer and machine operator are communicating the location of offsets. In this part, the center of rotation is set as the WCS, **Figure 12-16**. The important piece is consistency of programming, so program all subsequent multiaxis projects from the center of rotation.

Analyze the program line-by-line:

- O3312 is the program number.
- M6 T1 calls for a tool change to tool #1, a 1/4″ end mill.
- G0 G90 G54 X–1.5 Y0. A0. M3 S7640 is a standard positioning line with the A0. as a start angle. Turns on spindle to the correct rpm based on tool diameter and surface footage of material.
- G43 H1 Z1.5 turns on the tool height offset for tool #1.
- G1 Z.85 F10. changes to feed rate and goes to the bottom of the slot. Notice the Z position is a positive number because the WCS is coming off the centerline of the part.
- X–3. A10. F76.4 makes the feed rate move to our final X position and rotates the A axis simultaneously. This is the same as any other 2-axis movement in which the machine controller will calculate the direct motion to that position.
- Lines 7 through 10 make a second, deeper pass to finish the slot.
- Lines 11 through 13 finish the program and return to the machine Z home.

This is a very simple example of 4-axis programming. Multiaxis programming and part creation can be very complicated, and most programming would be performed in a CAM system. The CAM system has options for *Rotary* and *Rotary Advanced* toolpaths for completing advanced rotary programming, **Figure 12-17**.

12.3.3 3 + 2 Milling

The term *3 + 2 milling* refers to adding an additional two axes of movement, regardless of direction or function, to a 3-axis machining center. Versatile 5-axis-capable machining centers are rapidly becoming the standard in milling machine operations due to the availability of machines and CAM systems and the necessity of creating more complex machined parts. Earlier generations were responsible for putting the first cars on the road, the first aircraft in the sky, and men on the moon. Today's generation will be responsible for putting humankind on Mars, electric high-speed cars, and building aircraft capable of speeds up to 3,500 miles per hour. As the requirements for higher technological advances are continuously thrust upon the machining industry, our ability to create these complex designs requires more diversity and technology.

The 5-axis machine evolved separately from normal CNC equipment. Initially, standard 3-axis CNC machining centers were built by only a couple of different machine builders, with controllers made by only one company, FANUC. That led to very little diversity. Imagine if there were only two car makers and one engine maker. All cars would be very similar, and all the engines would be exactly the same. It also caused an issue with buying CNC machines, as delivery times were long and costs were high. And so, many companies, such as Haas, Cincinnati, Mazak, and Mitsubishi, started building their own 3-axis machining centers. These 3-axis machines were fairly similar to the originals, but when the need for multiaxis machines began to increase, all of these companies started to research and develop their own multiaxis equipment. This equipment varies tremendously from different machine builders. As previously discussed, a 5-axis machine can be a trunnion attachment to an existing 3-axis machine, or it can be a fully integrated machine built specifically for 5-axis machining. Before diving into 5-axis programming, it is important to understand the many variables associated with all of these different designs and machine configurations.

12.3.4 Trunnions

Figure 12-18 shows a *trunnion* unit mounted on a 3-axis machining center table. The platter or table rotates, and the saddle that encases the platter also rotates from a horizontal to a vertical position. This is a fairly common system, but it has the advantage of mounting and unmounting from the machine as needed. The disadvantage to this system is that it limits the working envelope of the machine. It is also very heavy and difficult to mount and unmount.

Another solution is a compact trunnion, **Figure 12-19**. These units take up less table space or work envelope; however, these smaller units do limit the size of the part or fixture that can be mounted to the trunnion.

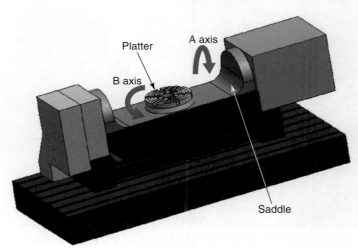

Goodheart-Willcox Publisher

Figure 12-18. A 5-axis trunnion mounted to a conventional machining center table.

Goodheart-Willcox Publisher

Figure 12-19. Axis of rotation on a compact trunnion.

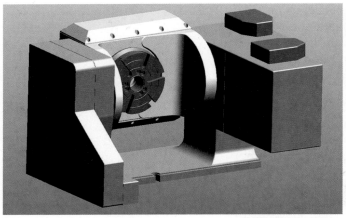

Goodheart-Willcox Publisher

Figure 12-20. Haas TR160 compact trunnion. A space-saving alternative to a full trunnion setup.

Goodheart-Willcox Publisher

Figure 12-21. Compact 5-axis trunnions can be light and easy to install and remove as required.

Figure 12-20 shows another Haas product, a compact TR160. Although this is a different design, it is similar in function to other compact trunnions.

All of these trunnions are supported on both ends through bearing blocks that give them added rigidity but also tend to slow rotation slightly. **Figure 12-21** shows another trunnion-style design that is only supported by one arm but is much lighter and faster.

12.3.5 Fully Integrated 5-Axis Milling

Rather than adding a separate piece of equipment to a 3-axis machine, machining centers can be built fully integrated as 5-axis machines. This is the best possible solution as they are very rigid, very fast, and do not influence the work envelope of the machine. Of course, there are multiple ways to accomplish this, and machine designers and builders continue to push the envelope of design and creativity. **Figure 12-22** shows a 3D model of one such machine in which the trunnion sits on the machine saddle and is driven by the ball screws.

Figure 12-23 shows an example of a horizontal machining center where the worktable is rotated and integrated into the machine base. All of

Goodheart-Willcox Publisher

Figure 12-22. A fully integrated 5-axis machine has the trunnion built into the machine frame and it cannot be removed.

Goodheart-Willcox Publisher

Figure 12-23. A horizontal machining center that is fully integrated for 5-axis machining. The arrows indicate where the rotation occurs.

the external sheet metal was left off the 3D model to expose the integration of rotation into the actual machine base. This machine is quite different from the add-on trunnion setups.

There are even more variable setups in 5-axis machines. **Figure 12-24** shows a machine setup known as a *head-table design*, with a close-up of the spindle and table design. The orange solid is the material workpiece. In these machines, the table rotates and also the entire spindle cartridge rotates around a fixed location. This is a very low-profile machine used in larger part manufacturing for its added clearance and extended reach.

12.3.6 Is This 5-Axis Work?

Most machining on a 5-axis machine is not true 5-axis machining. True *full 5-axis milling* would mean that the A or B axes were moving while the tool was engaged in an X-, Y-, or Z-axis movement. Usually we are positioning a workpiece in a set of specified angles and then doing traditional 3-axis-type toolpaths. But that comes with some challenges with a 5-axis machine.

Examine a TR210 trunnion and how it operates to see some differences from a 3-axis machining center. Using **Figure 12-25**, the block from the previous 4-axis example (**Figure 12-14**), compare how it functions as a 5-axis or 3 + 2 project.

If this were a 3-axis milling operation, a work coordinate (G54) would be set in a top corner, all of the tools would be set off the top of the part, and each side would be machined one at a time. With a 5-axis trunnion, only one setup is required and all of the sides can be machined in one operation. **Figure 12-26** shows the part with the trunnion oriented to its home position, or A0., B0. The 3D model also shows where the conventional X, Y, and Z WCS setting would be set.

That setup works perfectly for the first side of machining, but look what happens when the part is rotated to A axis 90° to begin machining

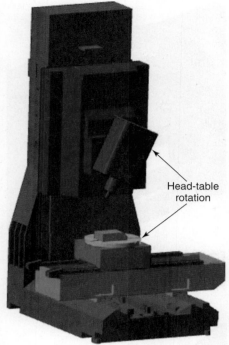

Head-table rotation

Figure 12-24. 5-axis machines can come in multiple setups, including this head-table design.

Figure 12-25. The sample rectangular block with the same machined features on multiple sides.

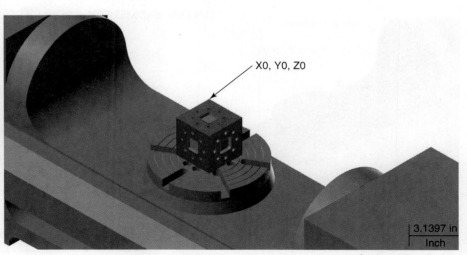

X0, Y0, Z0

3.1397 in

Inch

Figure 12-26. The original work coordinate home position used for programming in the top left corner.

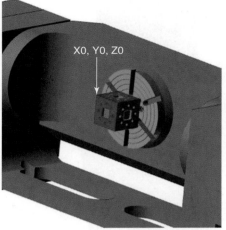

Goodheart-Willcox Publisher

Figure 12-27. As our sample part is rotated, the WCS position moves from its original position.

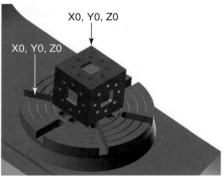

Goodheart-Willcox Publisher

Figure 12-28. The original WCS and the rotated position will vary depending on the center of rotation.

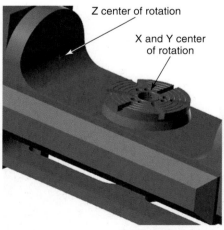

Goodheart-Willcox Publisher

Figure 12-29. The center of rotation is the point where all the axes intersect in X, Y, Z, A, and B.

the second side, **Figure 12-27**. The original WCS is no longer in the same location, not on the part or on the machine. To solve this, a WCS can be set up at each rotation, or trunnion angle. The machine's WCS—or G54, G55, G56, and so on—might look like this (with the blanks filled in with the part's exact position in X and Y):

| G54 X–____ Y–____ Z0. A0. B0. |
| G55 X–____ Y–____ Z0. A90. B0. |
| G56 X–____ Y–____ Z0. A90. B90. |
| G57 X–____ Y–____ Z0. A90. B180. |

For a second view of the changing WCS locations during rotation, look at the two different points in the 3D model, **Figure 12-28**. That is the same exact point on the part at two different rotations.

If programing this part manually, this is the only way to successfully use the WCS while rotating the part into the correct orientation. But it would be extremely rare to find a manufacturing facility that has multi-axis equipment and did not utilize all of the benefits of a high-end CAM system. There is a different strategy used in a CAM system that lets the computer do all the work.

On any 5-axis machine there is a point in space that is called the ***center of rotation***. It is a physical point where all rotation centerlines intersect. On the trunnion in **Figure 12-29**, the X- and Y-axis center of rotation is the center of the platter or tabletop. The Z-axis center of rotation is located at the center of the hub where the table is mounted to the trunnion gear and bearing blocks.

In a side view of the wire-framed trunnion, **Figure 12-30**, the point where the X, Y, and Z axes all merge is the true center of rotation. In a CAM system, this is the point from which to program, and by establishing work planes, a program can be generated that forces rotation. This CAM system program looks like this:

| O8258 (TR-210) |
| (T2\|1/4 FLAT ENDMILL\|H2\|D2\| DIA. .25) |
| G0 G28 G91 Z0. |
| (1/4 FLAT ENDMILL\|TOOL 2\|DIA OFF 2\|LEN 2\| DIA .25) |

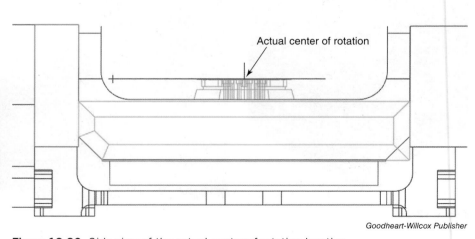

Goodheart-Willcox Publisher

Figure 12-30. Side view of the actual center of rotation location.

```
M11
M13
T2 M6
G0 G54 G90 X-.1935 Y-1.5294 B0. A90. S6112 M3
M10
M12
G43 H2 Z3.625 M8
Z2.125
G1 Z1.725 F12.22
X.265 Z1.701
X-.235 Z1.6748
X.265 Z1.6486
X-.235 Z1.6224
X.265 Z1.5962

...(Lines removed for clarity)

G1 Y-1.625
Z1.75 F12.22
G0 Z3.625
M9
M5
G0 G28 G91 Z0.
M30
```

Notice that the X, Y, and Z positions are odd in this code. The Z axis moves are positive, the X axis is going from positive to negative, and the Y is not where we would expect it. This is because the part has been rotated, but the center of rotation as our part home, or WCS, has been maintained.

 From the Shop

Dynamic Work Offsets

As machine builders have worked to create more 5-axis machinery, they have also made programming easier through improved controller technology. Most new controllers have *dynamic work offsets*. Dynamic work offsets allow you to set the center of rotation in a machine parameter and store it for machine calculations. You can then enter the material work coordinate location the same as you would on a 3-axis machine. The controller will do the calculations from center of rotation to part piece without any additional programming or offsetting—programming made simple.

12.3.7 Full-Axis Engagement

The process of full 4-axis or full 5-axis machining occurs when the material or workpiece is being simultaneously rotated while the tool is engaged in the material, forcing movement in another axis. This is referred to as

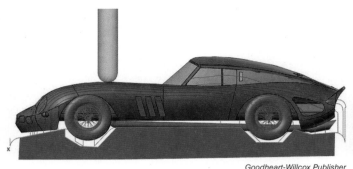

Goodheart-Willcox Publisher

Figure 12-31. Tool oriented to 90° of the machined surface.

Figure 12-32. As the tool climbs angled surfaces, the end mill side cuts the surface.

Goodheart-Willcox Publisher

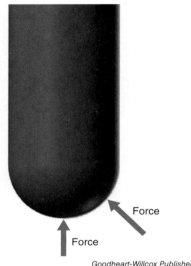

Force

Force

Goodheart-Willcox Publisher

Figure 12-33. Direction of cutting forces when the end mill is normal to cut surface and when it is side cutting the surface.

Goodheart-Willcox Publisher

Figure 12-34. In 5-axis machining, the part is rotated to machine in the normal cut direction.

full-axis engagement. For example, if a spherical object is being machined, the trunnion or table will be rotating, forcing a Z and X axis or Z and Y axis simultaneous movement. How much is it moving? The beauty of 5-axis engagement is that the centerline of the tool will stay oriented through the centerline of the part or the surface being machined. This is called machining "normal" to a surface.

Figure 12-31 shows an example of normal machining with a 3-dimensional model of a scaled-down car project. The tool is perpendicular to the surface being machined, or normal to the surface. The toolpath created is a *surfacing toolpath* and not a true 5-axis toolpath.

This is a typical example of 3 + 2 machining with surfacing paths, **Figure 12-32.** Farther along the path, as the tool encounters steeper angles, the tool is no longer cutting with the end of the ball end mill—now, it is side cutting. Side cutting with a ball end mill is done all the time, but it does have consequences. **Figure 12-33** shows the radial forces exerted on the tool when it cuts normal to the surface compared to side cuts to that surface. The forces in a normal cut are driven through the tool centerline and ultimately through the toolholder and spindle. This gives the best surface finish and the most rigidity. When the tool is side cutting, the forces are transmitted only through the tool. The tool will begin to chatter and vibrate with what are often called *harmonics.* This creates excessive tool wear, inferior surface finish, and can lead to a catastrophic failure.

The correct *tool engagement* should resemble **Figure 12-34**, with the tool normal to the surface being machined. Tool engagement is the angle and depth of the tool as it makes a cut. This engagement is directly related to chip thickness.

The only way to create a cutting profile that remains normal to the surface being machined is with a fully engaged 5-axis toolpath. The number of calculations required to create a full 5-axis path is staggering and will require a high-end computer system with advanced CAM software. These programs tend to be tens of thousands of lines long and vary greatly from one CAM system to another. A typical, small section of full 5-axis machining code might look like this:

```
N10 (3/8 BALL ENDMILL 0.06R|TOOL – 1/DIA. - .375)
M11
M13
T1 M6
```

```
G0 G54 G90 X1.1119 Y-3.4055 B149.492 A90. S10000 M3
G43 H1 Z2.7232
X1.002 Y-3.0688
Z2.454
Z2.054
G1 Z1.654 F50.
G93 X.9895 Y-3.0631 Z1.6618 B149.912 F950.
X.9624 Y-3.0511 Z1.6783 B150.814
X.9426 Y-3.0426 Z1.6901 B151.47
X.9203 Y-3.0333 Z1.7029 B152.203
X.9002 Y-3.0251 Z1.7141 B152.86
X.8738 Y-3.0148 Z1.7283 B153.715
X.8509 Y-3.0061 Z1.7402 B154.453
X.8235 Y-2.9961 Z1.754 B155.332
X.802 Y-2.9886 Z1.7644 B156.016
X.777 Y-2.9801 Z1.7761 B156.805
X.7513 Y-2.9717 Z1.7876 B157.613
X.731 Y-2.9653 Z1.7964 B158.248
X.7051 Y-2.9574 Z1.8073 B159.054
X.683 Y-2.9509 Z1.8161 B159.741
X.6593 Y-2.9443 Z1.8253 B160.469
X.637 Y-2.9382 Z1.8336 B161.154
X.6168 Y-2.933 Z1.8408 B161.771
X.5895 Y-2.9261 Z1.8502 B162.602
X.5598 Y-2.9191 Z1.8598 B163.503
X.5324 Y-2.913 Z1.8682 B164.33
X.511 Y-2.9084 Z1.8745 B164.973
X.4833 Y-2.9028 Z1.8822 B165.802
X.4562 Y-2.8976 Z1.8893 B166.612
X.4318 Y-2.8933 Z1.8953 B167.338
X.4066 Y-2.889 Z1.9012 B168.086
```

Five-axis toolpaths are created in the CAM software through the *Multiaxis* toolbar, **Figure 12-35**. There are multiple types of strategies, and the selection of the toolpath will depend on the type of path or the part being machined. Often, more than one strategy or toolpath is required to get the desired results.

 Safety Note

Know the sequence of operations executed by any program run by the machine you are operating. This allows you to recognize problems with the machine if it does something out of the ordinary.

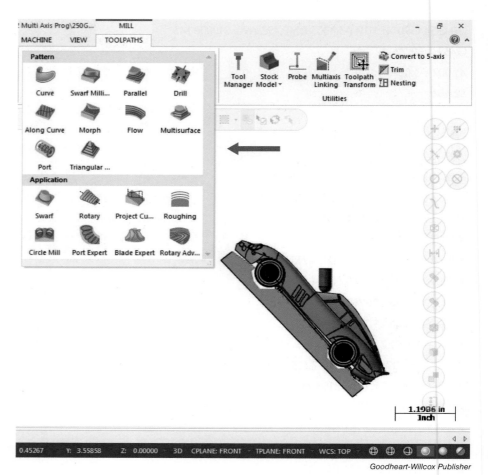

Figure 12-35. The *Multiaxis* toolbar is used to create multiple types of 5-axis toolpaths. The selection of the toolpath will depend on the type of feature being machined.

12.4 Mill-Turn Machines

Some milling machines can make round parts, and some lathes can do some milling, but these machines are either milling machines with attachments, or turning centers with additional axes or attachments. For example, a live tooling 3-axis lathe does have the ability to do some milling, but it is limited. A 3-axis lathe has an attachment in the tool turret that is either belt driven or, in higher-quality machines, gear driven. These attachments are usually limited in spindle speed (between 1,500 and 3,500 rpm) and significantly lower in horsepower (HP). A small, standard CNC machining center has a 30 HP motor, and most live tooling lathes have less than 5 HP in the milling attachment.

Machinists appreciate the versatility of a CNC turning center that also has a live tooling attachment, allowing for cross hole drilling or simple slotting. An indexer also offers versatility, allowing for more extensive milling work on a lathed part using a 4th-axis rotary in a CNC machining center. But of course, CNC machinists want full turning capability along with full milling capability all in one machine. That is why the ***mill-turn machine*** was developed.

Most mill-turn machines look a lot like a CNC turning center. They come with at least one spindle, or hydraulic chuck, but most often have

Pixel B/Shutterstock.com

Figure 12-36. A mill-turn machine performing finish milling machining operations in a yoke.

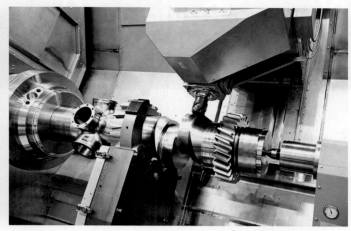

Sergey Ryzhov/Shutterstock.com

Figure 12-37. A mill-turn machine performing complex lobe and gear machining in a single operation.

two spindles to handle larger work and provide more rigidity. The real difference is in the milling. Unlike turning centers with additional live tooling features, these have full-capacity milling heads. They come out as a separate milling head that has the ability to change tools. The real advantages are travel range, horsepower, and a fully articulating arm, making it a true milling machine.

Figure 12-36 and **Figure 12-37** demonstrate some of the capabilities of the mill-turn machine. These machines can cut cams, gears, slots, ports, and almost any other part that would normally take a turned part with significant milling. As our engineering designs have become more complex and demanding, the advent of multifunctional machines like the mill-turn are becoming much more prevalent. This is technology that we should all keep our eyes on, as it will soon replace the 3-axis lathe setup. The complexity of machining in a very tight work envelope will require a strong CAM system built specifically for this type of machine. All of these machines are being developed individually by machinery makers, and they will vary greatly, especially on the programming side.

 **Thinking Green**

Leadership in Energy and Environmental Design (LEED)
Established by the US Green Building Council, LEED is the industry standard in green construction. LEED rates green building practices in such areas as energy conservation, building materials, water use, and air quality. Green practices are evaluated from the construction phase through the life of the building project. Buildings must meet certain criteria to achieve LEED certification, which offers numerous benefits. LEED certification can boost public image and qualify buildings for certain tax benefits and zoning allowances. Green practices are becoming more prevalent in factories, where smart choices for the environment can also promote efficiency and thereby reduce costs. Green machining practices and construction methods are an important part of LEED, alongside recycling and environmentally friendly cleaning methods.

Chapter Review

Summary

- Multiaxis machining can provide alternatives to machining complex parts. The use of multiaxis machines reduces setup times and additional machine operations.

- A CNC turning center becomes multiaxis when the spindle can be oriented and locked into place to create the C axis, or when the tooling moves off the part centerline to create the Y axis.

- A live tooling lathe is one that has the capability to do light milling or drilling operations.

- Live tooling lathes have limited horsepower and range of motion, impeding their ability to do full milling or heavy drilling-type operations.

- 2 + 1 turning is the use of an additional axes of movement from the standard X and Z axes of a normal lathe operation.

- Full 3-axis turning is the process of moving a primary axis simultaneously with a secondary axis. In this process, the X and Z axes are moving while the C axis is also rotating.

- Multiaxis turning requires additional G and M codes than does standard turning. Milling or drilling operations must be programmed in G98 (inches per minute), and M codes like M133, M135, and M155 are used to control the live tooling and C-axis engagement.

- The term *3 + 1 milling* refers to adding one additional axis of movement to a standard 3-axis machining center. It refers to positioning the one alternate axis in place and performing 3-axis machining programs. This alternate axis can be the A or B axis, depending on orientation or machine controller configuration.

- Additional equipment can be added to a machining center to achieve multiaxis machining. These are called trunnions or rotaries, based on their configuration.

- Full 4-axis machining is accomplished when all four axes of a machining center are moving simultaneously during a toolpath.

- The term *3 + 2 machining* refers to an additional two axes of movement being added to a standard 3-axis machining center. These axes are most often referred to as the A and B axes, but it can vary in certain machine controller configurations.

- Full 5-axis machining is where all five axes of a machining center are being simultaneously moved during a toolpath operation.

- Full-axis engagement is when all available machine axes are being used simultaneously.

- The mill-turn machine is a hybrid of a CNC turning center and a CNC machining center. It is capable of turning operations and full milling operations.

Review Questions

Answer the following questions using the information provided in this chapter.

Know and Understand

1. Any time there are more axes of movement than standard, the machining is considered _____.
   A. 3-axis
   B. multiaxis
   C. milling
   D. mill-turning

2. Many shops have turned to multiaxis equipment like _____, _____, and even fully integrated mill-turn machines to solve complex machining issues.
   A. end mills, lathes
   B. CAD, CAM
   C. rotaries, trunnions
   D. ball end mills, indexers

3. The lathe spindle can be oriented and locked into place. This is called the _____.
   A. C axis
   B. B axis
   C. A axis
   D. mill-turn machine

4. _____ is the process of moving a primary lathe axis simultaneously with a secondary axis.

   A. 3-axis milling
   B. 4-axis milling
   C. 4-axis turning
   D. 3-axis turning

5. 3 + 1 milling is the process of adding _____ axis/axes to a 3-axis machining center.

   A. a rotary
   B. a trunnion
   C. one additional
   D. two additional

6. *True or False?* Rotary axis units can be mounted in a vertical or horizontal position.

7. In _____, we are simultaneously moving one or all of the primary axes while also utilizing one additional axis.

   A. 5-axis milling
   B. 4-axis milling
   C. 3 + 1 milling
   D. 3 + 2 milling

8. Most programmers doing multiaxis machining will set the Z or _____ at the center of rotation.

   A. part
   B. X axis
   C. Y axis
   D. tool height offset

9. *True or False?* The use of a 3-axis-only machining center is rapidly becoming the standard in milling machine operations.

10. A 5-axis machine can be an attachment to an existing _____, or it can be a fully integrated machine built specifically for 5-axis machining.

    A. 3-axis machine
    B. 4-axis machine
    C. 2-axis machine
    D. multiaxis machine

11. *True or False?* Machining centers can be built fully integrated as 5-axis machines.

12. In a head-table design machine, the _____ rotates and also the entire _____ rotates around a fixed location.

    A. table, trunnion
    B. table, rotary
    C. table, programmer
    D. table, spindle cartridge

13. True 5-axis machining would mean that the _____ and _____ axes were moving while the tool was engaged in an X-, Y-, or Z-axis movement.

    A. G, M
    B. U, W
    C. A, B
    D. A, C

14. *True or False?* On any 5-axis machine the intersecting point of all the axes is called the center of origin.

15. The process of full 4- or 5-axis machining occurs when the material or workpiece is being _____ while the tool is engaged in the material.

    A. simultaneously rotated
    B. machined
    C. rotationally inverted
    D. horizontally rotated

16. When the tool is perpendicular to the surface being machined, it is _____ to the surface.

    A. vertical
    B. normal
    C. horizontal
    D. actual

17. The tool forces in a normal cut are driven through the tool centerline and ultimately through the _____ and _____.

    A. spindle, chuck
    B. vise, table
    C. turret, table
    D. toolholder, spindle

18. As a ball end mill encounters steeper angles, the tool is no longer cutting with the end, rather it is _____.

    A. side loading
    B. edge milling
    C. side cutting
    D. center cutting

19. *True or False?* The only way to create a cutting profile that remains normal to the surface being machined is with a fully engaged 5-axis toolpath.

20. The mill-turn machine gives us full _____ capability along with full _____ capability, all in one machine.

    A. turning, milling
    B. trunnion, rotary
    C. collet, chuck
    D. drilling, milling

Apply and Analyze

1. What makes a machine a *multiaxis* machine?

2. What is meant by the term *live tooling* lathe?

3. Name two possible additional axes of a lathe and describe the difference.

4. Describe the tool movement in full 3-axis lathe machining.

5. Name two different pieces of equipment that can be mounted on a CNC machining center to convert it into a multiaxis machine.

6. Name at least three different G and M codes that are specifically used for multiaxis turning operations and explain their use.

7. Define the center of rotation on a 5-axis machining center and briefly describe why that point is important to the machining or programming process.

8. Explain the term *full-axis engagement*.

9. What is the difference between a live tooling lathe and a mill-turn machine?

10. Why is the mill-turn machine considered a multiaxis machine, and what is its impact on manufacturing?

Critical Thinking

1. Multiaxis machining is becoming the standard in manufacturing. What are some of the industries that are most impacted by multiaxis machining and complex part manufacturing?

2. The ability to create multiaxis parts has also had a tremendous impact on CAM programs and engineering design. Why is it becoming increasingly important to utilize CAM software? How have engineering designs and capabilities increased with the availability of multiaxis machining?

3. Now that we have explored some of the capabilities of multiaxis machining and programming, how has your perception of technology and machining changed?

A 5-axis CNC machine cutting an automotive part.

13 Subprogramming and Probing

Chapter Outline

13.1 Programming Efficiency

13.2 Main Programs and Subprograms

13.3 Lathe Subprogramming

13.4 Implementation of Probing Systems

    13.4.1 Setting Work Coordinates with a Probe

    13.4.2 Bore Probing

    13.4.3 Boss Probing

    13.4.4 Outside Corner Probing

    13.4.5 Inside Corner Probing

    13.4.6 Single Surface Measurement

    13.4.7 Probing Inside the Program

Learning Objectives

After reading this chapter, you should be able to:

- Describe the difference between main programs and subprograms.
- Explain how a subprogram is distinguished by name from a main program.
- Explain the difference between M30 and M99.
- Describe the application of a subprogram.
- Identify the unique codes required to subprogram.
- Define how multiple subprograms can be used in the body of a main program.
- Explain the different applications for turning subprograms.
- Identify a probe head.
- Explain the benefits of probing on a CNC machine.
- Perform multiple WIPS probing operations.
- Set WCSs from probing routines.
- Describe the basic application of in-program probing.

Key Terms

contact measurement

G65

M99

M98

macro variables

main program

probe

probe head

probing system

safe move

single surface measurement

stylus

subprogram

wireless intuitive probing system (WIPS)

work shift

13.1 Programming Efficiency

CNC machinists are creative and want faster, more efficient ways to machine their parts. *Efficiency is directly correlated to time.* Every step and decision in manufacturing factors in time. We use multiaxis and mill-turn machines to reduce *setup times* and CAM systems to reduce *programming time.* We quote production costs by adding *material cost plus time.* The time it takes to machine a part determines how efficient and profitable we are in manufacturing. The customer will demand that every part is 100% in compliance with the print and will award contracts to the facility that can produce the part in the least amount of time, for the least cost.

As a machinist, you will be judged by how efficiently you create parts to print specification. Simply changing a finished part out of a machine with the next piece of raw stock adds time to the process and has to be done efficiently. As you progress in your career, you may have the opportunity to become a CNC programmer. Most companies have only a few programmers who are responsible for multiple machines. So, you will be asked to create efficient programs and processes for 5, 10, or 15 CNC machines, or maybe even more. Although it is a good start, efficiency is not just about running faster feed rates. It also comes into play in programming, fixturing, and inspection. The use of subprogramming CNC and in-process spindle probing can make programming and setup times more efficient, **Figure 13-1.**

Haas Automation, Inc.

Figure 13-1. CNC machining center using rotary indexers and subprograms to create complicated part geometry.

 Safety Note

To some degree, safety can be built into the program. Default lines at the beginning of programs cancel modal commands and add safety. Safe tool change positions can be programmed. Safety stops can be added with M00 and M01 where needed. Remember that safety is everyone's responsibility, including the programmer.

13.2 Main Programs and Subprograms

Recall that a program is the specific set of commands—G and M codes along with axis positions—that control a machine's movement and all of its functions. CNC machines will not do anything unless they are told to by a program. For example, the spindle will not turn on clockwise unless it has an M03 command. Up until this point in the text, we have dealt with these commands one program at a time, with a definite beginning and end. But it is possible to "leave" a program, perform a new function, and then reenter at a specific place in the original program. The initial program is called the ***main program***, and the secondary program is known as the ***subprogram***. This process is diagrammed in **Figure 13-2.**

A few rules govern this technique. First, a programmer must begin and end within the main program, not in the subprogram. Second, the subprogram must have a program number entered in the machine control,

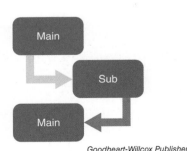

Goodheart-Willcox Publisher

Figure 13-2. A main program branches out to a subprogram and the subprogram must return back to the main program.

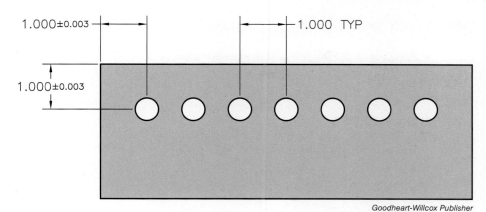

1.000±0.003

1.000 TYP

1.000±0.003

Figure 13-3. A possible use of a subprogram is for a part that has a repetitive pattern of features.

which will be "called" by the main program. In other words, the machine has to have direction on where to find the subprogram. Third, the subprogram cannot end with an M30 command (program end), rather it must be looped back to the main program with an *M99* command (end of subprogram/program restart). This is because the program cannot stop in the subprogram. It must reenter the main program.

Figure 13-3 shows the plans for a simple milled part. Focus on just the spot drill and drilling of the 1/2″ diameter holes for an example of a subprogram.

This part is 1″ thick with 1/2″ holes through the part. Looking at the print dimensions, it makes sense to use the top-left corner as the work coordinate system (G54). All of the holes are spaced 1″ apart. So, the program looks like this:

```
O1425
(SUBPROGRAM SAMPLE)
(T1|3/4 SPOTDRILL|H1|D1|DIA. .75)
(T2|1/2 DRILL|H2|D2|TOOL DIA. .5)
G20
G0 G17 G40 G49 G80 G90
T1 M6
G0 G90 G54 X1. Y–1. S2037 M3
G43 H1 Z.1
M8
G99 G81 Z–.26 R.1 F32.
X2.
X3.
X4.
X5.
X6.
X7.
G80
M5
G91 G28 Z0. M9
```

```
M01
T2 M6
G0 G90 G54 X1. Y–1. S1910 M3
G43 H2 Z.1
M8
G99 G81 Z–1.15 R.1 F15.3
X2.
X3.
X4.
X5.
X6.
X7.
G80
M5
G91 G28 Z0. M9
G28 X0. Y0.
M30
```

 **Tech Tip**

Tool 1 is a 3/4″ spot drill. It goes down to Z–.26 to create a small edge break. Using a 90° spot drill to create that edge break eliminates a secondary operation to deburr the hole after it is manufactured. This saves both time and money. Tool 2 is the drill, and it travels down to Z–1.15 to drill completely through the 1″ part. Although the part is only 1″ thick, the angle on the drill needs to completely travel through the part to create a through-hole to print specifications.

This is a fairly simple, straightforward program. But as a programmer, you may realize that your machine is large enough to create six parts at a time, which would reduce cycle time in a large order, **Figure 13-4.** This is a great application of a subprogram. To do so, set a unique work coordinate system, or WCS, for each part, allowing for individual changes to part hole locations. But be sure to set that WCS home on the same upper-left corner of each part so that one program can be used for every workpiece.

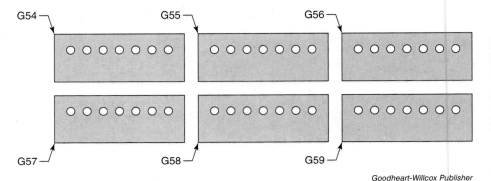

Figure 13-4. Subprograms can be used to create multiple parts in a single setup.

For this example, the single-part program (O1425) becomes the sub-program, and a separate main program is created to shift from part to part. A couple of small changes need to be made to program 1425, and a main program needs to be written that calls the subprogram multiple times. The main program should look like this:

O1111 (MAIN PROGRAM SAMPLE)	
G91 G28 Z0.	Safe move to send Z home at start of program
G20	
G0 G17 G40 G49 G80 G90	
G90 G54	Setting G54 for first part
M98 P1425	M98 "calls" a subprogram: Program #1425
G90 G55	Setting G55 for 2nd part
M98 P1425	
G90 G56	Setting G56 for 3rd part
M98 P1425	
G90 G57	Setting G57 for 4th part
M98 P1425	
G90 G58	Setting G58 for 5th part
M98 P1425	
G90 G59	Setting G59 for 6th part
M98 P1425	
M30	End main program

This program starts with a different program number, O1111. A G91 G28 Z0. is performed in the first line to make sure the spindle is in a safe position before starting. The fifth line is a G90 G54 by itself. This sets the WCS for the first part. The next line states an *M98*, which tells the machine to go to a subprogram. It also states P1425. The "P" stands for "program," and 1425 is the program number for the subprogram.

There are also a couple of changes needed in the subprogram. It is very important to consider what the main program is stating and what the subprogram is stating. Remember that many G codes are *modal*, meaning once they are turned on, they stay on until they are turned off. Remove all of the G54 commands inside program O1425 (now the subprogram) because the WCSs are set in the main program (G54 through G59) and should not be overridden in the subprogram. The subprogram needs to "rewind" and not stop. For that function, replace the M30 at the end with an M99. The only M30 in either program needs to be at the end of the main program. The altered subprogram looks like this:

O1425
(SUBPROGRAM SAMPLE)
(T1|3/4 SPOTDRILL|H1|D1|DIA. .75)
(T2|1/2 DRILL|H2|D2|TOOL DIA. .5)
G20
G0 G17 G40 G49 G80 G90
T1 M6

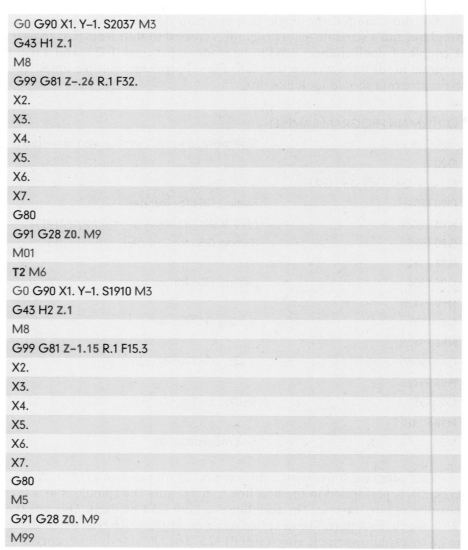

```
G0 G90 X1. Y–1. S2037 M3
G43 H1 Z.1
M8
G99 G81 Z–.26 R.1 F32.
X2.
X3.
X4.
X5.
X6.
X7.
G80
G91 G28 Z0. M9
M01
T2 M6
G0 G90 X1. Y–1. S1910 M3
G43 H2 Z.1
M8
G99 G81 Z–1.15 R.1 F15.3
X2.
X3.
X4.
X5.
X6.
X7.
G80
M5
G91 G28 Z0. M9
M99
```

Figure 13-5 maps the process path for these two programs, illustrating how they interact with one another, traveling from the main program to the subprogram and back multiple times.

It is also possible to go from a main program to a subprogram and then into a different subprogram. Depending on the controller manufacturer and the control model, it is also possible to go from subprogram to subprogram multiple times without going back to the main program. Programmers sometimes use this format to create subprograms for certain repetitive operations or for similar families of parts. For example, consecutive, simple operational subprograms can be used to spot drill, drill, and tap a 1/2″-13 hole in **Figure 13-3**. The advantage of this is the ability to save each operation and use them for any other program that has a spot drill, drill, or tap.

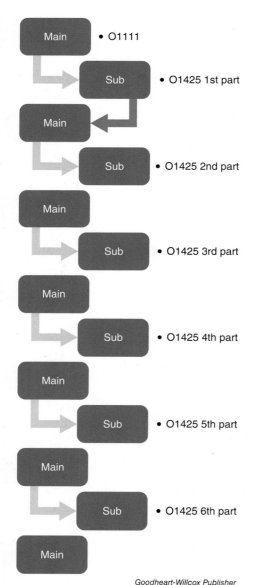

Main • O1111

Sub • O1425 1st part

Main

Sub • O1425 2nd part

Main

Sub • O1425 3rd part

Main

Sub • O1425 4th part

Main

Sub • O1425 5th part

Main

Sub • O1425 6th part

Main

Goodheart-Willcox Publisher

Figure 13-5. When creating multiple parts, the main program will call out each additional part made with a subprogram.

 Tech Tip

Taps are timed with spindle speed and feed for threads per inch (tpi), so each tap size would need its own program for use as saved subprograms.

The main program would look like this:

O1111 (MAIN PROGRAM DRILL AND TAP)	
G91 G28 Z0.	
G20	
G0 G17 G40 G49 G80 G90	
T1 M6	
G43 H1	
G90 G54 X1. Y–1.	Locates at X1. Y–1.
M98 P2222	M98 Spot Drill subprogram: Program #2222
G90 G54 X2. Y–1.	Locates at X2. Y–1.
M98 P2222	Runs Spot Drill subprogram at this location
G90 G54 X3. Y–1.	Locates at X3. Y–1.
M98 P2222	Runs Spot Drill subprogram at this location
G90 G54 X4. Y–1.	
M98 P2222	
G90 G54 X5. Y–1.	
M98 P2222	
G90 G54 X6. Y–1.	
M98 P2222	
G90 G54 X1. Y–1.	Locates at X1. Y–1.
T2 M6	
G43 H2	
M98 P2223	M98 Drill subprogram: Program #2223
G90 G54 X2. Y–1.	Locates at X2. Y–1.
M98 P2223	Runs Drill subprogram at this location
G90 G54 X3. Y–1.	
M98 P2223	
G90 G54 X4. Y–1.	
M98 P2223	
G90 G54 X5. Y–1.	
M98 P2223	
G90 G54 X6. Y–1.	
M98 P2223	
T3 M6	
G43 H3	
G90 G54 X1. Y–1.	Locates at X1. Y–1.
M98 P2224	M98 Tap subprogram: Program #2224
G90 G54 X2. Y–1.	Locates at X2. Y–1.
M98 P2224	Runs Tap subprogram at this location

```
G90 G54 X3. Y–1.
M98 P2224
G90 G54 X4. Y–1.
M98 P2224
G90 G54 X5. Y–1.
M98 P2224
G90 G54 X6. Y–1.
M98 P2224
M30
```

These three subprograms can be stored in the controller memory and used for any other program that has similar operations. The only changes might be the final Z depth. Remember that all programs begin with the letter "O," and the "P" is only used to call that program after an M98. These subprograms would look like this:

```
O2222 (3/4 SPOTDRILL)
G0 S2037 M3
M8
Z1.
G99 G81 Z–.26 R.1 F32.
G80
M99
```

```
O2223 (27/64 DRILL FOR 1/2-13 TAP)
G0 S1910 M3
M8
Z1.
G99 G81 Z–1.15 R.1 F15.3
G80
M99
```

```
O2224 (1/2-13 TAP RH TOOL)
G0 S534 M3
M8
G98 G84 Z–1.15 R.1 F41.1
G80
M99
```

Tech Tip

Although most computer-aided manufacturing (CAM) software does not have a specific path to enter subprograms into your CAM program, there are some workarounds. CAM programs do have a way to enter notes or additional code into the program. You can enter the M98 and corresponding program (P) number into the CAM file as a note, and it will post that note to the main program. But often the subprograms are completed in the CAM software and the main program is written by hand. With a little trial and error, you can get the CAM system to post the program right where you need it.

Although the main program is a little longer, these subprograms are very short. In machine shops where there are many similar operations, like drilling and tapping, the use of subprograms can save a lot of time. Subprograms are also useful in multiaxis mill programming, where multiple sides of a part have the same features.

13.3 Lathe Subprogramming

Similar techniques can be used on the CNC turning center. Drilling operations, simple OD turning operations with variable sizes, threading, and grooving operations are often simplified through subprograms.

The most common application of subprogramming in a turning center is in making multiple parts from bar stock. For example, **Figure 13-6** shows a part that might be a good candidate for subprogramming. This is a 3″ diameter part, but it is relatively short, at only 1.75″ long. With the right machine, more than one part can be made at a time.

The following program makes this part. Tool 1 (CNMG 432 insert) is a rougher tool, tool 2 (CNMG 432) is a finish tool, and tool 3 is a .250″ wide cut-off tool.

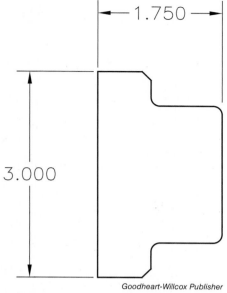

Goodheart-Willcox Publisher

Figure 13-6. A typical part that can be programmed in a turning center. Using subprograms allows for multiple parts in one setup. Note: for ease of viewing, this part is not fully dimensioned.

```
O2875 (LATHE SUB EXAMPLE)
(TOOL: 1 OFFSET: 1)
(OD ROUGH RIGHT - CNMG-432)
(ROUGH PASS)
G0 T0101
M8
G97 S246 M03
G0 X3.1 Z.05
G50 S3000
G96 S200
G71 U.1 R0.
G71 P100 Q102 U.02 W.01 F.01
N100 G0 X-.2 S200
G1 Z0.
X1.6875
G3 X2. Z-.1563 K-.1563
G1 Z-.9063
G2 X2.1875 Z-1. I.0938
G1 X2.6875
G3 X2.7317 Z-1.0092 K-.0313
G1 X2.9817 Z-1.1342
G3 X3. Z-1.1563 I-.0221 K-.0221
G1 Z-1.85
N102 X3.1
G0 Z.05
M9
G28 U0. W0. M05
T0100
M01
(OD FINISH RIGHT - CNMG-432)
(FINISH PASS)
G0 T0202
M8
```

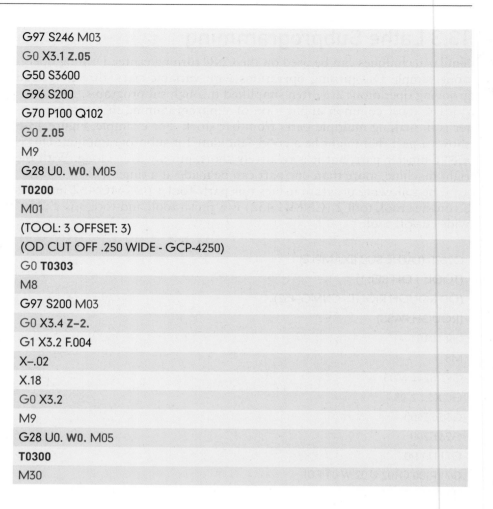

```
G97 S246 M03
G0 X3.1 Z.05
G50 S3600
G96 S200
G70 P100 Q102
G0 Z.05
M9
G28 U0. W0. M05
T0200
M01
(TOOL: 3 OFFSET: 3)
(OD CUT OFF .250 WIDE - GCP-4250)
G0 T0303
M8
G97 S200 M03
G0 X3.4 Z-2.
G1 X3.2 F.004
X-.02
X.18
G0 X3.2
M9
G28 U0. W0. M05
T0300
M30
```

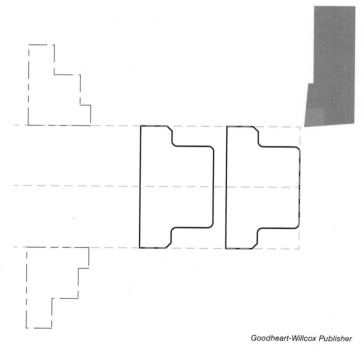

Goodheart-Willcox Publisher

Figure 13-7. A model illustration of creating two parts using a subprogram. Consider the location of workholding when programming.

A creative programmer, however, might see that two parts can be made in one setup by extending the bar stock out a little farther. So, every time the operator opens the door, two parts are complete, decreasing the time per piece and increasing machining productivity. **Figure 13-7** shows the stock and chuck jaw model for this setup. It appears that the stock is extended far from the chuck jaws, but this is 3″ diameter material sticking out only about 5″ from the grip of the jaws.

The sample part finishes at 1.75″ long and is cut off with a .250″ wide cut-off tool. That means that the second part must be more than 2″ behind the first part. Some stock should also be left on the second part to machine that face clean and flat, so start the second part at Z−2.050 from the face of the first part.

Until recently, there were no G54–G59 work coordinates in a lathe. There was no need since the offsets were set from the tool tip and the X0. position was always the spindle centerline. Earlier machines with older-style controllers use a different method to create multiple parts called a G10 *work shift*. Adding a G10 to a program shifts all of the tool offsets in a specified Z direction. It would be highly uncommon to shift the

X axis, usually it is only the Z axis. In our sample program, this line of code would be G10 W–2.050. Use the alternate axis name of W for the Z axis. The work shift must be "undone" at the end of program. The main program for the G10 work shift looks like this:

O1111 (LATHE MAIN PROGRAM EXAMPLE)
G28 U0. **W0.**
M98 **P2875**
G10 W–2.050
M98 **P2875**
G10 W2.050
M30

To use this as a subprogram, the only change needed is changing the M30 to an M99.

The more common way to accomplish this shift in newer controllers is the use of G54–G59 work coordinates. Once the tool offsets are set for all three tools, the G54 is set to X0. Z0. and the G55 is set to X0. Z–2.05. The G54 stays at 0 because the offsets should not be shifted on the first part, but on the second part with the G55. Using work coordinates, there is no need to "undo" the move. The main program would look like this:

O1111 (LATHE MAIN PROGRAM EXAMPLE)	
G28 U0. **W0.**	
G54	
M98 **P2875**	**subprogram O2875**
G55	
M98 **P2875**	**subprogram O2875**

This main-program-to-subprogram technique helps programmers avoid having to calculate all of the positions needed for multiple parts, and it also provides repeatability.

 Thinking Green

Proactive Maintenance
There are two kinds of maintenance—proactive and reactive. Reactive maintenance is maintenance done after a problem already exists (a *reaction* to the problem). Proactive maintenance means taking steps to *prevent* a problem before it happens, so it is sometimes called preventive maintenance. Proactive maintenance saves time, money, and other resources in a variety of ways. Performing routine maintenance on machines and replacing parts regularly can help prevent major breakdowns, which often require expensive parts and take extended time to repair. Proactive maintenance keeps machines running efficiently and accurately, preventing wasted work hours and wasted materials.

Figure 13-8. A probe head mounted in a vertical machining center.

13.4 Implementation of Probing Systems

In the Chapter 6 discussion on gaging and inspection tools, we discussed a CMM, or coordinate measuring machine. This is an inspection device that is similar to a CNC in that it moves through a Cartesian coordinate system. The difference is that a CMM does not have a cutting tool on it; instead, it has a *probe* that touches a surface of a part and records its position. This is not new technology, but CMMs have advanced rapidly and become the standard in most machine shops for inspection.

The most widely used *probing system* was developed by Renishaw™, who initially created the probe to solve complex metrology problems for Rolls Royce Olympus engines. Calibrated and used correctly, probes maintain an accuracy of .00005″ (50 millionths of an inch). The heart of the system is in the *probe head*, **Figure 13-8.** The probe head is the electronic instrument that takes a *contact measurement*, meaning that it has to physically touch the surface it is measuring. As the ruby-tipped stylus touches the part, it triggers a switch in the probe body and records the event, or position, in the machine controller.

The same basic probe head is adaptable to a CNC machine. There are applications for the turning center, but probes are most widely used in the machining center. They are directly mounted into a specialized toolholder and stored as any other tool in the automatic tool changer. Probing in machining centers has rapidly advanced and is quickly becoming standard equipment for any facility doing multiaxis work, a lot of setups, or close tolerance parts.

13.4.1 Setting Work Coordinates with a Probe

One of the tasks that can be completed with a probe is the setting of work coordinate offsets (G54–G59). Locating the correct corner or part edge accurately is critical in the setup process. When using a "wiggler"-style edge finder, the accuracy is no better than .002″ and is very dependent on the user correctly finding the edge. A fixed or programmable stop can be more accurate, but once the material is against a stop and the part is clamped, it tends to move slightly. If the correct position of the material cannot be identified, all programmed features on a part will be inaccurate by the same amount as the entered work coordinates. Fixture repeatability is another problem. It is possible to put a piece of material in a fixture and find it accurately—but can that same accuracy be repeated 100 times in a row?

To solve these problems, Renishaw has partnered with many machine controller manufacturers and integrated rapid part setup in their software. The best example of this has been with Haas Automation. Haas has called it the *wireless intuitive probing system (WIPS)*. The following sections look at the WIPS system and how to best utilize this tool.

 Tech Tip

Probes must be properly installed and calibrated before use. The probe tip must be indicated as accurately as possible and a calibration cycle performed to record all of the probe data onto the machine. Refer to the probing manual for more detailed instructions for a specific machine type.

To enter the *WIPS* page on a Haas controller, press *MDI* and then *Prog/Conv*, **Figure 13-9**. That enters you into the *Visual Quick Code (VQC)* pages. Cursor over to *Set Up* on the top of the page, and push the *Write/Enter* key, **Figure 13-10**. This activates the bottom tabs. Select the *Work* tab to enter the *WIPS* pages.

Once you have entered that section, there will be a series of pages and options that allow you to use the probing system to set work coordinates. Before using those, the probing tool must be staged into the spindle. The tabs are visual and display the approximate starting point of the probe before probing. Use the *Hand Jog* tab to manually position the probe before starting. The following sections examine a few of those frequently used tabs and how the pages are filled out in order to set work coordinates.

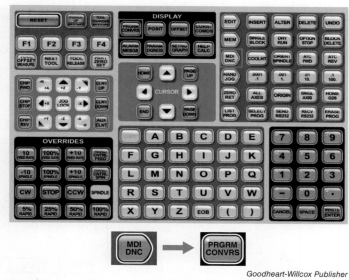

Goodheart-Willcox Publisher

Figure 13-9. Haas control panel showing the location of push buttons to enter into the WIPS page.

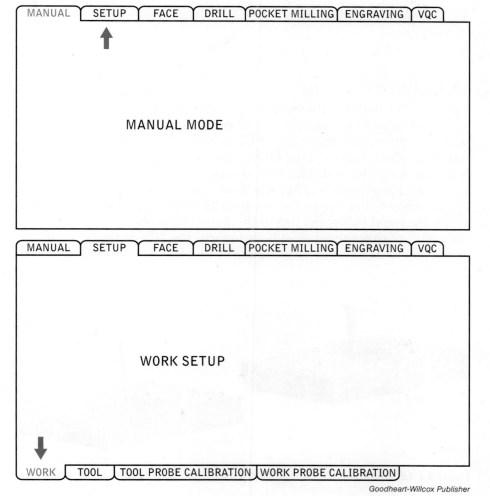

Goodheart-Willcox Publisher

Figure 13-10. To enter the *WIPS* page, arrow over to the *Setup* page and press *Write/Enter* to display the *Work* page.

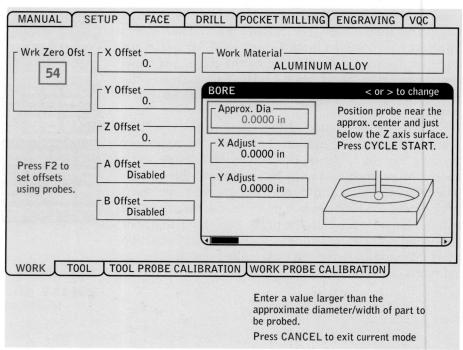

Figure 13-11. From the *Bore Setup* page, you can enter the bore diameter and WCS setting.

13.4.2 Bore Probing

Figure 13-11 displays the bore setup page. The work coordinate being set is displayed in the upper-left corner of the page. This can be altered with the arrow keys to set any WCS number. In the middle of the page is a box called *Approx. Dia*. Here, type in the approximate size of the bore you want checked. This sets parameters for the machine controller to know approximately what size hole is in the workpiece. The visual display shows that the probe needs to be approximately in the center of the bore and below the top of the part. The probe tip is a sphere, so the center of that sphere must be below the top face of the bore you are measuring to achieve an accurate probe setting, **Figure 13-12**. Once the

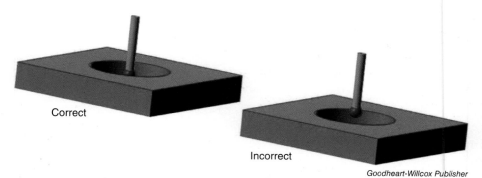

Correct

Incorrect

Figure 13-12. To use the bore measurement cycle, the probe centerline must be below the top of the part to be measured.

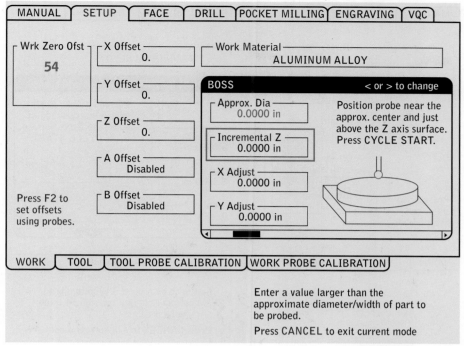

Figure 13-13. From the *Boss Setup* page, you can enter the boss diameter and WCS setting.

probe is in position and all the settings are correct, press *Cycle Start*, and the machine will automatically probe and measure the bore diameter while simultaneously setting the G54 work coordinate. You can proceed with machining from there.

13.4.3 Boss Probing

The next page over on the WIPS *Work* tab shows a similar probing routine. Instead of a bore measurement, this one produces a boss measurement. Because this is a boss measurement, the probe must go in the Z– direction to find the boss diameter, so one extra dialog box is needed, **Figure 13-13**.

The *Incremental Z* dialog box requires a Z height. This number is the amount the probe will travel down in the Z direction to find the outside diameter. The visual display shows that the probe is initially positioned above the center of the boss. The probe will travel outside of the *Approximate Diameter* size and then down in Z before it measures. This probing routine can measure very small diameter pins or very large diameter hubs.

13.4.4 Outside Corner Probing

WIPS can also set work coordinates for outside corners. The advantage of this option is that it sets both the X- and Y-axis work coordinates simultaneously. This is a very effective way to set the coordinates for a part set in a vise. The probe will be positioned above the corner of the part's top surface.

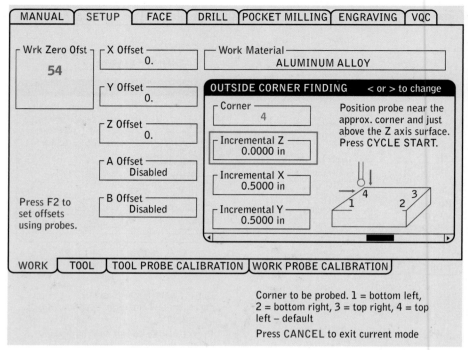

Goodheart-Willcox Publisher

Figure 13-14. In the outside corner finding routine, the corresponding corner must be noted per the illustration. The correct position for a corner measuring routine is centered on the corner and above the part surface.

The dialog box includes a couple of unique options, **Figure 13-14.** First, the graphic display shows a number in each corner of the part. The controller must know which corner it is searching for, as this dictates the direction of travel for the probe head. You must designate which corner you want to probe. Next, the *Incremental X, Incremental Y,* and *Incremental Z* dialog boxes dictate the amount of travel required in each axis. Since the correct corner is already indicated, these numbers do not require a + or – direction. Just like the bore probe routine, the Z axis number dictates how far down the Z travels before measurement.

13.4.5 Inside Corner Probing

The inside corner finding probe is identical to the outside corner routine with the exception of the graphic showing the location of the inside corner, **Figure 13-15.** This routine will find the inside corner and set the work coordinates at the intersection of those two surfaces. Even though this is an inside corner, the probe is located above the part's top surface.

13.4.6 Single Surface Measurement

The *single surface measurement* probe is probably the most commonly used probing routine. It is exactly what its name implies: a single surface measurement. If you only want to set the X-axis coordinate, or the Y-axis coordinate, this is the probing routine you would use. Use caution when setting the Z-axis surface, because it sets the WCS Z position and it will add additional Z-axis movement to your tool height offsets, **Figure 13-16.**

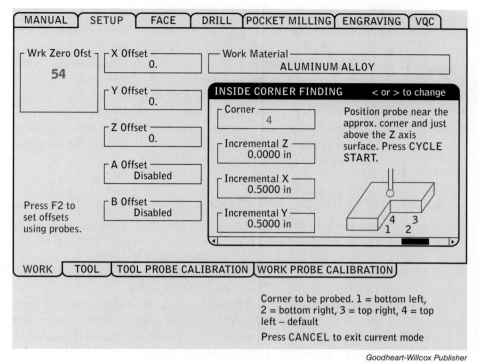

Figure 13-15. In the inside corner finding routine, the corresponding corner must be noted per the illustration. Also, an X and Y shift amount can be noted.

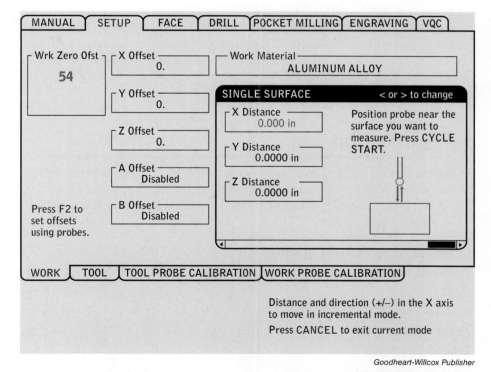

Figure 13-16. The single surface measurement routine can only be used to measure one axis at a time.

In this routine, you can only set and use one axis at a time. If you want to measure the X axis, fill in the direction of travel with the approximate distance to move. For example, if the measured surface is the right-hand edge of the part, place the probe in the correct Z-axis height to the right of the desired edge. In the *X Distance* dialog box, enter .500 (distance to search) and hit *Cycle Start*. This will cue the machine to move in the X positive direction. An X– entry will move the machine in an X– direction. If the probe cannot find a surface to measure, it will alarm out and stop. The probe will find a surface that is closer than .500″ and as far away as .750″, so the .500″ input is only an estimate of distance.

13.4.7 Probing Inside the Program

These probing routines provide accuracy and can save a lot of setup time. But would it not be helpful if we could put these routines right inside our main program and let the machine do all the work for setting work coordinates? How much time could we save if the probe could be programmed to find the part edges automatically? This is all possible.

The probing routines inside the WIPS system are nothing more than subprograms stored in the machine. We can call those subprograms and use them directly in our own programs. When you use the WIPS system, you will actually see a program running on the *MDI* screen. These are special programs that use stored machine information with a group of *macro variables* to perform special tasks. But, once again, our machining partners have simplified this system and made it possible for every machinist to utilize these special features. Because each machine controller defines macro variables distinctly for that machine, this section will once again only discuss the Haas machine controller. This can work on any controller, but the subprograms will be different and will be outlined in the user manual.

Figure 13-17 provides a sample part for which we will examine the subprograms and write a complete program. The probe can be treated just like any other milling tool. The special probe macros will be explained in some detail. This will be a useful program to save and call as a subprogram in any part with a top-left G54 WCS.

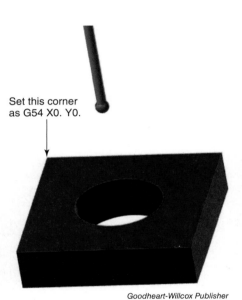

Set this corner as G54 X0. Y0.

Goodheart-Willcox Publisher

Figure 13-17. A sample part where probing is used to establish the work coordinate.

 Tech Tip

Before starting the program, set a "rough" G54 X0. Y0. and set the probe tool height offset, as with any programmed tool. The probe needs an approximate starting location in the G54 to find the material and a known tool height offset.

O3271 (PROBING TOP LEFT CORNER)	
(T10|6 MM PROBE HEAD)	
G20	
G0 G17 G40 G49 G80 G90	
T10 M6	
G0 G90 G54 X–1. Y–1.	
G43 H10 Z2.	
G65 P9832	PROBE ON

The program starts with a program number, the default block, and puts tool #10 (probe) into the spindle. Do not turn on the spindle. Position the spindle off the part at X–1. Y–1., and then turn on the tool height offset (G43) for tool #10. Notice the tool is kept 2″ away from the part in Z. The next line contains a **G65**. *Every* line used in probing will start with a G65; it is the code that tells the machine it is in probing mode. That line also contains a P9832, which powers on the probe head.

O3271 (PROBING TOP LEFT CORNER)	
(T10|6 MM PROBE HEAD)	
G20	
G0 G17 G40 G49 G80 G90	
T10 M6	
G0 G90 G54 X–1. Y–1.	
G43 H10 Z2.	
G65 P9832	PROBE ON
G65 P9810 Z–.5	SAFE MOVE

The next line is G65 P9810 Z–.5. The P9810 program is called a *safe move*. Every nonmeasurement move in the program should use this. It allows the probe *stylus* to move safely around the part, fixture, and clamps, protecting it from collision. If the stylus gets triggered or hits an unexpected object, the probe will stop moving so as not to damage the probe head.

O3271 (PROBING TOP LEFT CORNER)	
(T10|6 MM PROBE HEAD)	
G20	
G0 G17 G40 G49 G80 G90	
T10 M6	
G0 G90 G54 X–1. Y–1.	
G43 H10 Z2.	
G65 P9832	PROBE ON
G65 P9810 Z–.5	SAFE MOVE
G65 P9811 X0. S1.	MEASURE

The next line is G65 P9811 X0. S1. The P9811 is a measurement move. It says, "move to a position and when you touch a part, remember its position." The X0. tells the machine which axis and in which direction to move. Since the machine started at X–1. and is travelling to X0., it is moving in an X positive direction. This code also says that wherever the part edge is in the machine, that will be X0. Once the probe has travelled and found the part, S1. tells the program what to do with that information. The letter "S" represents the WCS, and the first one, S1., is the G54. Then, S2. represents G55, S3. represents G56, and so on.

With the X0. coordinate for G54 set, the probe needs to safely move over to the Y axis.

O3271 (PROBING TOP LEFT CORNER)	
(T10|6 MM PROBE HEAD)	
G20	
G0 G17 G40 G49 G80 G90	
T10 M6	
G0 G90 G54 X–1. Y–1.	
G43 H10 Z2.	
G65 P9832	PROBE ON
G65 P9810 Z–.5	SAFE MOVE
G65 P9811 X0. S1.	MEASURE
G65 P9810 Z1.	SAFE MOVE

The next line is G65 P9810 Z1. It is a safe move to raise the probe above the top of the part.

O3271 (PROBING TOP LEFT CORNER)	
(T10|6 MM PROBE HEAD)	
G20	
G0 G17 G40 G49 G80 G90	
T10 M6	
G0 G90 G54 X–1. Y–1.	
G43 H10 Z2.	
G65 P9832	PROBE ON
G65 P9810 Z–.5	SAFE MOVE
G65 P9811 X0. S1.	MEASURE
G65 P9810 Z1.	SAFE MOVE
G65 P9810 X1. Y1.	

The next line is G65 P9810 X1. Y1. It is a safe move to reposition the probe to the next location of measurement. Now the probe just needs to move down safely in Z and measure the Y axis.

O3271 (PROBING TOP LEFT CORNER)	
(T10|6 MM PROBE HEAD)	
G20	
G0 G17 G40 G49 G80 G90	
T10 M6	
G0 G90 G54 X–1. Y–1.	
G43 H10 Z2.	
G65 P9832	PROBE ON

G65 P9810 Z–.5	SAFE MOVE
G65 P9811 X0. S1.	MEASURE
G65 P9810 Z1.	SAFE MOVE
G65 P9810 X1. Y1.	SAFE MOVE TO X1. Y1.
G65 P9810 Z–.5	SAFE MOVE TO Z–.5
G65 P9811 Y0. S1.	MEASURE IN Y DIRECTION AND SET G54
G65 P9833	TURN OFF PROBE

The last three lines reposition the Z axis in a safe location, perform the measurement in the Y-axis direction, and turn the probe off. The S1. sets the G54 Y-axis origin. From here, we can return the Z axis home, turn off the probe, and start the machining process. The result of this probing code is two single-axis measurement routines built directly into your CNC program.

Examine one more sample program for bore measurement for the part in **Figure 13-18**:

Set G55 at center of 1.5″ diameter bore

Goodheart-Willcox Publisher

Figure 13-18. A sample part where probing is used on an established bore to set the work coordinate.

O3275 (PROBING BORE CENTER)	
(T10|6 MM PROBE HEAD)	
G20	
G0 G17 G40 G49 G80 G90	
T10 M6	
G0 G90 G54 X0. Y0.	
G43 H10 Z2.	
G65 P9832	PROBE ON
G65 P9810 Z–.5	SAFE MOVE
G65 P9814 D1.50 S1.	MEASURE 1.5″ BORE
G65 P9810 Z1.	SAFE MOVE
G65 P9833	PROBE OFF
G91 G28 Z0.	
M30	

A P9814 is included to use the bore measurement cycle. This line also needs to contain an approximate hole diameter. This diameter does not need to be accurate, as the probe cycle will calculate the center point. The program is completed with a G65 P9833 to turn off the probe head, and the machine is sent to Z home position. To use this as a subprogram, change the M30 to an M99, as outlined in Section 13.2, Main Programs and Subprograms.

There are multiple feature types that can be measured with a probe, including a boss, a web, internal corners, and external corners. There are also multiple options inside of each probing cycle that allow customization. This is a tremendously powerful tool that improves the accuracy and speed of any machining program. Manufacturers can no longer afford to do things the way they used to be done. The next chapter discusses automated manufacturing, and subprogramming and the use of probes makes much of those processes available. Embrace the latest technology and every tool available as a CNC machinist. Spend some time exploring all of the features available in subprogramming and probing, and be creative in their applications.

Chapter Review

Summary

- A subprogram is an auxiliary program that can be executed from the main program.
- In a main program, the subprogram is called, or executed, by using an M98 P number. The P number is the program number of the subprogram.
- An M30 code will stop the program and rewind it to its beginning. An M99 code used in subprogramming will loop the subprogram back into the main program.
- A subprogram is used to execute a secondary program from the main program. This can be used for multiple part machining, commonly used toolpath operations, or high-production variable machining.
- An M98 is used to designate the subprogram from the main program. An M99 is used to loop the subprogram back into the main program.
- A main program can call a subprogram that goes directly back into the main program, or a subprogram can execute other subprograms in succession.
- Subprograms are commonly used in a CNC turning center to make multiple parts from one piece of material or to create toolpaths on both sides of the workpiece.
- The probe head is the electronic instrument that takes a contact measurement from the material or surface.
- Probing can make programming and setup times more efficient. Also in-process part inspection can be performed using a probing system.
- Internal program probing can be used to find a part's work coordinate system or to perform on-machine inspection.

Review Questions

Answer the following questions using the information provided in this chapter.

Know and Understand

1. Efficiency is directly correlated to _____.
   A. more machinists
   B. quality of stock
   C. horsepower
   D. time

2. *True or False?* As a machinist, you will be judged by how efficient you are at creating parts to print specification.

3. The initial program in which we start is called the _____.
   A. main program
   B. subprogram
   C. probe program
   D. lathe program

4. The secondary program we enter is known as the _____.
   A. main program
   B. subprogram
   C. probe program
   D. lathe program

5. The subprogram cannot end with an M30 command, but must be looped with an _____ command.
   A. M98   C. M30
   B. M01   D. M99

6. We want to set a unique _____, or WCS, for each part.
   A. work Cartesian system
   B. work calibration system
   C. work coordinate system
   D. work changing system

7. An M98 in the program line tells the machine to go to a _____.

A. main program
B. subprogram
C. probe program
D. loop program

8. *True or False?* Drilling operations and simple OD turning operations with variable sizes, threading, and grooving operations are often simplified through subprograms.

9. Earlier machines with older-style controllers used a different method called a _____ work shift to create multiple parts.

A. G10
B. G01
C. G00
D. G98

10. The same basic probe head used in the _____ is adaptable to a CNC machine.

A. MMC
B. MCM
C. CCM
D. CMM

11. *True or False?* The probe is a non-contact measurement instrument, meaning that it does not have to physically touch the surface it is measuring.

12. Haas calls their built-in probing cycles the _____, or WIPS.

A. wireless intuitive probing system
B. work-in-progress system
C. wireless integrated programming system
D. working integrated probing system

13. The advantage of the corner finding probe cycle is that it sets both the _____ and _____ axis work coordinates simultaneously.

A. X, Z
B. Z, Y
C. X, Y
D. A, B

14. *True or False?* The single surface measurement is probably the most commonly used probing routine.

15. The probing routines inside the WIPS system are nothing more than _____ that are stored in the machine.

A. programs
B. subprograms
C. work coordinates
D. canned cycles

16. Every line we use in probing will start with a _____, the code that tells the machine we are in probing mode.

A. G95
B. G65
C. G30
D. G98

17. The P9810 program is called a _____.

A. boss measurement cycle
B. bore measurement cycle
C. measurement move
D. safe move

18. The _____ is a probing measurement move.

A. P9811
B. P9814
C. P9810
D. P9813

19. P9814 is used for the _____.

A. boss measurement cycle
B. bore measurement cycle
C. measurement move
D. safe move

20. Use a G65 P9833 to _____ the probe head.

A. turn on
B. initiate
C. turn off
D. return

Apply and Analyze

1. Write a main program, #O4628, that runs the subprogram, #3901, at a G54 WCS and G55 WCS.

2. Write a main program, #O1381, that runs three subprograms at G54 X3. Y–4. Program #O2319 is a spot drill program, program #O4567 is a 27/64″ drill program, and program #3922 is a 1/2″-13 tap program.

3. Write a turning center program, #O6565. Use subprogram #O3128 to create the first part and a G10 work shift to make the second part at Z–2.6″ using the same subprogram.

4. Write a turning center program, #O0425. Use subprogram #O1020 to create three parts using work coordinates G54, G55, and G56.

5. Write a probing program, #O0615, that sets the G54 X-axis work coordinate using a single-axis measurement probe program. The start position of the probe is X1. and Y–2. and the probe depth is Z–.5. The probe is set in tool position 10.

6. Write a probing program, #O7170, that sets the G54 X-axis work coordinate using a bore cycle measurement probe program. The start position of the probe is X0. and Y0. The probe depth is Z–.5. The probe is set in tool position 10. The bore diameter is 2.5″.

7. Write a probing program, #O0824, that sets the G54 X- and Y-axis work coordinate home position in the top-left corner of the workpiece. The probe depth is Z–.5. The probe is set in tool position 25.

Critical Thinking

1. Subprogramming offers the opportunity to CNC programmers to create multiple generic or repetitive programs and store them for future use. What are some of the long-term benefits of having these programs to reduce programming time in the future? How will having these skills increase your value to your company?

2. What are some of the ways that using probing programs will increase productivity in a CNC machine shop? Does the increase of accuracy with probing create more efficiency in the setup and operation of a CNC machine?

3. Probing programs can be used inside of any CNC program to inspect parts that are being produced. In your opinion, is it more or less efficient to measure the part inside the machine or outside the machine? Can you state examples of where it is more efficient in the machine and where it is less efficient?

This engineer is entering the parameters for a part into a CNC machine.

14

Automated and Additive Manufacturing

Chapter Outline

14.1 Introduction to Automated Manufacturing

14.2 Robotics

  14.2.1 Who Programs the Robot?

  14.2.2 Types of Robots

  14.2.3 Using Robots to Our Advantage

14.3 Cellular Manufacturing

14.4 Flexible Manufacturing Systems

14.5 Additive Manufacturing

  14.5.1 Vat Photopolymerization

  14.5.2 Binder Jetting Process

  14.5.3 Directed Energy Deposition

  14.5.4 Material Extrusion

  14.5.5 Material Jetting

  14.5.6 Powder Bed Fusion

  14.5.7 Sheet Lamination

14.6 The Future

Learning Objectives

After reading this chapter, you should be able to:

- Define *automated manufacturing*.
- Outline the benefits of automation.
- Explain the purpose of robotics in manufacturing.
- Identify the six types of robots.
- List some industries that have been influenced by robotics.
- Describe cellular manufacturing.
- Outline the benefits of improving process control and workflow.
- Explain a flexible manufacturing system and its components.
- Define *additive manufacturing* and contrast it to subtractive manufacturing.
- Identify the benefits and shortcomings of the seven additive manufacturing techniques.
- Discuss possible future technology and machine integration.

Key Terms

additive manufacturing

automated manufacturing

binder jetting process

cellular manufacturing

directed energy deposition

end-piece manipulator

extruder

fixed sequence robot

flexible manufacturing system (FMS)

fused deposition modeling (FDM)

integration

intelligent robot

just-in-time delivery

lean manufacturing

machine flexibility

manual robot

material extrusion

material jetting

numerically controlled robot

photopolymerization

playback robot

powder bed fusion

robotics

routing flexibility

sheet lamination

subtractive manufacturing

variable sequence robot

work cell

Chesky/Shutterstock.com

Figure 14-1. An automated manufacturing cell using robots, conveyors, and advanced workflow.

14.1 Introduction to Automated Manufacturing

Automated manufacturing is the integration of machinery and technology to create a process that performs manufacturing independently through automation. Automation includes not just machining, but also assembly and part movement. While most people think of Henry Ford's superior car design as his greatest innovation, it is actually his 1913 addition of a belt pulley system to the manufacturing process, which moved partially assembled cars through his factory at an amazing six feet per minute. It does not sound very impressive now, but this first moving assembly line was revolutionary at the time. Instead of making one car at a time, like everyone else, he started building hundreds of cars a month. By the 1940s, most production facilities were looking to automation to machine and build their product lines.

There are many pros and really only one con to automation. The benefits of automation include safety. Less hands-on and coordinated part movements create a safer workspace. Automated manufacturing can operate 24 hours a day and 7 days a week, providing more efficient use of time. Automation provides more consistent and higher-quality products. Repeatability in manufacturing creates higher yields. The only con is the initial cost to create automated systems.

There are many forms of automated manufacturing, with new advances made every day, **Figure 14-1**. Looking toward automation and the integration of computer programs and the many tools available into our manufacturing systems, today's manufacturers can compete on a global level with any system.

From the Shop

Our Robot Future

The use of robotics in manufacturing is not the future, it is the *now*. South Korea is leading the way in robot usage, but the rest of the world is running quickly to catch up. By 2035 it is estimated that 35%–50% of all jobs in manufacturing, transportation, and wholesale trade will be influenced by high automation. The cost of robots is less every year, and every forward-thinking manufacturing facility is trying to implement robotics. Embrace the technology and be excited by the great opportunities it can provide in your career journey.

14.2 Robotics

Robotics is the technology that deals with the design and production of robots. A robot is a machine that can help us do any job. It is designed to execute one or more tasks automatically with speed and precision. Robots can be remotely controlled by a human or programmed using computer software. The paradigm has shifted in manufacturing, and specifically in machining, around robots being considered a "bad thing." Fears that robots were going to take away all machinists' jobs have faded as an entirely new sector of manufacturing has opened up and allowed companies to compete at the highest levels. The reality is that robotics is here to stay, so machinists entering the field today should embrace it and hone their skills to become leaders in this field rather than be left behind.

14.2.1 Who Programs the Robot?

For robotic implementation in CNC machine shops, the CNC programmer will normally program the robot. The CNC programmer knows the machines being used, the process of machining, the best possible use of a robot, and has at least some knowledge of motion and *integration*,

Figure 14-2. Some machine shops with extensive robotic integration will have a full-time robotic programmer, but most shops will use another existing in-house resource to program and troubleshoot the robotics. It only makes sense for the CNC programmer to have the skills to tackle another programming challenge—the new shop robot. Programmers will need to gain new skills and training to tackle the robotics system. Fortunately, the technology associated with robotics programming is very advanced, making the use of robots fairly simple.

Figure 14-2. Three robots working simultaneously in a manufacturing cell.

14.2.2 Types of Robots

There are multiple types of robots, classified by function. They range from simple and "non-intelligent" all the way up to artificial intelligence (AI) units that can learn and handle many diverse operations. The six basic robot types are:

- **Manual robot.** This robot is driven strictly by a manual operator. Remote operated vehicles (ROVs) are good examples of these. They usually have an on-board camera and can be driven remotely by an operator offsite.

- **Fixed sequence robot.** A robot that performs predetermined, repetitive operations. This robot's sequence cannot be changed. They are usually programmed through a programmable logic controller (PLC). They simply step through a series of commands based on operations, position, or input signals.

- **Variable sequence robot.** Similar to a fixed sequence robot, but the command sequence can be easily altered. It follows a given set of commands repetitively, until altered by the operator.

- **Playback robot.** One of the two most common robots in machining, these are often referred to as "teach-programmed" robots. A human operator steps the robot through a series of positions or commands and at each step records that command, or "teaches" it. Those commands are stored in the memory of the controller.

- **Numerically controlled robot.** This is the second type of robot that is common in machining. Picture a cube of space that contains the extent of the robot's reach, **Figure 14-3.** Any place that the robot can reach can be defined by a Cartesian coordinate. Just as in CNC machining, that coordinate can be remotely programmed through software, and the robot can then perform a series of movements and commands based on that program.

- **Intelligent robot.** This robot is not completely AI, but it is making decisions on its own based on changes in the work environment through vision (camera), sensors, or even voice commands.

14.2.3 Using Robots to Our Advantage

Since robots can serve to help us in the machine shop, what are some operations where automation makes sense? First, identify where there are repetitive motions in the shop and what is happening to prevent spindles

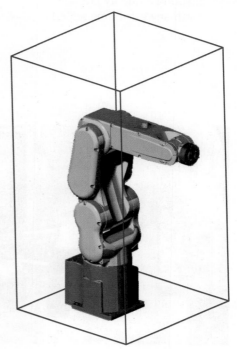

Figure 14-3. A 3D model of a robot. The box shows the limits of the robot's reach.

from turning for the maximum amount of time. Part loading and unloading comes to mind. This seems simple enough, but examine everything that happens in a simple turning center operation to safely load and unload a finished part with a new piece of material:

1. The cycle ends.
2. Open the door.
3. Open the hydraulic chuck while holding the finished part.
4. Remove the finished part and place it somewhere to prevent damage to it.
5. Secure a new piece of raw stock.
6. Center the new piece of stock in the hydraulic chuck.
7. Close the hydraulic chuck.
8. Release the piece of stock.
9. Close the door safely.
10. Activate the *Cycle Start* button.

Of these ten steps, some are machine functions and others are human operations that could be handled robotically. Opening the door, securing the part, and placing new stock are potential robotic functions. Opening and closing the chuck and starting the cycle are machine functions. The fact that we have independent machine functions and independent robot functions is exactly why we need to integrate the robot and machine. Both machine and robot makers have included systems that enable communication between the controllers, allowing them to work together, **Figure 14-4**.

Robots and integration can be expensive, so often a single robot is programmed to operate more than one machine at a time or perform multiple jobs, synchronized with multiple processes, **Figure 14-5**. The configurations of these integrated machines are only limited by the needs and imagination of the operators.

One industry that has been most influenced and positively impacted by the advent of robotics is the sheet metal industry. Shearing and forming sheet metal into a multitude of shapes and sizes has always been a very labor-intensive job. Safety has always been a major concern in braking and forming, with heavy equipment bending and forming metal into precision

Tech Tip

Robots can also be used to do on-machine inspection. Robots can be equipped with non-contact laser measuring devices or photo-driven inspection. Any on-machine inspection will slow the manufacturing process, but ensuring a quality product is worth any extra time added.

Suwin/Shutterstock.com

Figure 14-4. A robot working in conjunction with a CNC turning center to load and unload material.

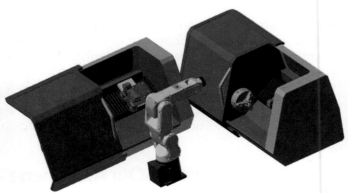

Goodheart-Willcox Publisher

Figure 14-5. Illustration of how a robot can be placed to perform multiple operations in a work cell.

Figure 14-6. A robot using vacuum manipulators to operate a CNC brake press.

Figure 14-7. An automotive manufacturing and assembly line using multiple robotic welders.

shapes. The edges of sheet metal are sharp and can easily cut someone. Robotic arms can use vacuum manipulators to place sheet metal into precision positions safely and accurately, **Figure 14-6.**

Another popular application is the robotic welding station. The art of precision welding is best performed with a steady hand and consistent travel speed—the perfect job for a robot. A robot can perform a weld over and over with exact repetition and precision. Robots can perform repetitive small welds or very long, continuous welds with the same consistency and performance time after time, **Figure 14-7.**

Adding to robots' versatility and endless possibilities of application, robotic "hands" or *end-piece manipulators*, **Figure 14-8**, have become more diverse and tactile in their functions. They can now pick up pallets full of steel parts or objects as delicate as fine glass.

Figure 14-8. An assortment of end-piece manipulators that can be installed on a robotic arm to perform various functions.

Robotics has numerous industrial uses, and it ranks as one of the fastest-growing technological fields in the world. Soon, robots will be built directly into machining and turning centers from the factory as optional equipment with the ability to program directly from the controller or computer-aided manufacturing (CAM) system. The consistent push toward more automation and extended efficiencies in machining will continue to drive manufacturing to automated machining and robotic implementation. Robots offer the ability to handle any work setting, from hazardous environments to fine inspection.

14.3 Cellular Manufacturing

Cellular manufacturing, or cell manufacturing, is a workflow that uses principles of just-in-time delivery and lean manufacturing to reduce costs and improve efficiency. The goal of cellular manufacturing is to reduce the wasted time of part movement and storage, while giving manufacturers the flexibility to make as many different parts as possible. Previous chapters have covered techniques for machining faster and more efficiently as well as using technology to improve the quality and quantity of manufactured products. But is there more efficiency to be gained in shop setup?

Let's start our own machine shop and see if setup makes any difference. **Figure 14-9** is our new place of business, CNC Machine Works, Inc. We purchase seven lathes and six mills and include an inspection department. We also create areas for finished goods and raw material storage.

This is a traditional machine shop setup. The large area for raw goods storage needs many racks and containers to store all the material we will buy, such as steel and castings. We also have a milling department with a supervisor who specializes in CNC mills, and we leave a lot of room between machines for tables and part movement. We have a lathe department with another supervisor who specializes in turning work. Of course, we need an

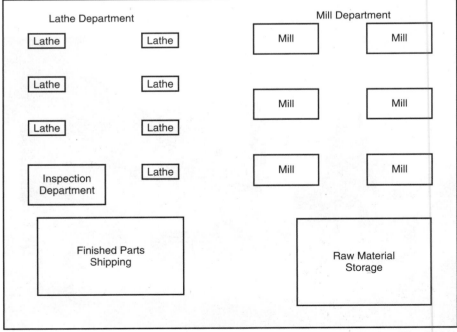

Figure 14-9. A traditional machine shop floor plan where turning and milling departments are utilized.

inspection department where we send all parts between operations and all finished work for certification.

Figure 14-10 is a model of our first part. It is fairly complicated, and we decide this part needs a lathe operation, three milling operations, and then one finish lathe operation. Of course, at each operation we will need a first article inspection and then a final inspection when the part is complete. The customer would like 500 parts, shipped at 100 pieces per month over the next five months. The piece is 4.8″ in diameter and 7.5″ long. Examine the workflow through our new shop and see what happens. We will use our shop layout to trace the steps required to complete this job.

Figure 14-11 shows a map of the part's travel, minus travel to and from inspection between each operation for inspection sign-off. Because of the shop's layout and machine time, we are probably making all 500 parts in each machine and then sending them to the next operation for completion. Notice the machine layout dictates that every machine needs one operator. For maximum spindle time, we need 13 operators, two supervisors, a raw storage handler, a shipping clerk, and an inspector. Let's try a different shop setup.

The practices of *just-in-time delivery* and *lean manufacturing* started to affect American manufacturing in the 1980s. Although they had been around before that and tested in other countries, the United States did not move toward these principles until the late 1970s. **Lean manufacturing** describes a process that utilizes workflow and process control to limit the resources used in creating a finished product. For example, in machining we create **work cells**, or U-shaped machining groups, to limit part movement and allow a single operator to operate more than one piece of equipment. We also utilize quick-change fixtures and fast turnover times to reduce the impact of setup times between jobs. **Just-in-time delivery** is customer driven and allows the customer to receive product only when they need it. For example, instead of buying 1,000 products and storing them in a warehouse, the customer wants 20 parts per week, as they are

Goodheart-Willcox Publisher

Figure 14-10. A part that requires multiple operations in both milling and turning.

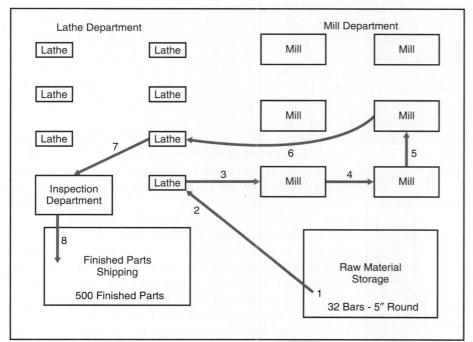
Goodheart-Willcox Publisher

Figure 14-11. The part-flow diagram of the sample part in our traditional shop floor plan.

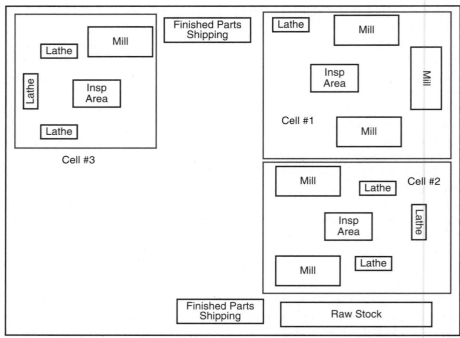

Goodheart-Willcox Publisher

Figure 14-12. A possible shop floor plan that has considered just-in-time delivery and lean manufacturing. This layout uses cellular manufacturing and improved workflow.

sold. Let's look at our new shop setup with some of these cellular manufacturing ideas in mind, **Figure 14-12.**

First, note that we still have 13 machines. We got rid of our inspection department and created an "inspection area" in each cell. This allows for quick inspection near each machine and no backup in inspection waiting for part approval. We have also created two smaller shipping areas so that we can ship directly from our cells. This is a factor of just-in-time delivery, because we are no longer going to create stockpiles of extra parts and store them for extended periods of time. It makes sense to do this job in cell #1, but all of these cells were created to be versatile and capable of handling many different job types. Examine the new part flow in cell #1, **Figure 14-13.**

In this model, we are only going to make 100 parts at a time. The customer only needs 100 parts per month, so that is all we need at one time. That gives us more time to run different parts in our cell and allows us to buy less material to be stored in our warehouse. Also, we do not need to store the 500 parts, because as soon as we finish 100 parts, they ship to the customer. We have rearranged our machines to fit into a U-shaped cell. This allows one operator to operate more than one machine at a time. With our inspection area inside our cell, the parts are actually receiving more inspection than they were before. Our costs per part are significantly reduced, our quality is better, and we free up money for future growth. This model only needs six operators, three inspectors, and one supervisor. We have reduced our workforce from 18 to 10, with additional floor space for expansion. The cell setups and functions will be determined by the types of parts and customers our company might service.

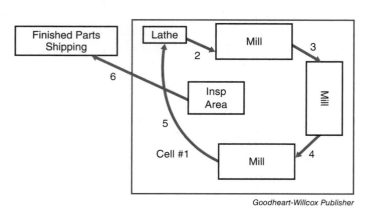

Goodheart-Willcox Publisher

Figure 14-13. The new part-flow diagram utilizing a single cell and improved productivity.

This is a simple example of cell manufacturing benefits. Cells can be U-shaped or linear with parts being moved down a line with robots or conveyors, **Figure 14-14.** The key is versatility and a good understanding of product development. Companies are investing more and more resources into cellular line manufacturing and quick-change setup modules.

14.4 Flexible Manufacturing Systems

A *flexible manufacturing system (FMS)* is a manufacturing process in which there is some amount of flexibility that allows the system to react in case of changes. This flexibility generally falls into two categories: routing flexibility and machine flexibility.

The first category, *routing flexibility*, refers to the system's ability to be altered to produce new product types and its ability to change the order of operations executed on any part. The second category is called *machine flexibility*, which is the ability to use multiple machines to perform the same operation on a part or the ability to mount a work fixture in many machines. This system allows for the ability to absorb large-scale changes, such as in volume, capacity, or capability.

Most FMSs consist of three main subsystems: 1) The *machines*, which are often CNC machines, are connected by 2) a *material handling system*, like a palletizer or robot, to optimize parts flow and 3) the *central control computer*, which controls material movements and machine flow.

The main advantage of an FMS is its high flexibility in managing manufacturing resources, such as time and effort, in order to manufacture a new product. The best application of an FMS is found in the production of small runs of products or products of a similar family type.

The FMS is designed specifically for versatility. Product fixtures are designed for quick change-outs and very little machine downtime. Instead of looking at a part and deciding how to machine that part, manufacturers analyze their FMS and decide how that part fits into that system. Often the part movement features, like conveyors or robots, are put on tracks or wheels to allow for easy reconfiguration, **Figure 14-15.** In certain cases, the machines can even be somewhat mobile to allow for rapid changes. All of this movement and reconfiguration is controlled by a centralized computer system that analyzes current and future manufacturing needs.

Hamik/Shutterstock.com
Figure 14-14. An engine-building facility that utilizes conveyors and a cellular manufacturing solution to progress engines through the machining process.

ssguy/Shutterstock.com
Figure 14-15. A flexible manufacturing system that allows for mobility and easy process changes.

 Tech Tip

An Enterprise Resource Planning software, or ERP, is used to analyze current and future workflow and predicted completion times. It can use that data to schedule production into machining cells for the most efficient part production. An ERP system can keep machines running and product flowing to the customer.

14.5 Additive Manufacturing

Probably the most talked about subject for the last ten years in manufacturing is 3D printing. 3D printing is an exciting, upcoming technology, but it is still in its infancy. Up until this point, this book has discussed machining starting with a "block" or piece of material and removing the unwanted material to reveal a part inside. That process is called *subtractive manufacturing*. The machine is subtracting unwanted material and leaving the wanted material. *Additive manufacturing* is a process that starts with raw material and adds it only to the places in space where material is desired. Material is added to create a part, **Figure 14-16**.

Will this replace CNC machining? The short answer is no; but like computers, CNC machines, or robots, this technology is only going to grow. We must embrace it and use it to our advantage. Additive manufacturing does have its limits, but it also provides some great new opportunities for the machining world.

Most think of 3D-printed plastic parts as additive manufacturing, but it goes much further than that. Following are seven different forms of additive manufacturing and how they are used.

14.5.1 Vat Photopolymerization

Photopolymerization was first developed in the early 1980s and was patented in 1984, first by the French and shortly after that by an American company. In this process, a liquid photopolymer, or light-activated polymer, is cured to create a 3D part. A vat is filled with a photopolymer fluid. An ultraviolet light is used to solidify or cure the polymer fluid where finished material is desired. The main types of this technology are stereolithography (SLA), digital light processing (DLP), and continuous digital light processing. Only plastic can be printed using these technologies.

This process can be found in rapid prototyping of casting components, such as investment castings. Investment castings are made from plastic molds. A tool or mold is created to make a plastic part. That part is then dipped repetitively in a cobalt solution, creating a very hard outer shell. The outer shell is then heated to melt out the plastic inner core, and molten metal is poured into the cobalt shell. After cooling, the shell is broken away, leaving a high-quality cast part. Creating the tool or mold used to make the initial part can take several weeks and several thousand dollars to create. This mold can then be used for thousands of parts. But what if

Kornilov007/Shutterstock.com

Figure 14-16. There are many additive manufacturing methods, including fused deposition modeling, direct metal laser sintering, stereolithography, digital light processing, and polyjet (shown here, from left to right).

only one prototype part or just a few parts are needed? In such cases, an SLA 3D-printed model can be made in just a few hours.

14.5.2 Binder Jetting Process

The *binder jetting process* uses a technique to selectively deposit a bonding agent or binding liquid to join powder material together to form a 3D part. This process is different from any other additive manufacturing technology in that it does not employ any heat during the process to fuse the material.

The process deposits a layer of powder, then a small layer of binding agent (think of it like a glue), then redeposits a powder over the glue to bind the powder layers together. Although this creates a metal part, it is not homogenous and is subject to cracking and fracturing.

14.5.3 Directed Energy Deposition

Directed energy deposition uses focused thermal (heat) energy such as a laser, electron beam, or plasma arc to fuse materials by melting them as they are being deposited. This process deposits powdered metal on a surface while directing a beam, or multiple beams, on the surface to melt the metal into place, **Figure 14-17**. The size of the powdered crystals (1/2–10 microns) and the directed energy technique determine the outcome of the finished part. This requires expensive equipment with high-energy use.

14.5.4 Material Extrusion

Material extrusion is an additive manufacturing technique that uses a continuous filament of plastic or composite material to construct 3D parts. *Fused deposition modeling (FDM)* is the only type in this category, and both plastic and composite can be printed using this technology. This is the 3D printer that has become readily available in the hobby market. Tremendous strides have been made in this technology allowing for the use of multiple plastic types, giving more versatility in the characteristics of the finished product, **Figure 14-18**.

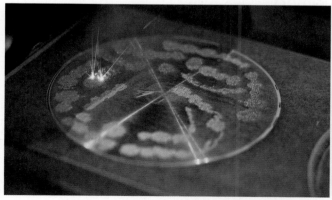

MarinaGrigorivna/Shutterstock.com

Figure 14-17. A part being 3D printed using direct energy deposition.

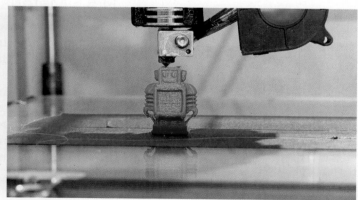

Kyrylo Glivin/Shutterstock.com

Figure 14-18. The most common 3D-printing technique using fused deposition modeling of plastic.

MarinaGrigorivna/Shutterstock.com

Figure 14-19. An example of a very complicated shape that can only be created using a 3D-printing process.

The technology in the plastic application is fairly simple. A spool of plastic line is threaded through an apparatus called an *extruder*. The extruder has a set of rollers that pulls the plastic through a heating element and deposits the liquid plastic through a nozzle. It looks somewhat like a hot glue gun melting the glue as you move the gun. The real key is in the software that can automatically form supporting features and "slice" the solid model into sections that designate the points of deposit. With this software and FDM technique, it is possible to create incredibly complicated designs that are virtually impossible to machine, **Figure 14-19.**

Although FDM cannot create close tolerance pieces, it can be used for fixtures, clamps, or stops in the modern machine shop. It is also too slow to use in a mass production application.

14.5.5 Material Jetting

In *material jetting*, material drops are deposited layer by layer into the build platform to form a 3D part. The powder is heated before the nozzle ejects and melts the powder as it is sprayed on a surface. This process can be followed by a UV-intensified light to create better bonding and quick drying. The nozzle tip is extremely small, and the preheated material bonds together better than in some other techniques, although it is still not completely homogenous. Powder material jetting includes the following commonly used printing technologies: UV-cured material jetting, drop on demand (DOD), and nanoparticle jetting (NPJ). It is a slower and less-accurate process.

14.5.6 Powder Bed Fusion

Powder bed fusion is an additive manufacturing technique that uses either a laser or electron beam to melt and fuse the material together to form a

MarinaGrigorivna/Shutterstock.com

Figure 14-20. A powder bed fusion technique of printing is used with a heat source to create printed parts.

3D-printed part. To build each new layer, the build platform is lowered in small increments. Simultaneously, a hopper containing material powder is raised, and a roller or vibrating blade spreads the next layer across the build platform. A laser, electron beam, or thermal printing head selectively melts the metal powder, which fuses to the preceding layer as it cools, **Figure 14-20.** The layering process continues until the object is fully printed. Although the vat will fill with powder, only the fused metal powder will create a solid shape. Powder bed fusion includes the following commonly used printing technologies: multi-jet fusion (MJF), direct metal laser sintering (DMLS), electron beam melting (EBM), selective heat sintering (SHS), selective laser melting (SLM), and selective laser sintering (SLS).

 **Thinking Green**

Additive Manufacturing

Additive manufacturing is considered a green technology in comparison to other metal-forming or machining processes because it typically produces less waste and requires less energy. For example, machining a part from a solid piece of metal might require cutting away up to 40% of the original material and discarding it. In contrast, additive manufacturing affords machinists the ability to shape any part they want without scrap. The metal powder used in additive manufacturing is created primarily from recycled materials. Furthermore, the energy required in the heating, or sintering, stage of powder metallurgy is less than the thermal energy required for melting processes such as forging.

BLACKDAY/Shutterstock.com

Figure 14-21. Often composite materials require a sheet lamination process to create complicated shapes. Sheets are bound together with adhesives and then the desired shape is cut away from the laminated structure.

Safety Note

Additive manufacturing can have many of the same safety hazards as subtractive manufacturing. Proper PPE should be worn. Be aware of respiratory hazards from ultrafine particulates and vapors. The materials used in the manufacturing process must be stored carefully due to fire hazard, and eye and skin exposure to materials may require eyewash or emergency shower access.

14.5.7 Sheet Lamination

In *sheet lamination*, material sheets are stacked and laminated together using either adhesives, ultrasonic welding, or brazing to form a 3D part, **Figure 14-21**. Once the object is built, unwanted sections are then cut layer by layer. This is not a normal 3D-printing process, but it is used in some composite material forming. Sheet lamination includes the following commonly used printing technologies: laminated object manufacturing (LOM), selective deposition lamination (SDL), and ultrasonic additive manufacturing (UAM). Because layers are formed and stacked, this process only provides strength in a single direction.

All of the additive manufacturing techniques discussed above have different applications and provide varying material properties. There is definitely a place for additive manufacturing in the design, engineering, and production of certain manufacturing sectors, but it also has its limitations. Notice some common characteristics: All but one of these processes uses heat in various forms. In machining, heating and then cooling material causes some issues. The first issue is warpage. As material is heated it expands and flows; when it cools, it solidifies and retracts. We can predict warpage, but we cannot control it. The second issue is surface finish. When metal is deposited in layers, the edges of the material are rough and uncontrolled. This creates a very rough outer edge, which is generally an issue for "finished" machined parts. After any 3D-printing application, parts often need to be remachined. The third issue may be the most problematic issue with 3D-printed parts, especially metal parts. Whenever metal is deposited in layers, porosity, or tiny microscopic gaps, is created in the material. These small holes create cracks, or what are referred to as micro-fractures. These micro-fractures become larger and larger under stress and can cause catastrophic failures in critical parts.

This defect or flaw in additive manufacturing prevents it from being used in critically engineered parts. However, not every part and not every feature is critical. Remember that machinists and engineers advance out of necessity and will no doubt develop a sector where additive manufacturing will fit and be cost effective.

 From the Shop

3D Printing

Additive manufacturing will not replace machining, but it definitely has a place in manufacturing. Industries like the medical field are 3D-printing joints, prosthetics, and even entire bone structures. Recently, human skin has been printed using stem cells and a bio printer. A 3D printer was sent to the International Space Station in 2014 to print tools and small components needed in space. NASA is exploring a recycling program where waste plastic is melted and reprinted into useful utensils like forks and spoons. As the world's plastic usage continues to grow, the repurposing of plastics might be the key to eliminating single-use plastics disposal. Research and development is underway to 3D-print homes using a large concrete depositor. The possibilities are truly endless as the technology expands.

14.6 The Future

In this text we have talked about the origins of machining, print reading, math, CNC programming, measurement, cutting tools, computer-aided manufacturing, robotics, and additive manufacturing. Some of these things, like additive manufacturing, are cutting edge and in their infancy, while some of these topics, like CAM and robotics, are not new but still developing.

So, what is the future of CNC machining? While there are always unknowns about what the future holds, here are some certainties: CNC machines will get faster with faster spindle speeds and faster feed rates. CNC machines are becoming more rigid, meaning bigger cuts are possible with less harmonics. CAM software is getting more user-friendly and will someday be the only way we program machines. CNC controls are going to continue to become more graphic and user-friendly. Engineering will require work with more exotic metals.

We can expect to see more integration of robotics and CNC. CNC machines are going to be shipped from the factory with robotic attachments. All machines are going to come with full inspection capability, either with a probe or laser non-contact inspection. Multiaxis machines, like mill-turn machines and fully integrated 5-axis milling, will completely take over the industry. CNC machine controllers will be full PCs with on-board CAM systems. Machinists will better utilize all the innovative software and hardware tools at our disposal.

Last but not least, the future of machining requires creative new minds that have grown up in the computer age and have new ideas. They will work with their hands to build machines their predecessors never imagined. They will push the very laws of physics with metal removal and redefine how we build every product our industry designs.

Chapter Review

Summary

- Automated manufacturing is the integration of machinery and technology.
- Automation reduces cost and improves product reliability.
- Robotics in manufacturing can be used for part loading and unloading, inspection, and material movement.
- There are six different types of robots in use today: the manual robot, the fixed sequence robot, the variable sequence robot, the playback robot, the numerically controlled robot, and the intelligent robot.
- Cellular manufacturing is a workflow that uses principles of just-in-time delivery and lean manufacturing to reduce costs and improve efficiency.
- Process control is the concept of improving part quality by controlling the process in which parts are made. Workflow is the path or direction that a part travels through the manufacturing process.
- A flexible manufacturing system (FMS) is one that can be adapted or physically moved to accommodate the products being made.
- Additive manufacturing is a process in which raw material is deposited to create a part in free space.
- The seven types of additive manufacturing are vat photopolymerization, binder jetting process, directed energy deposition, material extrusion, material jetting, powder bed fusion, and sheet lamination.
- Future technology involves machine-robot integration and further development of additive manufacturing techniques.

Review Questions

Answer the following questions using the information provided in this chapter.

Know and Understand

1. Automated manufacturing is the integration of _____ and _____ to create a process that performs manufacturing independently through automation.
   A. people, technology
   B. machines, robots
   C. machinery, technology
   D. programming, technology

2. The only con to automation is the _____ to create the automated systems.
   A. initial cost   C. loss of jobs
   B. time   D. process

3. _____ are machines that can help us do any job.
   A. Mills   C. 3D printers
   B. Lathes   D. Robots

4. *True or False?* Strictly speaking, for robotic implementation in CNC machine shops, normally the CNC programmer is going to program the robot.

5. Robots can be simple and _____ all the way up to artificial intelligence (AI).
   A. non-intelligent
   B. programmable
   C. integrated
   D. flexible

6. The term _____ refers to incorporating a robot and machine to function together.
   A. implementation
   B. integration
   C. diversification
   D. manipulation

7. A popular application of robotics is the robotic _____ station.
   A. milling   C. charging
   B. welding   D. turning

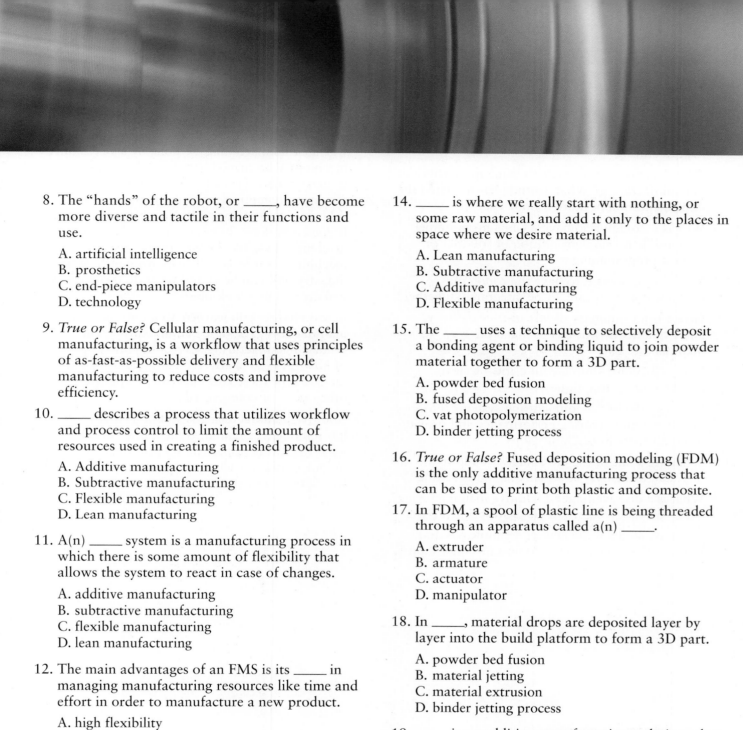

8. The "hands" of the robot, or _____, have become more diverse and tactile in their functions and use.

A. artificial intelligence
B. prosthetics
C. end-piece manipulators
D. technology

9. *True or False?* Cellular manufacturing, or cell manufacturing, is a workflow that uses principles of as-fast-as-possible delivery and flexible manufacturing to reduce costs and improve efficiency.

10. _____ describes a process that utilizes workflow and process control to limit the amount of resources used in creating a finished product.

A. Additive manufacturing
B. Subtractive manufacturing
C. Flexible manufacturing
D. Lean manufacturing

11. A(n) _____ system is a manufacturing process in which there is some amount of flexibility that allows the system to react in case of changes.

A. additive manufacturing
B. subtractive manufacturing
C. flexible manufacturing
D. lean manufacturing

12. The main advantages of an FMS is its _____ in managing manufacturing resources like time and effort in order to manufacture a new product.

A. high flexibility
B. rigidity
C. speed
D. process control

13. _____ is subtracting unwanted material and leaving the wanted material.

A. Lean manufacturing
B. Subtractive manufacturing
C. Additive manufacturing
D. Flexible manufacturing

14. _____ is where we really start with nothing, or some raw material, and add it only to the places in space where we desire material.

A. Lean manufacturing
B. Subtractive manufacturing
C. Additive manufacturing
D. Flexible manufacturing

15. The _____ uses a technique to selectively deposit a bonding agent or binding liquid to join powder material together to form a 3D part.

A. powder bed fusion
B. fused deposition modeling
C. vat photopolymerization
D. binder jetting process

16. *True or False?* Fused deposition modeling (FDM) is the only additive manufacturing process that can be used to print both plastic and composite.

17. In FDM, a spool of plastic line is being threaded through an apparatus called a(n) _____.

A. extruder
B. armature
C. actuator
D. manipulator

18. In _____, material drops are deposited layer by layer into the build platform to form a 3D part.

A. powder bed fusion
B. material jetting
C. material extrusion
D. binder jetting process

19. _____ is an additive manufacturing technique that uses either laser or electron beam to melt and fuse the material together to form a 3D-printed part.

A. Powder bed fusion
B. Material jetting
C. Material extrusion
D. Binder jetting process

20. *True or False?* Whenever we deposit metal in layers, we create a material that is just as structurally strong as raw steel.

Apply and Analyze

1. What is an early example of automated manufacturing?

2. What has been an obstacle to using more robotics in manufacturing? What attitude has hindered the implementation of more robotic use?

3. Why is it common to have a CNC programmer responsible for implementation of robotics or robotic programming?

4. Why is it important to have the robot and machine communicate, or be integrated, together?

5. Give a brief summary of *just-in-time delivery*, and explain why this has changed the way we manufacture parts.

6. Why is machine shop layout critical to lean manufacturing techniques?

7. How can flexible manufacturing systems (FMS) improve manufacturing performance and provide more diversity in machine shops?

8. What are some of the advantages of additive manufacturing over subtractive manufacturing?

9. What is a disadvantage of additive manufacturing? What are the structural issues with printed parts?

10. What are some of the newest technologies and how are they shaping the modern machine shop?

Critical Thinking

1. From 1775 until 1985, virtually all machining was manual machining, or done by a very skilled machinist who turned handles on a machine to make a part. From 1985 until today, CNC machines, computer technology, and now even robotics are dominating the machining industry. If you could look 20 years ahead, what will machining look like? Write a short paragraph on machining as it looks 20 years from now. What industry will you be in and what do the machines you are running look like?

2. In this chapter you learned about automated manufacturing, flexible manufacturing, and cellular manufacturing. Give a brief description of how these have affected manufacturing and include how these concepts can be applied to other areas of your everyday life.

3. Additive manufacturing is still in its infancy but it has unlimited potential. Write a brief summary of what the future impact of 3D printing could have on one industry, such as the medical field, space travel, or defense systems.

A model created in a laser sintering machine.

Reference Section

The following pages contain tables and charts that will be useful as reference in a variety of areas. To make locating information easier, the material in this section is listed below, along with the page number.

Common Shapes of Metals 333
Physical Properties of Metals 334
Rules for Determining Speeds and Feeds 335
Feeds and Speeds for HSS Drills, Reamers, and Taps 336
Decimal Equivalents: Number-Size Drills 337
Decimal Equivalents: Letter-Size Drills 337
60° V-Type Thread Dimensions: Fractional Sizes 338–339
60° V-Type Thread Dimensions: Metric Sizes 340
Tap Drill Sizes 341–342
Hardness Conversions 343
Conversion Table: US Customary to SI Metric 344
Conversion Table: SI Metric to US Customary 345
Decimal Conversion Chart 346
EIA and AIA National Codes for CNC Programming 347–349
Formulas 350–351

Common Shapes of Metals				
Shapes		**Length**	**How Measured**	**How Purchased**
	Sheet less than 1/4″ thick	Up to 144″	Thickness × width, widths to 72″	Weight, foot, or piece
	Plate more than 1/4″ thick	Up to 20′	Thickness × width	Weight, foot, or piece
	Band	Up to 20′	Thickness × width	Weight or piece
	Rod	12′ to 20′	Diameter	Weight, foot, or piece
	Square	12′ to 20′	Width	Weight, foot, or piece
	Flats	Hot rolled 20′–22′ Cold finished	Thickness × width	Weight, foot, or piece
	Hexagon	12′ to 20′	Distance across flats	Weight, foot, or piece
	Octagon	12′ to 20′	Distance across flats	Weight, foot, or piece
	Angle	Up to 40′	Leg length × leg length × thickness of legs	Weight, foot, or piece
	Channel	Up to 60′	Depth × web thickness × flange width	Weight, foot, or piece
	I-beam	Up to 60′	Height × web thickness × flange width	Weight, foot, or piece

Goodheart-Willcox Publisher

Physical Properties of Metals

Metal	Symbol	Specific Gravity	Specific Heat	Melting Point*		Lb per Cubic Inch
				°C	°F	
Aluminum (cast)	Al	2.56	.2185	658	1217	.0924
Aluminum (rolled)	Al	2.71	—	658	1217	.0978
Antimony	Sb	6.71	.051	630	1166	.2424
Bismuth	Bi	9.80	.031	271	520	.3540
Boron	B	2.30	.3091	2300	4172	.0831
Brass	—	8.51	.094	—	—	.3075
Cadmium	Cd	8.60	.057	321	610	.3107
Calcium	Ca	1.57	.170	810	1490	.0567
Chromium	Cr	6.80	.120	1510	2750	.2457
Cobalt	Co	8.50	.110	1490	2714	.3071
Copper	Cu	8.89	.094	1083	1982	.3212
Columbium	Cb	8.57	—	1950	3542	.3096
Gold	Au	19.32	.032	1063	1945	.6979
Iridium	Ir	22.42	.033	2300	4170	.8099
Iron	Fe	7.86	.110	1520	2768	.2634
Iron (cast)	Fe	7.218	.1298	1375	2507	.2605
Iron (wrought)	Fe	7.70	.1138	1500–1600	2732–2912	.2779
Lead	Pb	11.37	.031	327	621	.4108
Lithium	Li	.057	.941	186	367	.0213
Magnesium	Mg	1.74	.250	651	1204	.0629
Manganese	Mn	8.00	.120	1225	2237	.2890
Mercury	Hg	13.59	.032	−39	−38	.4909
Molybdenum	Mo	10.2	.0647	2620	47.48	.368
Monel metal	—	8.87	.127	1360	2480	.320
Nickel	Ni	8.80	.130	1452	2646	.319
Phosphorus	P	1.82	.177	43	111.4	.0657
Platinum	Pt	21.50	.033	1755	3191	.7767
Potassium	K	0.87	.170	62	144	.0314
Selenium	Se	4.81	.084	220	428	.174
Silicon	Si	2.40	.1762	1427	2600	.087
Silver	Ag	10.53	.056	961	1761	.3805
Sodium	Na	0.97	.290	97	207	.0350
Steel	—	7.858	.1175	1330–1378	2372–2532	.2839
Strontium	Sr	2.54	.074	769	1416	.0918
Tantalum	Ta	10.80	—	2850	5160	.3902
Tin	Sn	7.29	.056	232	450	.2634
Titanium	Ti	5.3	.130	1900	3450	.1915
Tungsten	W	19.10	.033	3000	5432	.6900
Uranium	U	18.70	—	1132	2070	.6755
Vanadium	V	5.50	—	1730	3146	.1987
Zinc	Zn	7.19	.094	419	786	.2598

*Circular of the Bureau of Standards No.35, Department of Commerce and Labor

Rules for Determining Speeds and Feeds

To Find	Having	Rule	Formula
Speed of cutter in feet per minute (fpm)	Diameter of cutter and revolutions per minute	Diameter of cutter (in inches), multiplied by 3.1416 (π), multiplied by revolutions per minute, divided by 12	$\text{fpm} = \dfrac{\pi D \times \text{rpm}}{12}$
Speed of cutter in meters per minute (mpm)	Diameter of cutter and revolutions per minute	Diameter of cutter, multiplied by 3.1416 (π), multiplied by revolutions per minute, divided by 1000	$\text{mpm} = \dfrac{D(\text{mm}) \times \pi \times \text{rpm}}{1000}$
Revolutions per minute (rpm)	Feet per minute and diameter of cutter	Feet per minute, multiplied by 12, divided by circumference of cutter (πD)	$\text{rpm} = \dfrac{\text{fpm} \times 12}{\pi D}$
Revolutions per minute (rpm)	Meters per minute and diameter of cutter in millimeters (mm)	Meters per minute, multiplied by 1000, divided by the circumference of cutter (πD)	$\text{rpm} = \dfrac{\text{mpm} \times 1000}{\pi D}$
Feed per revolution (FR)	Feed per minute and revolutions per minute	Feed per minute, divided by revolutions per minute	$\text{FR} = \dfrac{F}{\text{rpm}}$
Feed per tooth per revolution (ftr)	Feed per minute and number of teeth in cutter	Feed per minute (in inches or millimeters), divided by number of teeth in cutter $\times$ revolutions per minute	$\text{ftr} = \dfrac{F}{T \times \text{rpm}}$
Feed per minute (F)	Feed per tooth per revolution, number of teeth in cutter, and RPM	Feed per tooth per revolutions, multiplied by number of teeth in cutter, multiplied by revolutions per minute	$F = \text{ftr} \times T \times \text{rpm}$
Feed per minute (F)	Feed per revolution and revolutions per minute	Feed per revolution, multiplied by revolutions per minute	$F = \text{FR} \times \text{rpm}$
Number of teeth per minute (TM)	Number of teeth in cutter and revolutions per minute	Number of teeth in cutter, multiplied by revolutions per minute	$\text{TM} = T \times \text{rpm}$

rpm = Revolutions per minute
T = Teeth in cutter
D = Diameter of cutter
π = 3.1416 (pi)
fpm = Speed of cutter in feet per minute

TM = Teeth per minute
F = Feed per minute
FR = Feed per revolution
ftr = Feed per tooth per revolution
mpm = Speed of cutter in meters per minute

Goodheart-Willcox Publisher

		Drills			Reamers		Taps (sfm)			
							Threads per Inch			
Material	Brinell	(sfm)	Point	Feed	(sfm)	Feed	3–7 1/2	8–15	16–24	25–up
Aluminum	99–101	200–250	118°	M	150–160	M	50	100	150	200
Aluminum bronze	170–187	60	118°	M	40–45	M	12	25	45	60
Bakelite	...	80	60-90°	M	50–60	M	50	100	150	200
Brass	192–202	200–250	118°	H	150–160	H	50	100	150	200
Bronze, common	166–183	200–250	118°	H	150–160	H	40	80	100	150
Bronze, phosphor, 1/2 hard	187–202	175–180	118°	M	130–140	M	25	40	50	80
Bronze, phosphor, soft	149–163	200–250	118°	H	150–160	H	40	80	100	150
Cast iron, soft	126	140–150	90°	H	100–110	H	30	60	90	140
Cast iron, medium soft	196	80–110	118°	M	50–65	M	25	40	50	80
Cast iron, hard	293–302	45–50	118°	L	67–75	L	10	20	30	40
Cast iron, chilled*	402	15	150°	L	8–10	L	5	5	10	10
Cast steel	286–302	40–50*	118°	L	70–75	L	20	30	40	50
Celluloid	...	100	90°	M	75–80	M	50	100	150	200
Copper	80–85	70	100°	L	45–55	L	40	80	100	150
Drop forgings (steel)	170–196	60	118°	M	40–45	M	12	25	45	60
Duralumin	90–104	200	118°	M	150–160	M	50	100	150	200
Everdur	179–207	60	118°	L	40–45	L	20	30	40	50
Machinery steel	170–196	110	118°	H	67–75	H	35	50	60	85
Magnet steel, soft	241–302	35–40	118°	M	20–25	M	20	40	50	75
Magnet steel, hard*	321–512	15	150°	L	10	L	5	10	15	25
Manganese steel, 7% – 13%	187–217	15	150°	L	10	L	15	20	25	30
Manganese copper, 30% Mn.*	134	15	150°	L	10–12	L	...	...	...	...
Malleable iron	112–126	85–90	118°	H	...	H	20	30	40	50
Mild steel, .20 –.30 C	170–202	110–120	118°	H	75–85	H	40	55	70	90
Molybdenum steel	196–235	55	125°	M	35–45	M	20	30	35	45
Monel metal	149–170	50	118°	M	35–38	M	8	10	15	20
Nickel, pure*	187–202	75	118°	L	40	L	25	40	50	80
Nickel steel, 3 1/2%	196–241	60	118°	L	40–45	L	8	10	15	20
Rubber, hard	...	100	60-90°	L	70–80	L	50	100	150	200
Screw stock, C.R.	170–196	110	118°	H	75	H	20	30	40	50
Spring steel	402	20	150°	L	12–15	L	10	10	15	15
Stainless steel	146–149	50	118°	M	30	M	8	10	15	20
Stainless steel, C.R.*	460–477	20	118°	L	15	L	8	10	15	20
Steel, .40 to .50 C	170–196	80	118°	M	8–10	M	20	30	40	50
Tool, SAE, and forging steel	149	75	118°	H	35–40	H	25	35	45	55
Tool, SAE, and forging steel	241	50	125°	M	12	M	15	15	25	25
Tool, SAE, and forging steel*	402	15	150°	L	10	L	8	10	15	20
Zinc alloy	112–126	200–250	118°	M	150–175	M	50	100	150	200

Feeds and Speeds for HSS Drills, Reamers, and Taps

Goodheart-Willcox Publisher

*Use specially constructed heavy-duty drills.

Note: Carbon steel tools should be run at speeds 40% to 50% of those recommended for high-speed steel.

Spiral point taps may be run at speeds 15% to 20% faster than regular taps.

Decimal Equivalents: Number-Size Drills

Drill	Size of Drill in Inches	Drill	Size of Drill in Inches	Drill	Size of Drill in Inches	Drill	Size of Drill in Inches
1	.2280	21	.1590	41	.0960	61	.0390
2	.2210	22	.1570	42	.0935	62	.0380
3	.2130	23	.1540	43	.0890	63	.0370
4	.2090	24	.1520	44	.0860	64	.0360
5	.2055	25	.1495	45	.0820	65	.0350
6	.2040	26	.1470	46	.0810	66	.0330
7	.2010	27	.1440	47	.0785	67	.0320
8	.1990	28	.1405	48	.0760	68	.0310
9	.1960	29	.1360	49	.0730	69	.0292
10	.1935	30	.1285	50	.0700	70	.0280
11	.1910	31	.1200	51	.0670	71	.0260
12	.1890	32	.1160	52	.0635	72	.0250
13	.1850	33	.1130	53	.0595	73	.0240
14	.1820	34	.1110	54	.0550	74	.0225
15	.1800	35	.1100	55	.0520	75	.0210
16	.1770	36	.1065	56	.0465	76	.0200
17	.1730	37	.1040	57	.0430	77	.0180
18	.1695	38	.1015	58	.0420	78	.0160
19	.1660	39	.0995	59	.0410	79	.0145
20	.1610	40	.0980	60	.0400	80	.0135

Goodheart-Willcox Publisher

Decimal Equivalents: Letter-Size Drills

Drill	Size of Drill in Inches	Drill	Size of Drill in Inches	Drill	Size of Drill in Inches	Drill	Size of Drill in Inches
A	0.234	H	0.266	O	0.316	V	0.377
B	0.238	I	0.272	P	0.323	W	0.386
C	0.242	J	0.277	Q	0.332	X	0.397
D	0.246	K	0.281	R	0.339	Y	0.404
E	0.250	L	0.290	S	0.348	Z	0.413
F	0.257	M	0.295	T	0.358		
G	0.261	N	0.302	U	0.368		

Goodheart-Willcox Publisher

60° V–Type Thread Dimensions: Fractional Sizes

National Special Thread Series

Nominal Size	Threads per Inch	Major Diameter (Inches)	Minor Diameter (Inches)	Pitch Diameter (Inches)	Tap Drill for 75% Thread †	Clearance Drill Size*
1/16″	64	.0625	.0422	.0524	3/64″	51
5/64″	60	.0781	.0563	.0673	1/16″	45
3/32″	48	.0938	.0667	.0803	49	40
7/64″	48	.1094	.0823	.0959	43	32
1/8″	32	.1250	.0844	.1047	3/32″	29
9/64″	40	.1406	.1081	.1244	32	24
5/32″	32	.1563	.1157	.1360	1/8″	19
5/32″	36	.1563	.1202	.1382	30	19
11/64″	32	.1719	.1313	.1516	9/64″	14
3/16″	24	.1875	.1334	.1604	26	8
3/16″	32	.1875	.1469	.1672	22	8
13/64″	24	.2031	.1490	.1760	20	3
7/32″	24	.2188	.1646	.1917	16	1
7/32″	32	.2188	.1782	.1985	12	1
15/64″	24	.2344	.1806	.2073	10	1/4″
1/4″	24	.2500	.1959	.2229	4	17/64″
1/4″	27	.2500	.2019	.2260	3	17/64″
1/4″	32	.2500	.2094	.2297	7/32″	17/64″
5/16″	20	.3125	.2476	.2800	17/64″	21/64″
5/16″	27	.3125	.2644	.2884	J	21/64″
5/16″	32	.3125	.2719	.2922	9/32″	21/64″
3/8″	20	.3750	.3100	.3425	21/64″	25/64″
3/8″	27	.3750	.3269	.3509	R	25/64″

†Refer to tables "Decimal Equivalents: Number-Size Drills" and "Decimal Equivalents: Letter-Size Drills."

*Clearance drill makes hole with standard clearance for diameter of nominal size.

**Standard spark plug size.

(continued)

60° V–Type Thread Dimensions: Fractional Sizes *(continued)*

National Special Thread Series

Nominal Size	Threads per Inch	Major Diameter (Inches)	Minor Diameter (Inches)	Pitch Diameter (Inches)	Tap Drill for 75% Thread †	Clearance Drill Size*
7/16″	24	.4375	.3834	.4104	X	29/64″
7/16″	27	.4375	.3894	.4134	Y	29/64″
1/2″	12	.5000	.3918	.4459	27/64″	33/64″
1/2″	24	.5000	.4459	.4729	29/64″	33/64″
1/2″	27	.5000	.4519	.4759	15/32″	33/64″
9/16″	27	.5625	.5144	.5384	17/32″	37/64″
5/8″	12	.6250	.5168	.5709	35/64″	41/64″
5/8″	27	.6250	.5769	.6009	19/32″	41/64″
11/16″	11	.6875	.5694	.6285	19/32″	45/64″
11/16″	16	.6875	.6063	.6469	5/8″	45/64″
3/4″	12	.7500	.6418	.6959	43/64″	49/64″
3/4″	27	.7500	.7019	.7259	23/32″	49/64″
13/16″	10	.8125	.6826	.7476	23/32″	53/64″
7/8″	12	.8750	.7668	.8209	51/64″	57/64″
7/8″	18**	.8750	.8028	.8389	53/64″	57/64″
7/8″	27	.8750	.8269	.8509	27/32″	57/64″
15/16″	9	.9375	.7932	.8654	53/64″	61/64″
1″	12	1.0000	.8918	.9459	59/64″	1 1/64″
1″	27	1.0000	.9519	.9759	31/32″	1 1/64″
1 5/8″	5 1/2	1.6250	1.3888	1.5069	1 29/64″	1 41/64″
1 7/8″	5	1.8750	1.6152	1.7451	1 11/16″	1 57/64″
2 1/8″	4 1/2	2.1250	1.8363	1.9807	1 29/32″	2 5/32″
2 3/8″	4	2.3750	2.0502	2.2126	2 1/8″	2 13/32″

†Refer to tables "Decimal Equivalents: Number-Size Drills" and "Decimal Equivalents: Letter-Size Drills."

*Clearance drill makes hole with standard clearance for diameter of nominal size.

**Standard spark plug size.

Goodheart-Willcox Publisher

60° V–Type Thread Dimensions: Metric Sizes
International Standard

Major Diameter (mm)	Pitch (mm)	Minor Diameter (mm)	Pitch Diameter (mm)	Tap Drill for 75% Thread (mm)	Tap Drill for 75% Thread † (No. or Inches)	Clearance Drill Size*
2.0	.40	1.48	1.740	1.6	1/16″	41
2.3	.40	1.78	2.040	1.9	48	36
2.6	.45	2.02	2.308	2.1	45	31
3.0	.50	2.35	2.675	2.5	40	29
3.5	.60	2.72	3.110	2.9	33	23
4.0	.70	3.09	3.545	3.3	30	16
4.5	.75	3.53	4.013	3.75	26	10
5.0	.80	3.96	4.480	4.2	19	3
5.5	.90	4.33	4.915	4.6	14	15/64″
6.0	1.00	4.70	5.350	5.0	9	1/4″
7.0	1.00	5.70	6.350	6.0	15/64″	19/64″
8.0	1.25	6.38	7.188	6.8	H	11/32″
9.0	1.25	7.38	8.188	7.8	5/16″	3/8″
10.0	1.50	8.05	9.026	8.6	R	27/64″
11.0	1.50	9.05	10.026	9.6	V	29/64″
12.0	1.75	9.73	10.863	10.5	Z	1/2″
14.0**	1.25	12.38	13.188	13.0	33/64″	9/16″
14.0	2.00	11.40	12.701	12.0	15/32″	9/16″
16.0	2.00	13.40	14.701	14.0	35/64″	21/32″
18.0	1.50	16.05	17.026	16.5	41/64″	47/64″
18.0	2.50	14.75	16.376	15.5	39/64″	47/64″
20.0	2.50	16.75	18.376	17.5	11/16″	13/16″
22.0	2.50	18.75	20.376	19.5	49/64″	57/64″
24.0	3.00	20.10	22.051	21.0	53/64″	31/32″
27.0	3.00	23.10	25.051	24.0	15/16″	13/32″
30.0	3.50	25.45	27.727	26.5	1 3/64″	1 13/64″
33.0	3.50	28.45	30.727	29.5	1 11/64″	1 21/64″
36.0	4.00	30.80	33.402	32.0	1 17/64″	1 7/16″
39.0	4.00	33.80	36.402	35.0	1 3/8″	1 9/16″
42.0	4.50	36.15	39.077	37.0	1 29/64″	1 43/64″
45.0	4.50	39.15	42.077	40.0	1 37/64″	1 13/16″
48.0	5.00	41.50	44.752	43.0	1 11/16″	1 29/64″

†Refer to tables "Decimal Equivalents: Number-Size Drills" and "Decimal Equivalents: Letter-Size Drills."

*Clearance drill makes hole with standard clearance for diameter of nominal size.

**Standard spark plug size.

Tap Drill Sizes

Probable Percentage of Full Thread Produced in Tapped Hole Using Stock Sizes of Drill

Tap	Tap Drill	Decimal Equivalent of Tap Drill	Theoretical % of Thread	Probable Oversize (Mean)	Probable Hole Size	Percentage of Thread	Tap	Tap Drill	Decimal Equivalent of Tap Drill	Theoretical % of Thread	Probable Oversize (Mean)	Probable Hole Size	Percentage of Thread
0–80	56	.0465	83	.0015	.0480	74	8–32	29	.1360	69	.0029	.1389	62
	3/64	.0469	81	.0015	.0484	71		28	.1405	58	.0029	.1434	51
1–64	54	.0550	89	.0015	.0565	81	8–36	29	.1360	78	.0029	.1389	70
	53	.0595	67	.0015	.0610	59		28	.1405	68	.0029	.1434	57
1–72	53	.0595	75	.0015	.0610	67		9/64	.1406	68	.0029	.1435	57
	1/16	.0625	58	.0015	.0640	50	10–24	27	.1440	85	.0032	.1472	79
2–56	51	.0670	82	.0017	.0687	74		26	.1470	79	.0032	.1502	74
	50	.0700	69	.0017	.0717	62		25	.1495	75	.0032	.1527	69
	49	.0730	56	.0017	.0747	49		24	.1520	70	.0032	.1552	64
2–64	50	.0700	79	.0017	.0717	70		23	.1540	67	.0032	.1572	61
	49	.0730	64	.0017	.0747	56		5/32	.1563	62	.0032	.1595	56
3–48	48	.0760	85	.0019	.0779	78		22	.1570	61	.0032	.1602	55
	5/64	.0781	77	.0019	.0800	70	10–32	5/32	.1563	83	.0032	.1595	75
	47	.0785	76	.0019	.0804	69		22	.1570	81	.0032	.1602	73
	46	.0810	67	.0019	.0829	60		21	.1590	76	.0032	.1622	68
	45	.0820	63	.0019	.0839	56		20	.1610	71	.0032	.1642	64
3–56	46	.0810	78	.0019	.0829	69		19	.1660	59	.0032	.1692	51
	45	.0820	73	.0019	.0839	65	12–24	11/64	.1719	82	.0035	.1754	75
	44	.0860	56	.0019	.0879	48		17	.1730	79	.0035	.1765	73
4–40	44	.0860	80	.0020	.0880	74		16	.1770	72	.0035	.1805	66
	43	.0890	71	.0020	.0910	65		15	.1800	67	.0035	.1835	60
	42	.0935	57	.0020	.0955	51		14	.1820	63	.0035	.1855	56
	3/32	.0938	56	.0020	.0958	50	12–28	16	.1770	84	.0035	.1805	77
4–48	42	.0935	68	.0020	.0955	61		15	.1800	78	.0035	.1835	70
	3/32	.0938	68	.0020	.0958	60		14	.1820	73	.0035	.1855	66
	41	.0960	59	.0020	.0980	52		13	.1850	67	.0035	.1885	59
5–40	40	.0980	83	.0023	.1003	76		3/16	.1875	61	.0035	.1910	54
	39	.0995	79	.0023	.1018	71	1/4–20	9	.1960	83	.0038	.1998	77
	38	.1015	72	.0023	.1038	65		8	.1990	79	.0038	.2028	73
	37	.1040	65	.0023	.1063	58		7	.2010	75	.0038	.2048	70
5–44	38	.1015	79	.0023	.1038	72		13/64	.2031	72	.0038	.2069	66
	37	.1040	71	.0023	.1063	63		6	.2040	71	.0038	.2078	65
	36	.1065	63	.0023	.1088	55		5	.2055	69	.0038	.2093	63
6–32	37	.1040	84	.0023	.1063	78		4	.2090	63	.0038	.2128	57
	36	.1065	78	.0026	.1091	71	1/4–28	3	.2130	80	.0038	.2168	72
	7/64	.1094	70	.0026	.1120	64		7/32	.2188	67	.0038	.2226	59
	35	.1100	69	.0026	.1126	63		2	.2210	63	.0038	.2248	55
	34	.1110	67	.0026	.1136	60	5/16–18	F	.2570	77	.0038	.2608	72
	33	.1130	62	.0026	.1156	55		G	.2610	71	.0041	.2651	66
6–40	34	.1110	83	.0026	.1136	75		17/64	.2656	65	.0041	.2697	59
	33	.1130	77	.0026	.1156	69		H	.2660	64	.0041	.2701	59
	32	.1160	68	.0026	.1186	60							

(continued)

Tap Drill Sizes (continued)

Probable Percentage of Full Thread Produced in Tapped Hole Using Stock Sizes of Drill

Tap	Tap Drill	Decimal Equivalent of Tap Drill	Theoretical % of Thread	Probable Oversize (Mean)	Probable Hole Size	Percentage of Thread	Tap	Tap Drill	Decimal Equivalent of Tap Drill	Theoretical % of Thread	Probable Oversize (Mean)	Probable Hole Size	Percentage of Thread
5/16–24	H	.2660	86	.0041	.2701	78	1"–14	59/64	.9219	84	.0060	.9279	78
	I	.2720	75	.0041	.2761	67		15/16	.9375	67	.0060	.9435	61
	J	.2770	66	.0041	.2811	58	1 1/8–7	31/32	.9688	84	.0062	.9750	81
3/8–16	5/16	.3125	77	.0044	.3169	72		63/64	.9844	76	.0067	.9911	72
	O	.3160	73	.0044	.3204	68		1"	1.0000	67	.0070	1.0070	64
	P	.3230	64	.0044	.3274	59		1 1/64	1.0156	59	.0070	1.0226	55
3/8–24	21/64	.3281	87	.0044	.3325	79	1 1/8–12	1 1/32	1.0313	87	.0071	1.0384	80
	Q	.3320	79	.0044	.3364	71		1 3/64	1.0469	72	.0072	1.0541	66
	R	.3390	67	.0044	.3434	58	1 1/4–7	1 3/32	1.0938	84			
7/16–14	T	.3580	86	.0046	.3626	81		1 7/64	1.1094	76			
	23/64	.3594	84	.0046	.3640	79		1 1/8	1.1250	67			
	U	.3680	75	.0046	.3726	70	1 1/4–12	1 5/32	1.1563	87			
	3/8	.3750	67	.0046	.3796	62		1 11/64	1.1719	72			
	V	.3770	65	.0046	.3816	60	1 3/8–6	1 3/16	1.1875	87			
7/16–20	W	.3860	79	.0046	.3906	72		1 13/64	1.2031	79			
	25/64	.3906	72	.0046	.3952	65		1 7/32	1.2188	72	No		
	X	.3970	62	.0046	.4016	55		1 15/64	1.2344	65	test results		
1/2–13	27/64	.4219	78	.0047	.4266	73	1 3/8–12	1 9/32	1.2813	87	available		
	7/16	.4375	63	.0047	.4422	58		1 19/64	1.2969	72			
1/2–20	29/64	.4531	72	.0047	.4578	65	1 1/2–6	1 5/16	1.3125	87	Reaming		
9/16–12	15/32	.4688	87	.0048	.4736	82		1 21/64	1.3281	79	recommended		
	31/64	.4844	72	.0048	.4892	68		1 11/32	1.3438	72			
9/16–18	1/2	.5000	87	.0048	.5048	80		1 23/64	1.3594	65			
	33/64	.5156	65	.0048	.5204	58	1 1/2–12	1 13/32	1.4063	87			
5/8–11	17/32	.5313	79	.0049	.5362	75		1 27/64	1.4219	72			
	35/64	.5469	66	.0049	.5518	62							
5/8–18	9/16	.5625	87	.0049	.5674	80							
	37/64	.5781	65	.0049	.5831	58							
3/4–10	41/64	.6406	84	.0050	.6456	80							
	21/32	.6563	72	.0050	.6613	68							
3/4–16	11/16	.6875	77	.0050	.6925	71							
7/8–9	49/64	.7656	76	.0052	.7708	72							
	25/32	.7812	65	.0052	.7864	61							
7/8–14	51/64	.7969	84	.0052	.8021	79							
	13/16	.8125	67	.0052	.8177	62							
1"–8	55/64	.8594	87	.0059	.8653	83							
	7/8	.8750	77	.0059	.8809	73							
	57/64	.8906	67	.0059	.8965	64							
	29/32	.9063	58	.0059	.9122	54							
1"–12	29/32	.9063	87	.0060	.9123	81							
	59/64	.9219	72	.0060	.9279	67							
	15/16	.9375	58	.0060	.9435	52							

Taper Pipe		Straight Pipe	
Thread	**Drill**	**Thread**	**Drill**
1/8–27	R	1/8–27	S
1/4–18	7/16	1/4–18	29/64
3/8–18	37/64	3/8–18	19/32
1/2–14	23/32	1/2–14	47/64
3/4–14	59/64	3/4–14	15/16
1–11 1/2	1 5/32	1–11 1/2	1 3/16
1 1/4–11 1/2	1 1/2	1 1/4–11 1/2	1 33/64
1 1/2–11 1/2	1 47/64	1 1/2–11 1/2	1 3/4
2–11 1/2	2 7/32	2–11 1/2	2 7/32
2 1/2–8	2 5/8	2 1/2–8	2 21/32
3–8	3 1/4	3–8	3 9/32
3 1/2–8	3 3/4	3 1/2–8	3 25/32
4–8	4 1/4	4–8	4 9/32

Standard Tool Co.

Hardness Conversions

Brinell Indentation Diameter (mm)	Brinell Hardness Number		Rockwell Hardness Number		Rockwell Superficial Hardness Number (Superficial Diamond Penetrator)			Tensile Strength (Approximate) × 1000 psi
	Standard Ball	Tungsten-Carbide Ball	B Scale	C Scale	15 N Scale	30 N Scale	45 N Scale	
2.45	—	627	—	58.7	89.6	76.3	65.1	347
2.50	—	601	—	57.3	89.0	75.1	63.5	328
2.55	—	578	—	56.0	88.4	73.9	62.1	313
2.60	—	555	—	54.7	87.8	72.7	60.6	298
2.65	—	534	—	53.5	87.2	71.6	59.2	288
2.70	—	514	—	52.1	86.5	70.3	57.6	274
2.75	—	495	—	51.0	85.9	69.4	56.1	264
2.80	—	477	—	49.6	85.3	68.2	54.5	252
2.85	—	461	—	48.5	84.7	67.2	53.2	242
2.90	—	444	—	47.1	84.0	65.8	51.5	230
2.95	429	429	—	45.7	83.4	64.6	49.9	219
3.00	415	415	—	44.5	82.8	63.5	48.4	212
3.05	401	401	—	43.1	82.0	62.3	46.9	202
3.10	388	388	—	41.8	81.4	61.1	45.3	193
3.15	375	375	—	40.4	80.6	59.9	43.6	184
3.20	363	363	—	39.1	80.0	58.7	42.0	177
3.25	352	352	—	37.9	79.3	57.6	40.5	170
3.30	341	341	—	36.6	78.6	56.4	39.1	163
3.35	331	331	—	35.5	78.0	55.4	37.8	158
3.40	321	321	—	34.3	77.3	54.3	36.4	152
3.45	311	311	—	33.1	76.7	53.3	34.4	147
3.50	302	302	—	32.1	76.1	52.2	33.8	143
3.55	293	293	—	30.9	75.5	51.2	32.4	139
3.60	285	285	—	29.9	75.0	50.3	31.2	136
3.65	277	277	—	28.8	74.4	49.3	29.9	131
3.70	269	269	—	27.6	73.7	48.3	28.5	128
3.75	262	262	—	26.6	73.1	47.3	27.3	125
3.80	255	255	—	25.4	72.5	46.2	26.0	121
3.85	248	248	—	24.2	71.7	45.1	24.5	118
3.90	241	241	100.0	22.8	70.9	43.9	22.8	114
3.95	235	235	99.0	21.7	70.3	42.9	21.5	111
4.00	229	229	98.2	20.5	69.7	41.9	20.1	109
4.05	223	223	97.3	—	—	—	—	104
4.10	217	217	96.4	—	—	—	—	103
4.15	212	212	95.5	—	—	—	—	100
4.20	207	207	94.6	—	—	—	—	99
4.25	201	201	93.8	—	—	—	—	97
4.30	197	197	92.8	—	—	—	—	94
4.35	192	192	91.9	—	—	—	—	92
4.40	187	187	90.7	—	—	—	—	90
4.45	183	183	90.0	—	—	—	—	89
4.50	179	179	89.0	—	—	—	—	88
4.55	174	174	87.8	—	—	—	—	86
4.60	170	170	86.8	—	—	—	—	84
4.65	167	167	86.0	—	—	—	—	83
4.70	163	163	85.0	—	—	—	—	82
4.80	156	156	82.9	—	—	—	—	80
4.90	149	149	80.8	—	—	—	—	73
5.00	143	143	78.7	—	—	—	—	71
5.10	137	137	76.4	—	—	—	—	67
5.20	131	131	74.0	—	—	—	—	65
5.30	126	131	72.0	—	—	—	—	63
5.40	121	121	69.0	—	—	—	—	60
5.50	116	116	67.6	—	—	—	—	58
5.60	111	111	65.7	—	—	—	—	56

Carpenter Steel Co.

Conversion Table: US Customary to SI Metric

When You Know:	Multiply By:		To Find:
	Very Accurate	**Approximate**	
Length			
inches	* 25.4		millimeters
inches	* 2.54		centimeters
feet	* 0.3048		meters
feet	* 30.48		centimeters
yards	* 0.9144	0.9	meters
miles	* 1.609344	1.6	kilometers
Weight			
grains	15.43236	15.4	grams
ounces	* 28.349523125	28.0	grams
ounces	* 0.028349523125	0.028	kilograms
pounds	* 0.45359237	0.45	kilograms
short ton	* 0.90718474	0.9	tonnes
Volume			
teaspoons		5.0	milliliters
tablespoons		15.0	milliliters
fluid ounces	29.57353	30.0	milliliters
cups		0.24	liters
pints	* 0.473176473	0.47	liters
quarts	* 0.946352946	0.95	liters
gallons	* 3.785411784	3.8	liters
cubic inches	* 0.016387064	0.02	liters
cubic feet	* 0.028316846592	0.03	cubic meters
cubic yards	* 0.764554857984	0.76	cubic meters
Area			
square inches	* 6.4516	6.5	square centimeters
square feet	* 0.09290304	0.09	square meters
square yards	* 0.83612736	0.8	square meters
square miles		2.6	square kilometers
acres	* 0.40468564224	0.4	hectares
Temperature			
Fahrenheit	* 5/9 (after subtracting 32)		Celsius
Density			
pounds per cubic feet	1.602×10	16	kilograms per cubic meter
Force			
ounces (F)	2.780×10^{-1}		newtons
pounds (F)	4.448×10^{-3}		kilonewtons
kips	4.448		meganewtons
Stress			
pounds/square inch (psi)	6.895×10^{-3}		megapascals
kips/square inch (ksi)	6.895		megapascals
Torque			
ounce-inches	7.062×10^{3}		newton-meters
pound-inches	1.130×10^{-1}		newton-meters
pound-feet	1.356		newton-meters

* = Exact

Goodheart-Willcox Publisher

Conversion Table: SI Metric to US Customary

When You Know:	Multiply By:		To Find:
	Very Accurate	Approximate	
Length			
millimeters	0.0393701	0.04	inches
centimeters	0.3937008	0.4	inches
meters	3.280840	3.3	feet
meters	1.093613	1.1	yards
kilometers	0.621371	0.6	miles
Weight			
grains	0.00228571	0.0023	ounces
grams	0.03527396	0.035	ounces
kilograms	2.204623	2.2	pounds
tonnes	1.1023113	1.1	short tons
Volume			
milliliters		0.2	teaspoons
milliliters	0.06667	0.067	tablespoons
milliliters	0.03381402	0.03	fluid ounces
liters	61.02374	61.024	cubic inches
liters	2.113376	2.1	pints
liters	1.056688	1.06	quarts
liters	0.26417205	0.26	gallons
liters	0.03531467	0.035	cubic feet
cubic meters	61023.74	61023.7	cubic inches
cubic meters	35.31467	35.0	cubic feet
cubic meters	1.3079506	1.3	cubic yards
cubic meters	264.17205	264.0	gallons
Area			
square centimeters	0.1550003	0.16	square inches
square centimeters	0.00107639	0.001	square feet
square meters	10.76391	10.8	square feet
square meters	1.195990	1.2	square yards
square kilometers		0.4	square miles
hectares	2.471054	2.5	acres
Temperature			
Celsius	*9/5 (then add 32)		Fahrenheit

* = Exact

Goodheart-Willcox Publisher

Decimal Conversion Chart

Fraction		Inches	mm
	1/64	.01563	.397
1/32		.03125	.794
	3/64	.04688	1.191
1/16		.0625	1.588
	5/64	.07813	1.984
3/32		.09375	2.381
	7/64	.10938	2.778
1/8		.12500	3.175
	9/64	.14063	3.572
5/32		.15625	3.969
	11/64	.17188	4.366
3/16		.18750	4.763
	13/64	.20313	5.159
7/32		.21875	5.556
	15/64	.23438	5.953
1/4		.25000	6.350
	17/64	.26563	6.747
9/32		.28125	7.144
	19/64	.29688	7.541
5/16		.31250	7.938
	21/64	.32813	8.334
11/32		.34375	8.731
	23/64	.35938	9.128
3/8		.37500	9.525
	25/64	.39063	9.922
13/32		.40625	10.319
	27/64	.42188	10.716
7/16		.43750	11.113
	29/64	.45313	11.509
15/32		.46875	11.906
	31/64	.48438	12.303
1/2		.50000	12.700

Fraction		Inches	mm
	33/64	.51563	13.097
17/32		.53125	13.494
	35/64	.54688	13.891
9/16		.56250	14.288
	37/64	.57813	14.684
19/32		.59375	15.081
	39/64	.60938	15.478
5/8		.62500	15.875
	41/64	.64063	16.272
21/32		.65625	16.669
	43/64	.67188	17.066
11/16		.68750	17.463
	45/64	.70313	17.859
23/32		.71875	18.256
	47/64	.73438	18.653
3/4		.75000	19.050
	49/64	.76563	19.447
25/32		.78125	19.844
	51/64	.79688	20.241
13/16		.81250	20.638
	53/64	.82813	21.034
27/32		.84375	21.431
	55/64	.85938	21.828
7/8		.87500	22.225
	57/64	.89063	22.622
29/32		.90625	23.019
	59/64	.92188	23.416
15/16		.93750	23.813
	61/64	.95313	24.209
31/32		.96875	24.606
	63/64	.98438	25.003
1		1.00000	25.400

Goodheart-Willcox Publisher

EIA and AIA National Codes for CNC Programming
Preparatory (G) Functions

G word	Explanation
G00	Denotes a rapid traverse rate for point-to-point positioning.
G01	Describes linear interpolation blocks; reserved for contouring.
G02, G03	Used with circular interpolation.
G04	Sets a calculated time delay during which there is no machine motion (dwell).
G05, G07	Unassigned by the EIA. May be used at the discretion of the machine tool or system builder. Could also be standardized at a future date.
G06	Used with parabolic interpolation.
G08	Acceleration code. Causes the machine, assuming it is capable, to accelerate at a smooth exponential rate.
G09	Deceleration code. Causes the machine, assuming it is capable, to decelerate at a smooth exponential rate.
G10-G12	Normally unassigned for CNC systems. Used with some hard-wired systems to express blocks of abnormal dimensions.
G13-G16	Direct the control system to operate on a particular set of axes.
G17-G19	Identify or select a coordinate plane for such functions as circular interpolation or cutter compensation.
G20-G32	Unassigned according to EIA standards. May be assigned by the control system or machine tool builder.
G33-G35	Selected for machines equipped with thread-cutting capabilities (generally referring to lathes). G33 is used when a constant lead is sought, G34 is used when a constantly increasing lead is required, and G35 is used to designate a constantly decreasing lead.
G36-G39	Unassigned.
G40	Terminates any cutter compensation.
G41	Activates cutter compensation in which the cutter is on the left side of the work surface (relative to the direction of the cutter motion).
G42	Activates cutter compensation in which the cutter is on the right side of the work surface.
G43, G44	Used with cutter offset to adjust for the difference between the actual and programmed cutter radii or diameters.
G43	refers to an inside corner, and G44 refers to an outside corner.
G45-G49	Unassigned.
G50-G59	Reserved for adaptive control.
G60-G69	Unassigned.
G70	Selects inch programming.
G71	Selects metric programming.
G72	Selects three-dimensional CW circular interpolation.
G73	Selects three-dimensional CCW circular interpolation.
G74	Cancels multiquadrant circular interpolation.
G75	Activates multiquadrant circular interpolation.
G76-G79	Unassigned.
G80	Cancel cycle.
G81	Activates drill, or spotdrill, cycle.
G82	Activates drill with a dwell.
G83	Activates intermittent, or deep-hole, drilling.
G84	Activates tapping cycle.
G85-G89	Activates boring cycles.

(continued)

EIA and AIA National Codes for CNC Programming *(continued)*

Preparatory (G) Functions *(continued)*

G word	Explanation
G90	Selects absolute input. Input data is to be in absolute dimensional form.
G91	Selects incremental input. Input data is to be in incremental form.
G92	Preloads registers to desired values (for example, preloads axis position registers).
G93	Sets inverse time feed rate.
G94	Sets inches (or millimeters) per minute feed rate.
G95	Sets inches (or millimeters) per revolution feed rate.
G97	Sets spindle speed in revolutions per minute.
G98, G99	Unassigned.

Miscellaneous (M) Functions

M word	Explanation
M00	Program stop. Operator must cycle start in order to continue with the remainder of the program.
M01	Optional stop. Acted upon only when the operator has previously signaled for this command by pushing a button. When the control system senses the M01 code, machine will automatically stop.
M02	End of program. Stops the machine after completion of all commands in the block. May include rewinding of tape.
M03	Starts spindle rotation in a clockwise direction.
M04	Starts spindle rotation in a counterclockwise direction.
M05	Spindle stop.
M06	Executes the change of a tool (or tools) manually or automatically.
M07	Turns coolant on (flood).
M08	Turns coolant on (mist).
M09	Turns coolant off.
M10	Activates automatic clamping of the machine slides, workpiece, fixture, spindle, etc.
M11	Deactivates automatic clamping.
M12	Inhibiting code used to synchronize multiple set of axes, such as a four-axis lathe that has two independently operated heads or slides.
M13	Combines simultaneous clockwise spindle motion and coolant on.
M14	Combines simultaneous counterclockwise spindle motion and coolant on.
M15	Sets rapid traverse or feed motion in the + direction.
M16	Sets rapid traverse or feed motion in the − direction.
M17, M18	Unassigned.
M19	Oriented spindle stop. Stops spindle at a predetermined angular position.
M20-M29	Unassigned.
M30	End of data. Used to reset control and/or machine.
M31	Interlock bypass. Temporarily circumvents a normally provided interlock.
M32-M39	Unassigned.
M40-M46	Signals gear changes if required at the machine; otherwise, unassigned.
M47	Continues program execution from the start of the program, unless inhibited by an interlock signal.
M48	Cancels M49.
M49	Deactivates a manual spindle or feed-override and returns to the programmed value.

(continued)

EIA and AIA National Codes for CNC Programming *(continued)*

Miscellaneous (M) Functions *(continued)*

M word	Explanation
M50-M57	Unassigned.
M58	Cancels M59.
M59	Holds the rpm constant at its value.
M60-M99	Unassigned.

Other Address Characters

Address character	Explanation
A	Angular dimension about the X axis.
B	Angular dimension about the Y axis.
C	Angular dimension about the Z axis.
D	Can be used for an angular dimension around a special axis, for a third feed function, or for tool offset.
E	Used for angular dimension around a special axis or for a second feed function.
H	Fixture offset.
I, J, K	Centerpoint coordinates for circular interpolation.
L	Not used.
O	Used on some N/C controls in place of the customary sequence number word address N.
P	Third rapid traverse code—tertiary motion dimension parallel to the X axis.
Q	Second rapid traverse code—tertiary motion dimension parallel to the Y axis.
R	First rapid traverse code—tertiary motion dimension parallel to the Z axis (or to the radius) for constant surface speed calculation.
U	Secondary motion dimension parallel to the X axis.
V	Secondary motion dimension parallel to the Y axis.
W	Secondary motion dimension parallel to the Z axis.

Goodheart-Willcox Publisher

Formulas

Circle

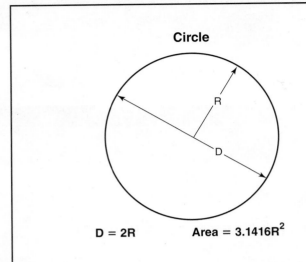

D = 2R Area = 3.1416R^2

Rectangle

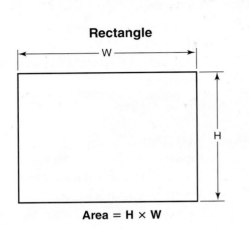

Area = H × W

Triangle

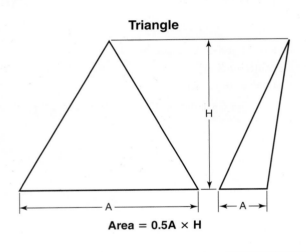

Area = 0.5A × H

Trapezoid

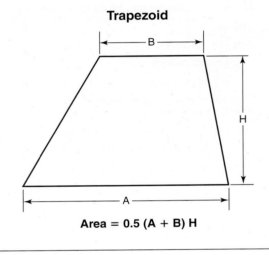

Area = 0.5 (A + B) H

Rectangular Prism

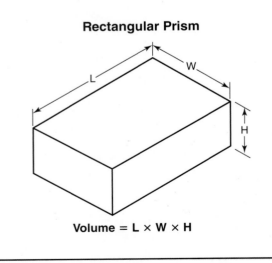

Volume = L × W × H

Cylinder

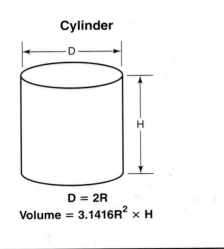

D = 2R
Volume = 3.1416R^2 × H

(continued)

Formulas *(continued)*

Cone

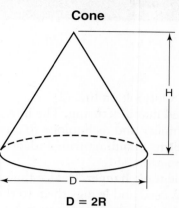

D = 2R

$$\text{Volume} = \frac{3.1416R^2 \times H}{3}$$

Sphere

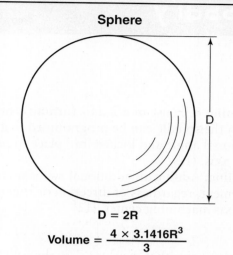

D = 2R

$$\text{Volume} = \frac{4 \times 3.1416R^3}{3}$$

Triangular Pyramid

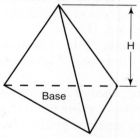

Base

$$\text{Volume} = \frac{\text{Area of base} \times H}{3}$$

Pythagorean Theorem

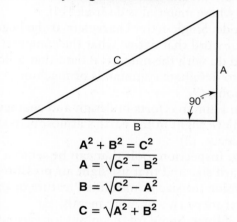

$$A^2 + B^2 = C^2$$
$$A = \sqrt{C^2 - B^2}$$
$$B = \sqrt{C^2 - A^2}$$
$$C = \sqrt{A^2 + B^2}$$

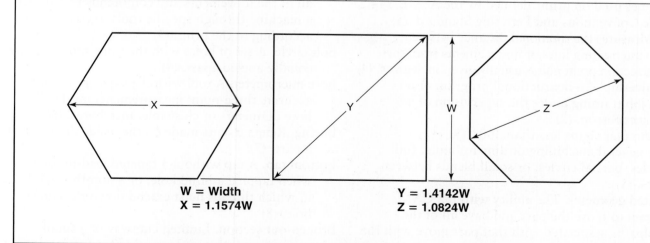

W = Width
X = 1.1574W

Y = 1.4142W
Z = 1.0824W

Glossary

2 + 1 turning. The use of a 2-axis turning center in which the spindle can be programmed to any rotational angle and locked into place to provide a third axis. (12)

3 + 2 milling. Adding an additional two axes of movement, regardless of direction or function, to a 3-axis machining center. (12)

A

acute angle. Angle that measures less than 90°. (4)

additive manufacturing. Machining that starts with raw material and adds it only to the places in space where material is desired. (14)

address code. Single-letter characters at the beginning of each word that define what the computer should do with the numerical data that follows. Used to designate commands or machine functions. (9)

aerospace. Human efforts in design and engineering to achieve flight in the earth's atmosphere as well as in space. (2)

air gaging. Inspection tools that can be set to a standard size and that use light air pressure to determine the size of machined feature or part. (6)

alloy. Mixture of two or more metals. (3)

alphabet of lines. Standard meanings for various types of lines used on prints defined by the ASME Y14.2 Line Conventions and Lettering Standard. (5)

angle. Measures the rotational distance between two intersecting lines or line segments from their intersection point and is usually given in degrees. (4)

apprenticeship. An instructional program of on-the-job training under the supervision of a journeyperson. (2)

arc. A circular shape less than 360°. (9)

arcs. In general machining or dimensioning, full circles, partial circles, or small blends between lines. (5)

associated geometry. The ability within a CAM system to move the part and have all of the toolpaths associated with that part move with the geometry, eliminating programming and setup errors. (11)

attitude. Outlook on life. (2)

automated manufacturing. The integration of machinery and technology to create a process that performs manufacturing independently through automation. (14)

automatic tool changer. Device that stores tools not being used and brings them to the ready position when a tool change is required. (7)

auxiliary view. A view in which a part is oriented and projected at an angle other than 90°. (5)

B

band saw. Cutting machine that uses a thin blade stretched over a set of wheels or pulleys. The band saw comes in a vertical or horizontal configuration. (1)

bilateral tolerance. Tolerance in both positive and negative directions. Also called plus-minus tolerance. (5)

binder jetting process. An additive manufacturing process which, using no heat, selectively deposits a bonding agent or binding liquid to join powder material together to form a 3D part. (14)

block threading method. See *longhand threading*. (10)

block-by-block threading. See *longhand threading*. (10)

body. The section of a program code that contains all of the movements and commands that direct a machine through specific tools to create a part according to the print. (9)

bolt circle. A set of holes with the same radii that are equally angled apart. (4)

bore micrometer. A tool with the capacity to make accurate three-point measurements inside large bore diameters or close tolerance bores. (6)

boring. Refers to cuts made on the inside diameter of a part. (7)

bottom tap. A tap without a tapered lead-in used when tapping a blind hole, or a non-through hole, in which threads must extend down the entire hole. (8)

broken-out section. Limited cutaway of a small section in a print to show a hidden feature or give more detail. (5)

Note: The number in parentheses following each definition indicates the chapter in which the term can be found.

C

calipers. A versatile measuring tool used to measure the outside of a part, the inside of a part, or a depth or step. (6)

canned cycle. Abbreviated multi-movement cycles designed to reduce repetitive amounts of machine code and make editing repetitive steps easy. (9)

Cartesian coordinate system. Specifies each point in a plane uniquely with a pair of alphanumeric coordinates. (4)

cellular manufacturing. A workflow that uses principles of just-in-time delivery and lean manufacturing to reduce costs and improve efficiency. (14)

center drill. A special drilling tool used to create a small starter hole prior to drilling, allowing a drill tip to enter the cut straight. Standard in manual machining but should not be used in CNC mills or lathes. (8)

center of rotation. A physical point where all rotation centerlines intersect. (12)

centerline. Line that designates the center of a hole, diameter, radius, or other symmetrical feature. (5)

chatter. Part vibration during machining. (7)

chemical vapor deposition (CVD). A tool-coating process in which tools are put into a furnace, reduced to a vacuum, and then coating materials are added as gases that bond to the insert. (8)

chip load. The actual thickness of a chip being cut or the depth of each cutting edge as it passes through a material. (4)

chips. Smaller pieces of material cut from a larger workpiece. (3)

chuck. A set of jaws used to secure the workpiece in a lathe. (1) Any workholding device, including multi-jaw chucks, collet closers, face plates, and any other fixtures or devices that secure a workpiece for machining. (7)

chuck key. Tool used to clamp material in a lathe chuck. (3)

circle. A closed plane curve that is an equal distance at all points from its center point. (4)

class A fire. Fire caused by normal combustible materials, such as wood, paper, and cloth. These types of fires require water and the cooling effect of a solution with high water content. This is the most common household fire, but class A fires are not as common in manufacturing. (3)

class B fire. Fire resulting from flammable liquids or grease. Class B fires can be put out by removing oxygen or smothering the fire, so class B extinguishers use dry chemicals or carbon dioxide. (3)

class C fire. Electrical fire that must be extinguished using nonconducting chemicals and extinguishing agents. (3)

class D fire. Fire involving flammable metals, such as lithium and magnesium. Class D extinguishers contain specialized heat-absorbing dry powders. (3)

closing statement. The conclusion, or ending lines, of a program code that prepares the machine to run the next part in a safe manner. (9)

CNC laser. A specialized CNC machine that uses a high-output laser cutter to engrave or precision cut metal. (2) A computer-controlled machine that directs a laser beam to cut sheet metal. (7)

CNC machining center. A CNC mill with an automatic or programmable tool changer. (7)

CNC machinist. A machinist with traditional machining skills as well as the skills to operate a technologically advanced computer-driven system. (1)

CNC mill. A milling machine with CNC controls but without an automatic tool changer. (7)

CNC programmer. CNC programmers write the instructions for the computing module that is used to run CNC machines. (2)

CNC router. A computer-controlled machine that directs a router to cut wood or plastic. (7)

CNC turning center. Computer-controlled lathe fitted with an automatic tool changer. Also called *CNC lathe*. (7)

CNC turning center spindle. Component of a CNC turning center made up of the chuck and headstock. (7)

coaxial. Having a common axis or two or more coincident axes. (6)

coaxial indicator. A special indicator used in machining that spins around the spindle centerline and is used to find bore or boss features. (6)

common denominator. Denominator shared by all fractions in a single expression. (4)

comparison gage. Tool that does not take an actual measurement but instead compares a feature or part to the known size of the gage or a hard fixed gage that can be compared to a machined feature. (6)

complementary angles. Two angles that equal 90°. (4)

computer numerically controlled (CNC) machines. Specialized machining equipment controlled by computer drives and servo motors. CNC machines require highly trained and skilled operators and are susceptible to failure in harsh environments. (2)

computer-aided drafting (CAD). The graphical process of creating true-to-form and scale parts, assemblies, or renderings, using lines, vectors, and arcs to create a digital recreation of a concept or part. (11)

computer-aided manufacturing (CAM). The process in which machining parameters are defined using computer software. (11)

contact measurement. Probe measurement in which the probe head must physically touch the surface it is measuring. (13)

contouring. A type of toolpath that is typically performed with an end mill and follows a joined path or single piece of geometry. (9)

coordinate measuring machine (CMM). An inspection tool similar in design to a milling machine. A CMM uses the Cartesian coordinate system to drive around an electronic probe to find part features. (6)

cosine. For a given angle, the ratio of the adjacent side to the hypotenuse of a right triangle. (4)

counterbore. Flat-bottomed hole used to mount hardware flush with or below surface level. (5) A tool designed to create clearance diameters for bolt heads and most socket head cap screws. (8)

countersink. Tapered angle, or chamfer, placed on the outermost edge of a hole. (5) A tool designed to create relief angles or clearance diameters for screw heads, typically cut at 82° or 100° depending on the type of screw being used in the assembly. (8)

cross threading. Misaligned threading due to the tool not synchronizing with the rotary encoder signal and not reentering the threads at the same starting position. (10)

cutter compensation. Accounting for the diameter of the tool in use when programming a cut. (9)

cutting plane line. Line on a print that defines where a cut occurs and in which direction it is shown in additional views. (5)

D

data control. The task of ensuring that various versions and revisions of a program and all associated files are properly named, maintained, and stored. (11)

default block. Code in the opening statement of a program that cancels any previous cycles that may still be active. (9)

denominator. The number on the bottom of a fraction that indicates the size of the fraction parts. (4)

depth micrometer. A common tool used when measuring from one face or surface of a part to another face. (6)

design intent. How to interpret a designer's idea when print reading. (9)

dial indicator. An accurate and versatile indicator with a dial face that can be attached to other accessories, such as a magnetic base, height gage, tramming bar, or mill indicator holder. (6)

diameter. The distance across a circle through the center point. (4)

diameter offset. The diameter of a tool referenced when using cutter compensation. (9)

dimension. Measurable extent of an object indicating length, width, depth, height, or other measurement. (5)

dimension line. Line that appears on a print between extension lines and indicates the extent of a part dimension. (5)

Direct Exchange Format (DXF). A file format that was developed specifically by a CAD company to convert its files out to other systems. (11)

directed energy deposition. An additive manufacturing process in which focused thermal (heat) energy—such as a laser, electron beam, or plasma arc—fuses materials by melting them as they are being deposited. (14)

drawing number. Print identifier that can be associated with a particular part number or appear as a stand-alone engineering number. (5)

drill fixture. Tool used to create multiple holes in a workpiece. A drill fixture contains stops to secure the workpiece and bushings to guide the drill to a precise location. (1)

drill press. Machine commonly used in secondary operations to drill holes in workpieces that do not require the accuracy of milling machines. (1)

drop indicator. An indicator with a vertically dropping stem controlled by a series of springs and gears that is generally capable of extended travel but is less accurate when compared with a dial indicator. (6)

E

electronic gaging. Fast, accurate, and increasingly cost-effective inspection method that uses sensitive electronic triggers to inspect parts. (6)

end mill. A cutting tool used in milling for roughing, finishing, slotting, contouring, and profiling. (8)

end-piece manipulator. Robotic "hands" with diverse, tactile functions like grasping, vacuum pick-up, or magnetic. (14)

equilateral triangle. Triangle with all equal sides and all equal angles. (4)

ethical behavior. Conforming to accepted standards of fairness and good conduct. Integrity, confidentiality, and honesty are crucial components. (2)

extension lines. Lines that extend from the edge of a feature on a print and display the specified dimensions. (5)

extruder. The apparatus in fused deposition modeling that consists of a set of rollers that pulls plastic filament through a heating element and deposits the liquid plastic through a nozzle during 3D printing. (14)

F

face mill. A cutting tool used in milling, designed to machine the top face of a workpiece or part. (8)

facing. Machining the front face of a part to remove any excess material on the front face or to establish a consistent part length through multiple parts. (7)

feature-based machining. An HSM CAD function that allows the programmer to pick a solid and let the software decide the appropriate toolpaths. (11)

feeler gage. A shim or thin piece of metal used to determine the size of a gap or space between two parts or features. Feeler gages come in sets with multiple sizes and are often used during setup or inspection. (6)

fillet gage. A simple comparison gage precut with radii or partial arcs to compare against a machined surface. (6)

fillets. In general machining or dimensioning, small arcs that blend two intersecting lines. (5)

five-second rule. Safety practice in which a worker takes five seconds before turning on or operating any piece of equipment to look around the workspace and equipment for any impending danger. (3)

fixed sequence robot. A robot that performs predetermined, repetitive operations, the sequence of which cannot be changed. (14)

fixturing/fixture. Any device used for holding material during machining. (7)

flexible manufacturing system (FMS). A manufacturing process in which there is some amount of routing and/or machine flexibility that allows the system to react in case of changes. (14)

flutes. Number of cutting edges on a cutting tool. (4)

full 3-axis turning. The process of moving a primary axis simultaneously with a secondary axis in a lathe. (12)

full 4-axis milling. Milling in which all of the primary axes are moving while one additional axis is utilized simultaneously. (12)

full 5-axis milling. Milling in which all of the primary axes are moving while the two additional axes are utilized simultaneously. (12)

full-axis engagement. Machining in which the material or workpiece is being simultaneously rotated while the tool is engaged in the material, forcing movement in another axis. (12)

fused deposition modeling (FDM). Material extrusion technique that can use plastic or composite to print 3D parts. (14)

G

G65. A macro call identifying a probing function. (13)

gage blocks. Precision-ground rectangular blocks used as comparison gages. Also called *Johansson blocks* or *Jo blocks*. (6)

geometric dimensioning and tolerancing (GD&T). Comprehensive system used to define nominal geometry and allowable tolerances. (5)

geometric shape. In mathematics, a 2-dimensional object, such as a circle, arc, triangle, polygon, or quadrilateral; or a 3-dimensional object, such as a sphere, pyramid, cube, or polyhedron. (4)

geometry creation. Drawing lines, arcs, or solids that define a part's shape. (11)

go/no-go gages. Internal or external gages manufactured at the high and low limits of thread specification to measure threads exactly. (5) A style of gage in which the gage either fits the feature or does not. (6)

grade. Rating that establishes and identifies the precision and quality of a gage block set. (6)

grinding. Wearing, smoothing, sharpening, or removing material by abrasion or friction. The process by which most sharp cutting tools are made. (1) An abrasive process where material is removed in small increments. (7)

gross domestic product (GDP). Monetary value of all goods and services produced by residents of a nation during a specific time period. (2)

H

half section. View that shows a part split into halves or quarter sections to show hidden or internal features. (5)

hard tool. A tool that only cuts on the sides and must enter through a preexisting hole. (9)

hardened steel. Materials that are extremely wear-resistant naturally, or made very wear-resistant through a process of heat treatment. (1)

harmonics. A chatter or vibration of the tool due to forces being transmitted only through the tool when side cutting. (12)

hazard alert label. Any additional label or warning that might be placed around a shop or classroom lab to identify potential risks. (3)

hazardous materials identification system (HMIS). Numerical hazard rating system that uses labels and colors to communicate chemical hazards. (3)

headstock. Includes all components that secure a chuck or open and close a chuck. (7)

head-table design. A type of 5-axis machining center in which the table rotates and also the entire spindle cartridge rotates around a fixed location. (12)

height gage. A vertical comparison gage used to compare a known height to an unknown height. (6)

hidden line. Line used to show a feature that may be visible in another view but is hidden or invisible in the current view. (5)

high-speed machining (HSM). CAM toolpaths that account for solid modeling to speed up the machining process without concern for violating any adjoining part feature. (11)

horizontal band saw. A band saw with a blade oriented parallel to the floor or base. (1)

horizontal machining center. A machining center in which the spindle is oriented on the side wall of the machine, not the top, and the table or workpiece moves into the spindle. (7)

hypotenuse. The longest side of a right triangle, which is always located across from the right angle. (4)

I

I, J, and K method. An alternative process for creating arcs using incremental coordinates from arc end to arc center or for creating full circles. (9)

improper fraction. Fraction whose numerator is larger than its denominator. (4)

incremental position. The distance from the start point to the next point. (9)

indexer. A precision work-positioning device that rotates parts to be machined. (12)

indicator. A widely used comparison gage that indicates the difference in measurement between a known and unknown surface. (6)

information transfer. Conveying all of the information gathered in the programming stages to the shop floor and inputting it in a CNC machine controller. (11)

Initial Graphics Exchange Specification (IGES). A file type that can be imported from CAD to CAM. IGES files will import all 2D geometry, but when importing solids, it will import them as surfaces. (11)

insert. A small, clamped-in cutting tool. (8)

inside diameter (ID) grinder. Machine with a mounted grinding wheel that reciprocates in and out of a part to remove material from the inside of the workpiece while it spins. (1)

inside micrometer. A tool used when measuring internal features such as slots, grooves, or large bore diameters. (6)

integration. The incorporation of a robot and machine to function together. (14)

intelligent robot. A robot that, while not completely AI, makes decisions on its own based on changes in the work environment through vision (camera), sensors, or voice commands. (14)

International Organization for Standardization (ISO). An international standard-setting body that promotes global industrial standards. (5)

isosceles triangle. Triangle with two equal sides and two equal angles. (4)

J

journeyperson. A skilled laborer who has completed an apprenticeship. (2)

just-in-time delivery. A customer-driven, cellular manufacturing concept that allows the customer to receive product only when they need it. (14)

L

lathe. A machine that spins or turns the workpiece while the cutter is fixed or driven through the material. (1)

layers/levels. A CAD tool that allows the operator to hide or show specific features within a drawing depending on the desired view or print. (11)

leader line. Line used to point to a specific location or note on a print. (5)

leadership. Ability to guide and motivate others to complete tasks or achieve goals. It involves communicating well with others, accepting responsibility, and making decisions with confidence. (2)

lean manufacturing. A cellular manufacturing concept that utilizes workflow and process control to limit the resources used in creating a finished product. (14)

limiting tolerance. Tolerance defined by given upper and lower dimensions. (5)

limits. Definitive maximum and minimum sizes of a part. (5)

line segment. Part of a line with a definite beginning and end. (4)

line. Continuous, straight, 1-dimensional geometric element with no end. (4)

live tool. An attachment that allows for drilling or minimal milling operations to be performed on a lathe. (12)

lock out/tag out (LOTO). System used to secure machines and prevent accidental operation during maintenance or other unsafe conditions. (3)

longhand threading. The single point block programming of a lathe threading tool that calculates every pass of the tool, allowing the programmer greater control of the programmed paths and depths of cut and the ability to cut tapered or straight threads in one cycle. (10)

M

M98. The M code that identifies a subprogram to be called. (13)

M99. The M code that returns a subprogram back to the main program. (13)

machine controller. The central processing area of a CNC machine where edits, program uploads, and all tooling and work offsets are made. (7)

machine flexibility. A flexible manufacturing system concept that refers to the system's ability to use multiple machines to perform the same operation on a part or the ability to mount a work fixture in many machines. (14)

machine function (M code). Program code that serves as a switch for various machine operations. (9)

machine home. The machine position at which X, Y, and Z axes are at their furthest limits. (9)

machine operator. An entry-level employee responsible for the day-to-day operation of a CNC machine, including pre-operation, start-up procedures, daily maintenance, and machine adjustment. (2)

machine setup operator. Worker responsible for ensuring CNC machine, tooling, and work holding accurately produce parts specified by programmers. (2)

machine vise. A heavy clamping system mounted on a mill table that secures material by means of a screw or lever. (1)

machining specialist. A highly technical and qualified machinist, or someone with advanced training who operates machinery for a specific industry. (1)

machinist. A person who operates a machine. (1)

macro variables. A stored parametric parameter that can be used to program defined variables. (13)

main program. Initial, overarching program that provides direction to the machine and may contain subprograms. (13)

manual machine. Machines that are operated by physically moving the machine, material, or cutter to shape a desired part. (1)

manual machinist. Highly skilled and dexterous operator of manual machines. (1)

manual robot. A type of robot driven strictly by a manual operator, usually including an on-board camera and the capability to be driven remotely by an offsite operator. (14)

manufacturing engineer. Work to improve the process of making finished products. Also called *tooling engineer*. (2)

material extrusion. An additive manufacturing process in which a continuous filament of plastic or composite material is melted and extruded to construct 3D parts. (14)

material jetting. An additive manufacturing process in which material drops are deposited layer by layer into the build platform to form a 3D part. (14)

mechanical engineer. Responsible for machine design or modification and often work on new designs or improving existing equipment. (2)

micromachining. The process of manufacturing extremely small parts. (2)

micrometer. An accurate measuring tool used to make small measurements. (1) A measuring device that uses a rotating scale on a high precision screw thread to precisely measure objects. Also called *micrometer screw gage*. (6)

mill vise. Milling tool that clamps a workpiece in place to keep it completely still while the cutter rotates through the material. (3)

milling machine. A machine that uses a rotating cutter to remove material from a stationary workpiece. (1)

mill-turn machine. A true hybrid of a machining center and turning center that is fully capable of both turning and milling. (7) A CNC machine that offers full turning capability along with full milling capability all in one machine. (12)

mixed number. A whole number with a fraction. (4)

modal command. Programmed commands that stay on until an alternate movement is commanded or they are turned off. (9)

multiaxis. The utilization of more axes of movement than standard in a lathe or mill. (12)

multiaxis machine. A machine with more axes of travel than a standard 3-axis machining center. (7)

multi-turret turning center. A CNC turning center with more than one turret used in operations where it is advantageous to machine from both sides of a part or perform multiple operations simultaneously. (7)

multiview drawing. A 2-dimensional drawing of a 3-dimensional part that includes multiple views of the part. (5)

N

National Tooling and Machining Association (NTMA). Organization that supports the growth of the manufacturing industry through online training. (2)

negotiation. Process of agreeing to an issue that requires all parties to give and take. The goal is a "win-win" solution. (2)

NIMS. Organization formed in 1995 to develop a set of credentials for metalworking professionals. NIMS offers credentials in a range of metalworking operations based on industry-established standards. (2)

nonverbal communication. Sending and receiving messages without using words. It involves body language, which includes the expression on your face and your body posture. (2)

note. Supplementary information needed for manufacturing and post-manufacturing processes that go beyond graphical information about part shape. (5)

numerator. The number on the top of a fraction that indicates the number of fraction parts. (4)

numerical control (NC). The automation of machine tools by use of a computer. (1)

numerically controlled robot. A robot that can be programmed remotely through software and uses Cartesian coordinates to perform a series of movements and commands based on that program. (14)

O

obtuse angle. Angle that measures more than 90°. (4)

on-the-job training (OJT). A common training practice in which new employees are given only the training required to do their immediate job. (2)

opening statement. The beginning, or starting lines, of a program code, which contains the program number and a default block that cancels any previous cycles that may still be active. (9)

optical comparator. A measuring device that uses the shadow of a part to magnify features and measure them onscreen. Optical comparators are especially useful for measuring irregularly shaped parts. (6)

origin. Zero point on a line or coordinate system. (4)

orthographic projection. Method used to create multiple views of a part in which each view is offset by 90°. (5)

outside diameter (OD) grinder. Machine that applies a precisely balanced grinding wheel to the outside of a spinning cylindrical part to slowly remove material from the outside of the workpiece. (1)

P

parallelogram. A quadrilateral in which opposite sides are both equal and parallel. (4)

parting off. Creating multiple parts from bar stock in a single operation by cutting off each part from the stock as it is completed. (10)

part number. Unique number assigned to a part to identify and distinguish it from similar parts. (5)

peck drilling. A technique for incrementally drilling a hole in which the tool drives down to a partial depth, fully retracts to remove chips and allow coolant to enter, and then re-enters the hole at a deeper depth. Used whenever the depth of a hole is four times greater, or more, than the diameter of the drill. (9)

personal protective equipment (PPE). Any equipment worn to minimize exposure to hazards that cause workplace injuries or illnesses. PPE is determined by the job being performed and commonly includes safety glasses, gloves, aprons, earplugs, steel-toed shoes, and respirators. (3)

photopolymerization. An additive manufacturing process in which a liquid photopolymer, or light-activated polymer, is cured using ultraviolet light to create a 3D part. (14)

physical vapor deposition (PVD). A tool-coating process in which negatively charged inserts have positively charged coating material applied to them electrically. (8)

pinch turning. Process performed on multi-turret turning centers where tools cutting on both sides of a part can prevent part vibration or chatter. (7)

pin gage. A precision round pin used to check the size of a machined feature, such as the size of a round hole or the width of a slot or groove. Pin gages come in sets, in a variety of grades and increments. (6)

pitch. Number of peaks or total turns in a 1″ span of a thread, measured in threads per inch. (5)

plasma cutting. Process in which electrically charged, superheated gas is forced through a nozzle to cut ferromagnetic materials. (7)

playback robot. A robot in which a human operator steps the robot through a series of positions or commands, records, or "teaches," those commands at each step, and then those commands are stored in the memory of the controller. (14)

pocket milling. Machining inside a closed area or boundary. A pocket can have a flat or tapered walls and bottom as well as islands or internal components that are not meant to be machined. (9)

polar coordinate. Dimension that references a distance and an angle. (5)

polycrystalline cubic boron nitride (PCBN). A material used in hard-turning inserts that provides superior edge integrity in materials above 45 Rockwell C scale hardness. (8)

polycrystalline diamond (PCD). A manufactured diamond material used in inserts that is especially effective in applications involving high speeds, high volumes, or certain nonferrous materials, such as aluminum or graphite. (8)

polygon. A 2-dimensional shape with straight sides. (4)

post-processing. The function within CAM systems that translates the generic machine code to the actual machine controller. (11)

powder bed fusion. An additive manufacturing process that uses either a laser or electron beam to melt and fuse the material together to form a 3D-printed part. (14)

precision measurement. Any measurement smaller than .005″ (five-thousandths of an inch) or less than 1° of accuracy. (6)

preparatory command (G code). Program code that puts the CNC control in a specific mode or state of operation. (9)

print. Plan that transfers information from the designer to the builder, allowing the builder to know exactly what the designer envisioned. (5)

probe. A precision measurement instrument that touches a surface of a part and records its position. (13)

probe head. The ruby-tipped stylus on the end of a probe that touches the part, triggers a switch in the probe body, takes a contact measurement, and records the event, or position, in the machine controller. (13)

probing system. A hardware addition to a CNC machine that provides accurate measurement feedback. (13)

profile cutting. CNC lathe toolpath to machine the outside diameter or inside diameter of a part. (10)

protractor. A measuring tool used in manufacturing to measure angles between two surfaces. (6)

punctual. Always prompt and on time. (2)

Pythagorean theorem. Mathematical property of right triangles that states that the square of one side plus the square of the other side is equal to the square of the hypotenuse or longest side. (4)

Q

quadrant. One of four distinct areas in a 2-axis coordinate system. (4)

quadrilateral. A 4-sided, 2-dimensional polygon. (4)

quality control (QC) inspector. Works with manufacturing team to ensure products meet form, fit, and function requirements. (2)

R

radius. The distance from the center point to the edge of a circle. (4)

rectangle. A quadrilateral with two pairs of equal sides and four right (90°) angles. (4)

reference. An individual who will provide important information about you to a prospective employer.

A reference can be a teacher, school official, previous employer, or any other adult outside your family who knows you well. (2)

résumé. A brief outline of your education, work experience, and other qualifications for work. (2)

rhombus. A quadrilateral with four equal sides but no right (90°) angles. (4)

right angle. Angle that measures exactly 90°. (4)

right triangle. Triangle with one 90° angle. (4)

robotics. The technology dealing with the design and production of robots. Robots are designed to replicate human movement and can be remotely controlled by a human or programmed using computer software. (14)

rotary. An additional axis that rotates around a single axis, that can be mounted to a 3-axis machining center. (12)

roughing end mill. An end mill with a special, serrated cutting-edge geometry that allows for varying chip formation and the removal of large volumes of material. (8)

routing flexibility. A flexible manufacturing system concept that refers to the system's ability to be altered to produce new product types and its ability to change the order of operations executed on any part. (14)

S

safe move. Programming that allows the probe stylus to move safely around the part, fixture, and clamps, protecting it from collision. (13)

safety data sheet (SDS). Detailed informational document prepared by the manufacturer of a hazardous chemical that describes the physical and chemical properties of the product. (3)

scalene triangle. Triangle with no equal sides or angles. (4)

screw pitch gage. Tool used to verify the pitch on a threaded part. (6)

section lines. Diagonal or crosshatched lines that indicate a feature is not on a normally external side of a part. (5)

sectional view. A view showing a part cut or sectioned to reveal internal or hidden features that cannot be shown in normal views. (5)

self-motivation. An inner urge to perform well. (2)

semiprecision measurement. Any measurement greater than .005″ or 1° of accuracy. (6)

setup sheet. A series of pictures and diagrams that include all machine setup information. (11)

sheet lamination. An additive manufacturing process in which material sheets are stacked and laminated together using either adhesives, ultrasonic welding, or brazing to form a 3D part. (14)

sheet metal fabrication. The process of cutting, forming, and bending pieces from flat sheet metal into usable shapes. (2)

sine. For a given angle, the ratio of the opposite side to the hypotenuse of a right triangle. (4)

sine bar. A tool used in conjunction with gage blocks to measure angles. (6)

single surface measurement. Probe routine that measures and sets the work coordinate for a single axis. (13)

Society of Manufacturing Engineers (SME). Organization that provides training for the manufacturing industry, traditionally for manufacturing engineers and technicians, but more recently for a range of skill levels through online courses. (2)

soft skills. Desirable qualities in an employee that are not acquired in school, such as common sense, the ability to deal with people, and a positive attitude. (2)

solid model. 3-dimentional objects that have mass and volume. (11)

spindle. Device that secures a tool holder in place and rotates cutting tools. (7)

spiral flute tap. A tap with a flute twist that allows chips to be pulled out of the existing hole, optimal for tapping holes that do not go completely through a piece of material. (8)

spot drill. A drilling tool that has a short flute with a large tip and a 90° point. Used to start holes in CNC applications. (8)

spotface. Shallow counterbore that typically has close tolerance. (5)

square. A quadrilateral with four equal sides and four right (90°) angles. (4)

steel rule. A precision-made measuring tool, available in several basic types and graduations, including fractional inch, decimal inch, and metric. (6)

straight flute tap. A tap with no twist in the clearance flutes and coated to prevent chip buildup, optimal for tapping holes that go through a part and do not need chip evacuation. (8)

stylus. The tip of the probe head that engages the surface being measured. (13)

subprogram. Secondary programs that can be called within a main program to perform specific tasks. (13)

subtractive manufacturing. Machining that starts with a "block" or piece of material and removes the unwanted material to reveal a part inside. (14)

subtractive process. Process that starts with a piece of stock and subtracts material to leave the finished part. (3)

supplementary angles. Two angles that equal 180°. (4)

surface footage. Number of linear feet a location on a rotating component travels in one minute. (4)

surface grinder. A precision grinding machine that is used to grind flat surfaces or make custom cutting tools by holding a workpiece stationary on a magnetic surface plate. (1)

surfaces. 3-dimentional objects that have no thickness or mass, like stretching a skin over a frame to form a shape. Surfaces can be "stitched" together to create complicated forms. (11)

surfacing toolpath. A machining path that follows a meshed surface. (12)

Swiss screw machine. An automatic lathe that has a sliding headstock and guide bushing. (7)

symmetry line. A line drawn on a print for a symmetric part that shows that the dimensions for one set of features are mirrored in the features on the other side of the part. (5)

T

tailstock. A moveable or programmable device that provides longitudinal support to a workpiece. (7)

tangent. For a given angle, the ratio of the opposite side to the adjacent side of a right triangle. (4)

tangent line. A line that touches an arc or circle at exactly one point. (4)

tap. A tool used for creating internal threads in a drilled hole. (8)

team. Small group of people working together for a common purpose. Teamwork often requires cooperation, flexibility, and a willingness to try new ways of doing things. (2)

telescoping gage. A type of comparison gage used to measure semiprecision bore diameters. (6)

thread gage. A go/no-go style gage used to measure threads. (6)

threads per inch (TPI). The amount of full threads in one inch, or pitch. (9)

title block. Part of a print that contains identifying information about the part and print as well as supplementary information, such as tolerances, scale, materials, and part designer. (5)

tolerance. Amount of acceptable deviation from a desired size. (5)

tolerance block. Part of a print that sets limits for machining dimensions. (5)

tool engagement. The angle and depth of the tool as it makes a cut. (12)

tool nose radius compensation. In lathes, cutter compensation for the radius on the nose of the tool. (10)

tool offset. The distance from the bottom of the tool to the programmed Z0. position. (9)

toolpath. The cutting strategy used to create a part. (11)

trigonometry. A branch of mathematics that deals with the relationships between the sides and angles of triangles and with functions of angles. (4)

trunnion. A multiaxis attachment mounted directly onto a machine table to allow for rotation in one (4-axis) or two (5-axis) directions beyond standard 3-axis machining. (7) (12)

turning. A machining operation performed on the outside or outside diameter of a part. (7)

turret. Tool holder used in a lathe. (1) Automatic tool changer in a CNC turning center where tools are fixed and stored in place when not in use and secured for cutting. (7)

twist drill. Any solid body drill bit. (8)

U

unilateral tolerance. Describes a dimension with tolerance in only one direction. (5)

unspecified tolerance. Default tolerance used when no other tolerance is specified on a print. (5)

V

variable sequence robot. A robot that performs a set of commands repetitively until the command sequence is altered by the operator. (14)

verbal communication. Speaking, listening, and writing. (2)

vernier height gage. Specialized height gage that can be fitted with a carbide attachment to scribe or mark a surface at a given dimension in addition to comparing one height to another. (6)

vernier scale. A scale made to slide along divisions of a larger graduated scale and take measurements between the larger divisions. (6)

vertical band saw. A band saw with a blade oriented perpendicular to the floor. (1)

vertical machining center. A machining center in which the spindle travels vertically on the Z axis. (7)

visible lines. Lines that define the part shape or surface of an object. Also called *object lines*. (5)

W

water jet. An industrial CNC tool capable of cutting a variety of materials using a high-pressure jet of water, or a mixture of water and an abrasive substance, such as silica. (2) Tool that uses a mixture of abrasive particles in high-pressure water to cut out intricate shapes in various materials. (7)

wireless intuitive probing system (WIPS). Haas Automotive controller software, developed in partnership with Renishaw™, that integrates a probing system and rapid part setup in CNC machines. (13)

work cell. U-shaped machining groups used in lean manufacturing to limit part movement and allow a single operator to operate more than one piece of equipment. (14)

work coordinate system (WCS). A program that defines where a part is on the table or fixture by telling the machine where the zero point of the part physically sits within the work envelope of the machine. (9)

work envelope. Any space within range of a machine's normal operation. (3)

workholding. Any device that is used to secure a workpiece against the forces of machining. (9)

workholding table. Machine table that secures fixtures, vises, and material in place for machining. (7)

workpiece. Material to be worked in a mill or lathe. (1)

work shift. Programming code that shifts all of the tool offsets in a specified direction. (13)

wringing. The process of stacking gage blocks together. (6)

Z

zero return. The movement of the machine to its machine home position, at which X, Y, and Z axes are set at zero. (9)

Index

2 + 1 turning, 267–268
2-dimensional (2D) geometry, 247–248, 251
3 + 1 milling, 270–271
3 + 2 milling, 272–273
3-axis turning, 268–269
3-dimensional geometry, 249, 251
3D printing, 327
5-axis milling, 274–277
60° V-type thread dimensions, 338–340

A

abbreviations, feeds and speeds, 335
abrasive process, 147
accidents, causes of, 40
acute angles, 73
additive manufacturing, 322–327
    definition, 322
    green technology, 325
    limitations of, 326–327
    safety, 326
address codes, 347–349
    lathe programming, 217–220
    mill programming, 184–187
adjustable squares, 6
aerospace, 25
air gaging, 127
alloys, 50
alphabet of lines, 95
American National Safety Institute (ANSI), 60
American Society of Mechanical Engineers (ASME), 91
angles, 73–74
    complementary, 74
    lathe programming, 229–230
    mill programming, 195
    supplementary, 74
anvils, micrometer parts, 116
apprenticeships, 41
arcs, 104–105, 249
    lathe programming, 230
    mill programming, 192
area, 344–345, 350

associated geometry, 256
attitude, 37–38
automated manufacturing, 314–327
automatic tool changers
    CNC machining centers, 136
    CNC turning centers, 142
automotive industry, machining and, 24
auxiliary views, 93

B

ball end mills, 278
band saws, 10–11
barrels, micrometer parts, 116
basic dimensions, 98–100
bench grinders, 12
bilateral tolerance, 101
binder jetting process, 323
block-by-block threading, 235
block threading method, 235
blocks, gage, 121–122
blueprints, 88–89
body language, 39
body (drills), 164
body (programming), 182–183
bolt circles, 76
bolt hole circles, 100
bore micrometers, 118
bore probing, 300–301
boring, 144
boring cycles, 201–202
boring machines, 5, 6
boss probing, 301
bottom taps, 166
Brinell hardness number, conversions, 343
broken-out sections, 94

C

C axis, 267
CAD (computer-aided drafting), 89, 244–246
calculating speeds and feeds, 79–81
calipers, 114–115

CAM (computer-aided manufacturing), 29, 244–259
CAM software
    5-axis toolpaths, 279
    lathe operations, 269
canned cycles, 184
    lathe programming, 231–235
    mill programming, 184, 197–202
carbide, 162–163
    drilling, 232
    inserts, 173f
careers, 28–41
    applying for jobs, 30–36
    CNC manufacturing, 28–30
    succeeding in workplace, 36–41
Cartesian coordinate system, 77–79
CAT taper, 169
cellular manufacturing, 318–321
center drills, 165
centerlines, 96
center of rotation, 276
central control computer, 321
ceramics, 164
chamfers, 103
chemical tagging, 58
chemical vapor deposition (CVD), 163
chip break peck drilling, 199
chip load, 80–81
chips, 50
    monitoring machine output, 51
    reducing, 141
    safety and removal, 50–52
chuck key, 52
chucks, 8, 141–142
circles, 70
    area, 350
    milling, 249
class A fires, 55
class B fires, 55
class C fires, 55
class D fires, 55f–56
class K fires, 55f
clearance diameters, 167
climb cut, 189–190

Note: Page numbers followed by *f* indicate figures.

closing statement
    lathe programming, 217
    mill programming, 182–184
    safety, 216
CMM (coordinate measuring machine), 30, 126–127, 298
CNC grinders, 147–148
CNC lasers, 28, 150–151
CNC lathes, 27, 78
CNC machines, 22–23, 134–151
    development of, 16–17
    careers operating, 28–29
    modern working environment, 22–23
CNC machining centers, 134–140
CNC machinists, 15, 16–17
CNC manufacturing
    careers, 28–30
    future of, 327
    progression to, 4–17
CNC mills, 79, 134
CNC plasma machines, 150
CNC programmers, 29
CNC programming codes, 347–349
CNC routers, 27, 151
CNC turning centers, 140–146
    axes of movement, 143–144
    components of, 141–143
    multi-spindle, 146
    multi-turret, 145
    operations, 144–145
coatings, 163
coaxial, 125
coaxial indicators, 125–126
codes
    CNC programming, 347–349
    G, 184–186, 218–219, 347–348
    M, 186–187, 219–220, 348–349
collet holders, 171
combustible metals, 50
combustion engine, 6
commands
    M30, 289
    M98, 291
    M99, 289
    modal, 192
    preparatory, 184
    See also functions
common denominator, 67
communication skills, 38–39
comparison gages, 118–120
compensation, cutter, 189–191
complementary angles, 74
computer-aided drafting. *See* CAD
computer-aided manufacturing. *See* CAM

computer numerically controlled machines. *See* CNC machines
computing industry, machining and, 26–27
cones, volume, 351
constructive criticism, 38
contact measurement, 298
contour toolpath, 253–254
contouring, 187–191
controllers
    CNC machining centers, 136–137
    CNC turning centers, 142–143
conversions
    decimals and fractions, 69, 346
    hardness, 343
    between SI Metric and US Customary, 344–345
coordinate measuring machine (CMM), 30, 126–127, 298
coordinates, polar, 100
cosine, 75
counterbore cycle, 200
counterbores, 104, 167
countersinks, 103, 167
created solids, 250
critical-thinking skills, 38
cross threading, 235
cutter compensation, 189–191
cutter entry, 195–196
cutters
    diamond, 163–164
    plasma, 27, 150
cutting edges, 81, 165
cutting fluids, green, 234
cutting plane lines, 93
cutting process, 12
cutting technique, 246
cutting tools, 158–173
    countersinks and counterbores, 167
    drills, 164–165
    end mills, 167–168
    face mills, 168–169
    machinability ratings, 173
    taps, 166
    toolholders, 169–172
    See also inserts
CVD (chemical vapor deposition), 163
cycles
    canned, 197–202
    drilling canned, 231–235
    threading, 235–237
cylinders, volume, 350

D

data control, 258
decimal dimensions, 97

decimal equivalents, drills, 337
decimals, conversions, 69, 346
deep hole drill, lathe programming, 232–233
default block, 182
defense, machining and, 25–26
denominators, 66–67
density, measurement conversions, 344
depth micrometers, 117
Descartes, René, 77
design intent, 97, 180
dial calipers, 114
dial indicators, 124
diameter, 70
diameter offset
    lathe programming, 220
    mill programming, 191, 204–205
diamond cutters, 163–164
digital prints, 89
dimension lines, 95
dimensions, 97–100
Direct Exchange Format (DXF), 251
directed energy deposition, 323
DOE (U.S. Department of Energy), 246
door interlocks, safety, 256
drawing numbers, 92
drawings and prints, 88–91
drill fixtures, 11
drill presses, 11–12
drilling cycle, 197–200
drills, 164–165
    components of, 164–165
    decimal equivalents, 337
    feeds and speeds, 336
    sizes, 337, 341–342
    types of, 164–165
drive motors, 16
drop indicators, 124–125
dynamic work offsets, 277

E

edge calipers, 114
edge cutting geometry, 167–168
education, traditional, 43
efficiency, 288
electrical process, 150
electronic CAD drawings, 89
electronic gaging, 127
emergency action plans, 53f
emergency stop, 217
employee tests, 36
end mills, 167–168
end-piece manipulators, 317
engineers, 30

Enterprise Resource Planning (ERP) software, 321
Environmental Protection Agency (EPA), 246
equilateral triangles, 71
ethical behavior, 39
exposure limits, common metals, 50f
extension lines, 95
Extrude, 250
extruder, 324
eye protection, 51f, 60–61

F

face mills, 168–169
facing, 144–145
FDM (fused deposition modeling), 323
feature-based machining (FBM), 256
feed rates, 79–81, 233
feeds and speeds
    calculating, 79–81
    common tools and materials, 336
    rules for determining, 335
feeler gages, 119
fillet gages, 119
fillets, 104, 105
fine boring cycles, 201–202
fire extinguishers, 55–56
fire safety, 53–56
five-second rule, 48
fixed sequence robots, 315
fixtures, drill, 11
fixturing, 140
flat, 14
flexible manufacturing systems (FMS), 321
floating taps, 9
flutes
    drills, 164
    end mills, 81
formulas, 335, 350–351
four-quadrant plotting, 78
fractional dimensions, 97
fractional sizes, 60° V-type thread dimensions, 338–339
fractions, 66–70, 346
frames, micrometer parts, 116
full 3-axis turning, 268–269
full 4-axis milling, 271–272
full 5-axis milling, 275
full-axis engagement, 277–279
full retract peck drilling, 197–199
fully integrated 5-axis milling, 274–275
functions, 347–349
    miscellaneous, 348–349
    other characters, 349

preparatory, 347–348
See also commands
fused deposition modeling (FDM), 323

G

G02 and G03 radial movements, 192–194
G41 left-hand compensation, 189
G73 drilling cycle, 199
G81 drilling cycle
    lathe programming, 231
    mill programming, 199–200
G82 drilling cycle
    lathe programming, 231–232
    mill programming, 200
G83 drilling cycle
    lathe programming, 232–233
    mill programming, 197–199
G84 tapping cycle
    lathe programming, 233–234
    mill programming, 200–201
G85 boring cycle, 201–202
G86 boring cycle, 202
G codes, 347–348
    lathe programming, 218–219
    mill programming, 184–186
gage blocks, 121–122
gages, 118–123, 127
    advanced, 126–127
    comparison, 118–120
    go/no-go, 105–106, 118
    height, 120–121
    telescoping, 123
gang tool, 172
GD&T (geometric dimensioning and tolerancing), 106–107
GDP (gross domestic product), 23–24
gearbox, 9f
generic machine code file, 259
geometric characteristics, 107f
geometric dimensioning and tolerancing (GD&T), 106–107
geometric shapes, 70–73
geometry, 247–251
    2-dimensional, 247–248
    3-dimensional, 249
    associated, 256
    creating, 247–250
    importing, 251
geometry creation, 247
gloves, safety, 51
go/no-go gages, 105–106, 118
grades, gage blocks, 121
granite plates, resurfacing, 125

green initiatives, 246
green manufacturing, 17, 325
greening, 17
grinders, 12–14, 147–148
grinding, 12–14, 147–148
    precision, 12–13
    process, 14
    safety, 61
    types of, 148
grinding wheels, safety, 148
grooving, 234–235
gross domestic product (GDP), 23–24

H

H numbers, 204
Haas, Gene, 17, 26, 266
Haas Machines, 26
half sections, 93–94
hard tools, 181
hardened steel, 12–13
hardness conversions, 343
harmonics, 278
hazard alert labels, 61
hazardous materials identification system (HMIS), 56–57
head-table design, 275
headstocks, 142
health and hygiene, 36–37
height gages, 120–121
helical entry, 196
hidden lines, 96
high-speed machining (HSM), 158, 255
high-speed steel (HSS), 158
HMIS (hazardous materials identification system), 56–57
holes, 102–104
horizontal band saws, 10–11
horizontal machining centers, 137–138
horizontal milling machines, 6, 7f
HSM (high-speed machining), 158, 255
HSS (high-speed steel), 158
hypotenuse, 72

I

I, J, and K method, 194–195
improper fractions, 68
inches per minute (ipm), 80
inches per revolution (ipr), 81
incremental position, 194
indexable drills, 165
indexer, 266
indicators, 123–126
industrial revolution, 5, 14–15

information transfer, 257
Initial Graphics Exchange Specification (IGES), 251
insertable drilling tools, 165
inserted end mills, 168
inserts, 158–164
    checking data, 158
    materials, 162–164
    naming, 158–162
inside corner probing, 302
inside diameter (ID) grinders, 13
inside diameter (ID) grinding, 148
inside micrometers, 117
integration, 314–315
intelligent robots, 315
International Organization for Standardization (ISO), 91, 158
interpersonal skills, 39–40
interviews, 34–36
ipm (inches per minute), 80
ipr (inches per revolution), 81
ISO naming conventions, 158–162
isosceles triangles, 71

J

Jo blocks, 121
jobs, 257–259
    post-processing, 259
    setup sheets, 257
Jobs, Steve, 26
Johansson blocks, 121
journeyperson, 41
just-in-time delivery, 319

L

laser beams, 151
lasers, CNC, 28
lathe programming, 212–237
    address codes, 217–220
    angles, 229–230
    arcs, 230
    canned cycles, 231–235
    formatting, 215–217
    offsets, 220–221
    parting off, 237
    planning, 212–214
    profile cutting, 222–230
    threading cycles, 235–237
    tool nose radius compensation, 221–222
    tool setting, 220
    work coordinate offsets, 221
lathes, 8–10
    calculating feeds and speeds, 81
    CNC, 27
    components of, 9

early, 4
subprogramming, 295–297
toolholders, 171–172
types of, 8–9
See also lathe programming
layers, CAD programs, 245
leader lines, 96
leadership, 40
lean manufacturing, 319
LEED (Leadership in Energy and Environmental Design), 281
left-hand compensation, 189
length offsets
    lathe programming, 220
    mill programming, 203–204
    safety, 204
length, measurement conversions, 344–345
letters of application, 31, 33f
letters of recommendation, 31
levels, CAD programs, 245
limiting tolerance, 101
limits, 101
line segments, 73
line usage, 95–97
lines, 73, 95–97, 249
    alphabet of, 95
    cutting plane, 93
    geometry creation, 249
    usage on prints, 95–97
    section, 93
live tools, 267
lock out/tag out (LOTO), 59–60
locknuts, micrometer parts, 116
longhand threading, 235

M

M30 command, 289
M98 command, 291
M99 command, 289
M codes, 348–349
    lathe programming, 219–220
    mill programming, 186–187
machinability ratings, 173
machine controllers, 136–137
machine flexibility, 321
machine function, 186
machine home, 205
machine operators, 28
machine service technicians, 15
machine setup operators, 29
machine shops
    layout, 319–321
    success working in, 36
machine tools, early, 4–5
machine vise, 6–7

machinery, CNC, 134–151
Machinery's Handbook, 166
machines
    boring, 5, 6
    gantry-style, 27
    mill-turn, 146–147, 280–281
    multiaxis, 138–140
    sewing, 6
    Swiss screw, 147
machining centers, 137–138
machining
    feature-based, 255
    high-speed, 158, 255
    modern work environment, 22–24
    multiaxis, 266–267
    skills to compete, 22–43
    solids, 255–256
    supplied industries, 24–28
    surfaces, 256
    trunnions, 273–274
machining math, 66–81
machining specialists, 15, 17
machinists, 14–17
    CNC, 15–17
    manual, 6, 15–16
macro variables, 304
magnetic plates, use of, 14
main programs, 288–294
maintenance, proactive, 297
managers, 30
manual machines, 6–14
    drill presses, 11–12
    grinders, 12–14
    lathes, 8–10
    milling, 6–8
    saws, 10–11
manual machinists, 6, 15–16
manual robots, 315
manufacturing engineers, 30
manufacturing techniques, 17, 314–327
material description, 92
material extrusion, 323–324
material handling system, 321
material jetting, 324
mathematics, 66–81
Maudslay, Henry, 4, 6
measuring tools, 112–127
    calipers, 114–115
    early, 5–6
    gages, 118–123, 127
    indicators, 123–126
    micrometers, 116–118
    protractors, 113
    steel rules, 112–113
mechanical engineers, 30

medical technology, machining and, 24–25
melting points, metals, 334
metals, 333–334
    common shapes, 333
    physical properties, 334
    symbols, 334
metric sizes, 60° V-type thread dimensions, 340
micro-fractures, 326
micromachining, 24
micrometer screw gages, 116
micrometers, 4–6, 106, 116–118
    bore, 118
    components of, 116
    depth, 117
    inside, 117
mill programming, 180–207
    address codes, 184–187
    angles, 195
    arcs, 192
    canned cycles, 184, 197–202
    contouring, 187–191
    formatting, 182–184
    offsets, 191, 203–205
    planning, 180–182
    pocket, 195
    radius and angle, 191–195
    tapping cycle, 200–201
    tool setting, 202–205
    work coordinate offsets, 205–207
mill vise, 52
milling
    3 + 1, 270–271
    3 + 2, 272–273
    5-axis, 274–277
    full 4-axis, 271–272
    multiaxis, 270–279
    toolholders, 169–171
milling machines, 6–8
mills, 6–8, 134
    calculating feeds and speeds, 80–81
    CNC, 134
    end, 167–168
    face, 168–169
mill-turn machines, 146–147, 280–281
mixed numbers, 68
modal commands, 192, 291
multiaxis, 266
    machines, 138–140
    machining, 266–267
    milling, 270–279
    programming, 266–281
    turning, 267–269

multi-spindle turning centers, 146
multi-turret turning centers, 145
multiview drawings, 90–91

N

National Tooling and Machining Association (NTMA), 42
negotiation, 40
nested, 150
NIMS, 42
nonverbal communication, 38–39
notes
    prints, 89–90*f*
    programming, 216
number lines, 77
numerators, 66
numerical control (NC), 16–17
numerically controlled robots, 315

O

object lines, 95
obtuse angles, 73
Occupational Safety and Health Administration (OSHA), 41
offsets
    diameter, 191, 220
    dynamic work, 277
    length, 220
    tool, 191
    tool diameter, 204–205
    tool length, 203–204
    work coordinate, 205–207
oil spills, 54
oily rag disposal, 53
on-the-job training (OJT), 41–42
opening statement
    lathe programming, 215–216
    mill programming, 182–183
operations, secondary, 11
optical comparator, 127
orthographic projection, 91
OSHA (Occupational Safety and Health Administration), 41
outside corner probing, 301–302
outside diameter (OD) grinders, 13
outside diameter (OD) grinding, 148

P

P9810 program, 305
P9811 program, 305
P9814 program, 307
parallel, 14
parallelograms, 72
Parameters page, 253
part numbers, 92
parting off, 237

peck drilling, 197–199
    chip break, 199
    full retract, 197–199
pedestal grinders, 12
personal protective equipment (PPE), 60–61
photopolymerization, 322
physical vapor deposition (PVD), 163
pilot holes, 196
pin gages, 120
pitch, thread, 105
plasma cutters, 27, 150
plasma cutting, 150
playback robots, 315
plotting, 78–79
plus-minus (+/–) tolerance, 101
pocket milling, 195
pocketing, 195–197
pockets, 195, 254–255
points, 165, 248
polar coordinates, 100
polycrystalline cubic boron nitride (PCBN), 164
polycrystalline diamond (PCD), 163–164
polygons, 71
post-processing, 259
powder bed fusion, 324–325
precision grinders, 12
precision measurements, 112
preparatory commands, 184
print reading, 88–107
print review
    lathe programming, 212–213
    mill programming, 180–181
prints, 88–89
proactive maintenance, 297
probe head, 298
probes, 298
    setting work coordinates, 298–299
    single surface measurement, 302–304
    stylus, 305
probing, 298–307
    bore, 300–301
    boss, 301
    inside corner, 302
    inside programs, 304–307
    outside corner, 301–302
    routines, 299–304
probing systems, 298
production machinists, 15
professional behavior, 38
profile cutting, 222–230
program body, 216

programming
    efficiency, 288
    multiaxis, 266–281
    robots, 314–315
    rules, 288–289
    safety, 288
    See also subprogramming
programs, 288–294
    main, 288
    probing, 305–307
    secondary, 288
    subprograms, 288–297
projection, orthographic, 91
protractors, 113
punctual, 37
Pythagorean theorem, 72, 351

Q

quadrants, 78
quadrilaterals, 72
quality control (QC) inspectors, 30

R

radial boring machines, 6
radial movements, 192–194
radius, 70, 105
radius and angle milling, 191–195
ramp angle entry, 196
ratchets, micrometer parts, 116
reamers, 181, 336
record keeping, 57–58
rectangles, 72, 350
rectangular prisms, volume, 350
recycling
    cutting tools, 167
    tungsten carbide drills, 200
reference dimensions, 97–98
references, 31
relief angles, 167
respirators, safety, 61
résumés, 31, 32*f*
retention knobs, 170
return, zero, 183
revolutions per minute (rpm), 80
Revolve, 250
rhombuses, 72
right angles, 73
right triangles, 71–72
robotics, 314–318
    sheet metal industry and, 316–317
    welding stations and, 317
robots, 314–318
    manufacturing applications, 315–318
    programming, 314–315

types of, 315
using for inspection, 316
Rockwell scale, 164, 343
rotary, 270
roughing end mills, 168
roughing method, 196–197
round inserts, 168
routers, CNC, 27, 151
routing flexibility, 321

S

safe move, 305
safety, 40–41, 48–61
    chip removal, 50–52
    fire, 53–56
    five-second rule, 48
    guards, 59
    hazard alert labels, 61
    hazardous materials, 49–50, 56–57
    lock out/tag out (LOTO), 59–60
    personal protective equipment (PPE), 60–61
    shop, 48–53
    tool, 52–53
safety data sheets (SDS), 50, 57–58
saws, 10–11
scale calipers, 114
scalene triangles, 71
screw pitch gages, 119
screw threads, standardization of, 4
SDS (safety data sheets), 50, 57–58
secondary operations, 11
section lines, 93
sectional views, 93–94
sections, 93–94
self-motivation, 37
semiprecision measurements, 112
sequence of operations, 279
servo control, 16
servo drives, 16
servomechanisms, 16
servo motors, 17
setup, 257–259
setup sheets, 257
sewing machines, 6
shears, 27
sheet lamination, 326–327
sheet metal fabrication, 27–28
sheet sizes, 91
shrink fit holders, 171
simple solids, 250
sine, 75
sine bars, 122–123
single surface measurement, 302–304
size, thread, 105

Society of Manufacturing Engineers (SME), 42
soft skills, 36
solid body holders, 171
solid body reamers, 103*f*
solid models, 249–250
solids, 249–250, 255–256
    creating, 249–250
    machining, 255–256
specific gravity, metals, 334
specific heat, metals, 334
speeds and feeds. *See* feeds and speeds
spheres, volume, 351
spindle stop boring cycle, 202
spindles
    CNC machining centers, 135
    CNC turning centers, 141–142
    micrometer parts, 116
spiral flute taps, 166
spot drilling, 199–200
spot drills, 165
spotface (SF), 104
squares, 72
start position, 188–189
steel rules, 112–113
straight flute taps, 166
stress, measurement conversions, 344
stylus, 305
subprogramming, 288–297
    lathes, 295–297
    workarounds in CAM software, 294
subprograms, 288
subtractive manufacturing, 322
subtractive process, 50
supplementary angles, 74
surface feet per minute (sfm), 80
surface footage, 80
surface grinders, 13–14
surfaces, 249
    imported, 251
    machining, 256
surfacing toolpath, 278
Swiss screw machines, 147
symmetry, 96
symmetry lines, 96–97

T

tabular dimensions, 98
tailstocks, 9, 142
tangent, 75
tangent lines, 73
tap-drill chart, 166
tap drill sizes, 341–342
tapered angles, 103
tapping cycles, 200–201

tapping speeds, 201
taps, 166
    feeds and speeds, 292, 336
team, 39–40
technical training, 41–43
technology industry, machining and, 26–27
telescoping gages, 123
temperature, measurement conversions, 344–345
templates, 118
thimbles, micrometer parts, 116
third-angle projection, 91
thousandths accuracy, 69–70
thread-cutting lathes, 4
thread dimensions, 338–340
thread gages, 118–119
threading, 235
threading cycles, 235–237
threads, 105–106
threads per inch (TPI), 201
three-dimensional plotting, 78–79
time management, 37
title blocks, 92–93
tolerance, 101–102
tolerance block, 93
tool engagement, 278
tool nose radius compensation, 221–222
tool offset, 191
tool selection
    lathe programming, 213–214
    mill programming, 182
tool setting
    lathe programming, 220
    mill programming, 202–205
tool start position, 188–189
toolholders, 169–172
    collet, 171
    lathe, 171–172
    milling, 169–171
    shrink fit, 171
    solid body, 171
tooling engineers, 30
toolpaths, 251–256
    associated geometry, 256
    contours, 253–254
    generating, 251–256
    machining, 255–256

pockets, 254–255
    selection for drilling, 252–253
    surfacing, 278
tools
    advanced gaging and inspection, 126–127
    diameter offsets, 204–205
    Extrude, 250
    forces acting on, 278
    gages, 118–126
    holders, 169–172
    length offsets, 203–204
    live, 267
    measuring, 112–118
    Revolve, 250
    See also cutting tools; inserts
torque, measurement conversions, 344
TPI (threads per inch), 201
trainers, 30
training, technical, 41–43
trapezoids, area, 350
triangles, 71
    area, 350
    types of, 71
triangular pyramids, volume, 351
trigonometry, 74–76
trunnions, 138, 273–274
tungsten carbide, 162–163
    recycling, 200
turning, 144
    2 + 1, 267–268
    full 3-axis, 268–269
    multiaxis, 267–269
    operations, 144–145
    See also CNC turning centers
turret lathes, 8
turrets, 8, 142
twist drills, 164–165

U

UNC (Unified National Coarse), 166
UNF (Unified National Fine), 166
unilateral tolerance, 101–102
unspecified tolerance, 102

V

variable sequence robots, 315
vat photopolymerization, 322–323

VDI tooling, 172
verbal communication, 38
vernier height gages, 120
vernier scale, 120
vertical band saws, 10
vertical comparison gages, 120
vertical machining centers, 137
vertical milling machines, 7
VF-1 CNC milling machines, 17
views, 93–94
vise
    machine, 6–7
    mill, 52
visible lines, 95
volume, 344–345, 350–351

W

walking the path, 222
water jets, 28, 149–150
weight, measurement conversions, 344–345
welding, robots and, 317
Whitney, Eli, 5, 6
Wilkinson, John, 4, 5
wiper inserts, 168
wireless intuitive probing system (WIPS), 298
woodworking, machining and, 27
work cells, 319–321
work coordinate offsets
    lathe programming, 221
    mill programming, 205–207
work coordinate system (WCS), 205–207
work envelope, 49
work habits, 37
work shift, 296
workholding, 8
    lathe programming, 213
    mill programming, 181–182
    part, 181–182, 213
    repeatability, 206
workholding tables, 135–136
workpiece, 6
wringing, 122

Z

zero return, 183